A Letter to Instructors

It would seem reasonable to begin a quantitative literacy book by defining what quantitative literacy means. It turns out that's quite a bit harder than you might think. For many people, quantitative literacy is kind of like love: you recognize it when you see it, but it's not so easy to write out a dictionary-style definition. In our view, it's important to focus on the "literacy" part, since we're all pretty good at defining what quantitative thinking is about. And like many words in math, it can be instructive to think about what the word literacy means in plain English.

When thinking of it from a language standpoint, everyone knows what literacy means, and has a natural understanding of why it's important. In short, literacy is the minimal skill one would need in terms of using written and spoken language to be successful in society. That my friends, can serve as a fine definition of what we feel like quantitative literacy is about. We live in the information age of a monetary-based society, and there are some basic skills in working with quantities, and (more importantly) of understanding how and why to apply those skills, that are just simply required to meet a level of success in that society that all college students should be striving for. And that's the goal of this book.

There are, of course, hundreds of books that do a fine job of teaching quantitative and algebraic skills from a traditional perspective. There are far fewer books that are completely designed to teach those skills from a literacy standpoint, and precious few that require students to develop the important skills and concepts on their own. The process of teaching students to read and write has always been treated as more self-guided than math education: the teacher provides some guidance and framework, of course, but ultimately the responsibility lies with the student to take those first few fragile steps on their own. In quantitative literacy, the goal is to follow this path, providing guidance and framework, but not simply demonstrating procedures, and *never* accepting mimicry of solved problems with no contextual framework.

The teaching of mathematics to college students has undergone some dramatic and long-overdue changes in the last five to ten years, mostly in the classes taken by students in non-STEM programs. The traditional curriculum has been need-based: what skills do students need for the next class? We're advocates for thinking about the type of thinking that is *useful* for students, rather than focusing on what students *need* to do. That paradigm shift allows us to move away from skill and drill learning and focus on critical thinking and problem-solving skills that form the backbone of this course.

So what type of student is ready for success in quantitative literacy? We don't feel like there's a single answer to that question. A certain comfort level with arithmetic skills is certainly helpful: students that are comfortable working with numbers in many different forms and settings will have a leg up, and experience working with algebraic formulas is always helpful. Beyond that, it's more a question of willingness to learn, persistence in the face of struggle, and openness to conceptual thinking that are the key prerequisites. Each lesson begins with a short list of prep skills problems designed to let both students and instructors know what core skills are called upon within that lesson.

For some students, quantitative literacy will serve as a terminal course, at least from a math standpoint: we've yet to have reports of this material actually being fatal for anyone. Others will move on to a course in statistics, or maybe further studies in liberal arts math or quantitative reasoning. But more important than the destination is the journey: the thinking skills honed in this course will serve students well across the curriculum and beyond, and if all goes well will finally answer the eternal question for many math students: "Why are we doing this?"

MATH IN OUR WORLD: A QUANTITATIVE LITERACY APPROACH
Published by McGraw-Hill Education, 2 Penn Plaza, New York, NY 10121.

Some ancillaries, including electronic and print components, may not be available to customers outside the United States.

This book is printed on acid-free paper.

1 2 3 4 5 6 7 8 9 DOW 21 20 19 18 17 16

ISBN 978-1-259-82796-9
MHID 1-259-82796-8

ISBN 978-1-259-82788-4 (Annotated Instructor's Edition)
MHID 1-259-82788-7

Chief Product Officer, SVP Products & Markets: *G.Scott Virkler*
Vice President, General Manager, Products & Markets: *Marty Lange*
Managing Director: *Ryan Blankenship*
Brand Manager: *Adam Rooke*
Director, Product Development: *Rose Koos*
Product Developer: *Vincent Bradshaw*
Director of Marketing: *Sally Yagan*
Marketing Coordinator: *Annie Clarke*
Director of Digital Content Development: *Robert Brieler*
Associate Digital Product Analyst: *Adam Fischer*
Director, Content Design & Delivery: *Linda Avenarius*
Program Manager: *Lora Neyens*
Content Project Manager: *Peggy J. Selle*
Assessment Project Manager: *Emily Windelborn*
Buyer: *Sandy Ludovissy*
Design: *Tara McDermott*
Content Licensing Specialist Photo: *Carrie Burger*
Content Licensing Specialist Text: *Lori Slattery*
Cover Image: © *Pixdeluxe/Getty Images (farmers);* © *Fuse/Getty Images (chopping vegetables);*
© *Lisovskaya Natalia/Getty Images (plate)*
Compositor: *SPI-Global*
Typeface: *10/12 Stix Mathjax MAIN*
Printer: *R. R. Donnelley*
All credits appearing on page or at the end of the book are considered to be an extension of the copyright page.

Library of Congress Cataloging-in-Publication Data

Names: Sobecki, Dave. | Mercer, Brian A.
Title: Quantitative literacy in our world / Dave Sobecki, professor of mathematics, Miami University Hamilton, Brian Mercer, professor of mathematics, Parkland College.
Description: New York, NY : McGraw-Hill Education, [2017]
Identifiers: LCCN 2016017048| ISBN 9781259827969 (alk. paper) | ISBN 9781259827884 (annotated instructor edition)
Subjects: LCSH: Mathematics—Textbooks. | Mathematics—Study and teaching (Higher) | Mathematical statistics—Study and teaching (Higher)
Classification: LCC QA135.6 .S5658 2017 | DDC 510.71/1—dc23 LC record available at https://lccn.loc.gov/2016017048

mheducation.com/highered

About the Authors

Dave Sobecki

Dave was born and raised in Cleveland, and in spite of starting college with a major in creative writing, he somehow ended up with a Doctorate degree in math. Go figure. Dave is a tenured professor of math at the Hamilton campus of Miami University in southwest Ohio. He has won a number of teaching awards in his career, and has written or co-authored either nine or sixteen books, depending on how you count them. He's also working on his first novel. When not teaching or writing, Dave's passions include Ohio State football, Cleveland Indians baseball, heavy metal music, travel, golf, collecting fine art, and most importantly spending time with his wife Cat and over-attached retrievers, Macleod and Tessa.

Courtesy of Dave Sobecki

Brian Mercer

Brian is a tenured professor at Parkland College in Champaign, Illinois, where he has taught developmental and transfer math courses for 18 years. He began writing in 1999, and has currently co-authored 6 textbooks, with others in the planning stages. Outside of the classroom and away from the computer, Brian is kept educated, entertained, and ever-busy by his wonderful wife Nikki, and their two children, Charlotte, 9, and Jake, 8. He is an avid St. Louis Cardinals fan and enjoys playing softball and golf in the summertime with colleagues and friends.

Courtesy of Brian Mercer

Table of Contents

Unit 1: THE IMPORTANCE OF BEING NUMERATE

Unit 2: CONTENDING WITH CHANGE

Unit 3: MANAGING YOUR MONEY

Unit 4: STATISTICALLY SPEAKING

A Letter to Students

This book has the term "Quantitative Literacy" in the title, so we probably ought to begin by describing what that means. But nothing is for free in this course, so before we do that, we're going to ask you to think about an answer to this question: Why are you taking this class? For that matter, why are you in college at all? It would be great if "To get an education" was your answer, or at the very least crossed your mind. It might be even better if you thought "To get smarter," that should be the goal of everyone that's going to school, really.

So can you actually get smarter? OF COURSE YOU CAN. You almost certainly know that you can get stronger and more fit than you currently are by exercising your body. Then why in the world wouldn't you think you can get smarter by exercising your brain? You can, but not if you occupy a chair and expect your professor to just magically fill your head with math. That's not mental exercise. True mental exercise requires challenging yourself to think, to solve problems, to make connections, and to explain your understanding of concepts. Which brings us back to our part of the bargain, a description of what quantitative literacy means.

At its core, this course is about exercising your brain by challenging you to really LEARN, not just to memorize formulas or mimic step-by-step procedures. When you hear the word "literacy," you probably think about reading and writing, and that's a good thing—you know that these are the basic skills that anyone should have if they're going to have a shot at being successful in twenty-first-century society. By including the phrases "Math in Our World" and "Quantitative Literacy" in the title of this book, we're simply pointing out the fact that the skills you'll develop in this course are also necessary for being successful in society.

Certainly being comfortable with numbers is important in a monetary society. But it goes beyond that: Someone once told us that nobody is going to pay people to solve problems that have already been solved, and that kind of became a framework for this course. We gathered as much information as we could about what employers want college graduates to be able to do these days, and three things consistently jumped out at us: being able to work effectively in groups, being able to solve problems and figure stuff out without always being told what to do, and being able to use technology well. It's those three things that became the key principles of this book.

The Approach: What truly makes this course different

The thing that makes this a course in quantitative literacy, rather than a survey of mathematical topics, is the way the material is presented, and more importantly, how we expect you to interact with the material. If you're dead-set on sitting in the back of the room and being talked at for an hour, you'll find out pretty quickly that you're in the wrong place: Active participation is required. You won't find a bunch of solved examples in this book, like you're probably used to seeing, what you will find is a series of activities designed to encourage (and by "encourage" we mean "force") you to take responsibility for your own learning. Discovery learning and productive struggle are the key phrases that describe our philosophy.

Discovery learning is the process of learning new ideas on your own, rather than having them presented to you. That doesn't mean that you're expected to "teach this course to yourself." Your professor is by far the most valuable resource you'll have on this journey. But his or her job is to help you to learn, not to show you exactly what to do. Going back to the literacy analogy, you can't teach someone to read by just reading to them: At some point you have to guide them through reading on their own, and that's the role of your professor in this course.

Productive struggle is described really well by the exercise analogy from earlier: The best way to really learn something, and to make your brain stronger in the process, is to have to work at it. If you get every question in this book right on the first try, then we've failed as writers, to be honest. Life can be a hard place, especially when you're responsible for your own care and feeding, and we can promise you that you will get NOWHERE in this world if you give up every time you encounter some resistance. In our view, the single most valuable skill you'll practice in this course is GRIT: the ability to overcome obstacles and keep trying until you reach a goal.

Because working effectively in groups is so important, the book has been designed as a series of in-class activities for you to work through in a group setting. There are two things you need to do in order to maximize your chance for success in this course: be willing to put in a lot of work, and be willing to learn with and learn from your group mates, as well as to help teach them from time to time. If we were pure evil and decided to kidnap you and dump you out in the wilderness somewhere, you surely would appreciate not being alone, right? As you wander the path toward an adult education, having peers there with you should be comforting, and should be a resource that you take advantage of whenever possible.

Once class is over, there will be a lot of work you're expected to do on your own, in a variety of different settings. If you turn ahead to one of the pages in the book marked "Portfolio," you can see the different categories of homework, and no doubt your professor will talk about them on the first day of class. If you keep thinking back to the exercise analogy, you'll have a much easier time getting motivated to keep up with the homework. Why are you in this class? To get smarter. That will happen if you work hard toward the goal, and are never, ever afraid to think on your own. You were smart enough to make the decision to further your education in college, so you're smart enough to think, learn and succeed in quantitative literacy. Set a goal, work toward it, and believe in yourself. And don't be afraid to drop us a line if you have questions about the book, suggestions for improvement, or just want to share your thoughts on how the course is going.

davesobecki@gmail.com
bmercer51@gmail.com

The Content

The backbone of any well-designed math course is content, so let's talk about the content that we've chosen as the vehicle for becoming quantitatively literate.

Unit 1: Numeracy

Trying to become quantitatively literate without first being comfortable with basic numeracy is like trying to become language literate without knowing the alphabet. So the first leg of our journey runs through a series of topics that are designed to foster solid numeracy skills in contexts that help students to grasp the significance of those skills in a wide variety of applied settings. While we're practicing skills that will be used in the remainder of the course, we're not doing so just to prepare for later material, but also setting up the frame of mind that defines a quantitative literacy course.

Unit 2: Algebraic Thinking

The single thing that distinguishes algebra from arithmetic is working with quantities that can vary (or as I like to call them, "variables"), so in Unit 2 we approach the importance of studying changing quantities in our world. By beginning with applied settings throughout, and building up to the use of symbols to represent variable quantities, we demonstrate the historically accurate view that algebra wasn't developed as an abstract subject and applied at some later date, but rather was a natural outgrowth of applying numerical thinking to a world where change is inevitable.

Unit 3: Consumer Math

Did we mention that we live in a monetary society? Consumer math is kind of the low-hanging fruit of quantitative literacy content: everyone who's ever spent their own money on anything has a natural understanding of how financial literacy is important. The focus of this unit isn't "here's how to do financial calculations:" it's "let's understand the concepts and situations that lead to important financial considerations." While we absolutely will study formulas that calculate interest, loan payments, and the like, we certainly will not just provide the formulas, ask students to plug a bunch of numbers in, and call that education. Instead, students are asked to think like consumers, and develop a true understanding of the math that serves us well in navigating the rocky financial waters of modern society.

Unit 4: Probability and Statistics

We wrap the course up by focusing on data gathered from all around us, studying the likelihood of events occurring, and different methods of gathering, organizing, and interpreting data. Anyone can find the average (whatever that means) of a list of numbers, but that's a hollow exercise without the ability to interpret the story told by that statistic. While we don't feel that basic quantitative literacy requires high-level statistical analysis, we do feel that our society will be much the better if everyone has experience with sifting through the mountain of information they're exposed to every day, and separating the gold nuggets from the cow pies, if you will. So a major focus of, and the culmination of the unit, is recognizing ways that statistics can be manipulated to tell a story that has at best a passing resemblance to the truth. While politicians and advertisers may not appreciate these efforts, we think that students interested in getting an education will.

One of the best things about a course of this nature is that the material is not completely linear. It's perfectly possible for instructors to choose the content that's most relevant for his or her students, and reordering of some of the topics is reasonable as well. Unit 1 was designed to set the tone for the course and refresh background skills, but depending on the prerequisite level of your students, it's pretty much up to you to decide on how to arrange the material.

The Approach: What truly makes this course different

Maybe you've noticed that the topic list for this book isn't exactly revolutionary: most if not all of the topics can be found scattered through a traditional math curriculum. The thing that makes this a course in quantitative literacy, rather than a survey of mathematical topics, is the way the material is presented, and more importantly, how we expect students to interact with the material. Those who are dead-set on sitting in the back of the room and being talked at for an hour will find out pretty quickly that they're in the wrong place: active participation is required. You won't find any traditional solved examples in this book. What you will find is a series of activities designed to encourage (and by "encourage" we mean "force") students to take responsibility for their own learning. Discovery learning and productive struggle are the key phrases that describe our philosophy.

Students will always learn and retain ideas more efficiently when they develop their own understanding of those ideas. And the importance of persistence and grit in the face of intellectual difficulty can't possibly be overstated. Nobody is going to pay someone to solve problems that they can solve on their own, so a willingness and an ability to work around the edges of a problem and persist to a solution is, in our opinion, the most valuable part of this course.

How the instructor chooses to facilitate that productive struggle is up to them. Some prefer to have students work through every lesson in groups, in which case the instructor becomes a consultant, roaming the class, providing their expertise on an as-needed basis, and occasionally bringing the class together for mini-lectures and summary of key points. When using this approach, there are two key things to keep in mind. First, the phrase "talk to each other, people" is particularly useful, as students tend to think group work in math means that everyone answers questions on their own, then compares their answers. That pretty much misses the point in this course. Second, even if you're a world-class lecturer, and better than anyone you know at showing students how to do problems, sometimes you just have to keep quiet and let them figure it out among themselves. When you see a group of students have the proverbial "light bulb going on over the head" moment, you'll be very glad you let that happen.

Others will prefer to use the activities within the book as in-class worksheets, with the instructor leading the class through discussions of each question. We've found that in some cases, those who prefer a more traditional approach will slowly transition into more group-oriented classes, which is fine: the important thing is to teach in a method that works for you. Interestingly, each of these approaches has been used successfully in an online setting, and the resources provided as part of our program are there to help.

The Components of Each Lesson

Every lesson is divided into distinct portions, so let's take a look at how each is intended to be used.

Group and Class

The material designed to be covered in a typical class is divided into two parts: Group and Class. The order varies from lesson to lesson depending on the content. Both Group and Class can be worked in the group-based format we discussed earlier. The difference between the two is level of accessibility. The portion labeled as Group is the material we feel is most accessible, and most effective, for students to discover while working with their group mates. Class, on the other hand, is the content we feel that students are likely to need a little more guidance on. Sometimes this is due to a higher level of math skills needed. Other times it's about concepts that are a bit more difficult to understand, or applications that require some higher-level thinking. It's completely up to the instructor to decide how to distinguish between these two portions in class.

Portfolio

This is the portion of each lesson that is meant for out-of-class work. Notice that once you get to the portfolio, there's a spot at the top of the page for students to write their name, as that portion of the lesson is meant to be turned in. First, you'll see a **Technology** assignment, in which we ask students to make meaningful use of spreadsheets, graphing calculators, or Internet research to enhance understanding, putting the computational burden on technology so that students are freer to focus on interpretation. Next, the **Online Practice** portion is where you'll find auto-graded problems **in Connect Hosted by ALEKS.** All of the skills problems are found there, as well as many additional applied problems. If you're not familiar with Connect Hosted by ALEKS, or CHBA as we say, we encourage you to take it for a test drive. We're really proud of the way this system was designed to encourage meaningful problem solving and discourage mimicry and system-gaming. There's a **Reflections** section, where students are asked to write their thoughts on the most important concepts in the lesson. This also provides a place where students can ask the instructor conceptual questions. We've also included a **Looking Ahead** feature, so that students know where to look each day for any work that must be done in preparation for the next lesson. The most important part of the Portfolio is so important that it actually gets its own segment of the lesson—Applications.

Applications

The cornerstone of quantitative literacy is also the cornerstone of each lesson's homework: Applications. That may bring to mind the classic "word problems" that appear in every math textbook written since . . . well, since forever. But for us, the Applications are much more. Certainly the main point is applying things learned in the lesson, but we also focus on conceptual understanding and interpretation of mathematical results. In an ideal world, the Applications portion would always be handed in on paper and hand graded, but we know this isn't always feasible for busy instructors. So we've also included versions of the Applications in Connect Math Hosted by ALEKS allows you to choose a machine-graded option without sacrificing too much of the interpretation that makes the Applications so important

Implementation, Training, and Support

Significant changes to a curriculum or course create a ripple effect on other activities that need to be accomplished for a successful implementation. We recognize that those activities can be the straw that breaks the camel's back. That's why we've put together a collection of resources to help you and your institution have a successful implementation of our materials in your Quantitative Literacy course. We have the following editable documents available for you to customize for your institution in Connect Math Hosted by ALEKS:

- Course description—for use in course catalogs/websites
- Sample course syllabi
- Transition guide from a traditional "Survey of Mathematics" course to a Quantitative Literacy Course
- Training Guide and videos—for faculty and adjuncts
- Transferability information—for establishing transfer guidelines between institutions
- Group projects designed to synthesize key ideas from within each unit

Of course, you also have us to help you through this process.

Acknowledgments

So many people were involved in bringing this project to life that it's tempting to just thank the whole world and move on, but some folks deserve special recognition. We sincerely hope we didn't omit anyone in this category. If we did, blame it on Dave and he owes you some beverages of your choice.

We spent the better part of two years traveling all over the country visiting conferences and campuses, and conducting focus groups and symposia. We thank everyone who attended and shared their ideas on improving college math for non-STEM students. You are the main reason this project exists.

Elizabeth Betzel
Columbus State Community College
Colleen Bye
Utah Valley University
Erin Cooke
Perimeter College at Georgia State University
Steven Cosares
LaGuardia Community College
Jodi Cotten-Konsur
Westchester Community College
Lisa Creighton
Central Piedmont Community College
Tim Delworth
Purdue University, West Lafayette
Raina J. Eckhardt
Manchester Community College
Jennifer Fitzgerald
Northern Essex Community College
Debra Hall
James Madison University
Julie Hartzler
Des Moines Area Community College
Lori Heymans
Northern Essex Community College
Barbie Hoag
Oakland Community College
Heather Howington
University of North Georgia
Gretta Johnson
Amarillo College
Rachel King
Northern Essex Community College
Kurt Klopstein
Bakersfield College
Bernadette Kocyba
J. Sargeant Reynolds Community College
Kelly Kohlmetz
University of Wisconsin - Milwaukee
Chris Mansfield
Durham Technical Community College
Susan McLoughlin
Union County College
Linda Murphy
Northern Essex Community College
Jonathan Oaks
Macomb Community College
Rebecca Rose
Northern Essex Community College
Robin Rufatto
Ball State University
Mike Sieve
Ridgewater College
Jim Sullivan
Northern Essex Community College
Prudence York-Hammons
Temple College

Our deepest appreciation goes to all of our friends in North Carolina who inspired and supported this project. In particular, Paula Savich from Mayland Community College was the first person who suggested a version of Math in Our World in a group project–based format. Chris Mansfield from Durham Tech went above and beyond, working with us on developing a content plan. Several instructors provided feedback on the various drafts of the manuscript.

Kari Arnoldsen
Snow College
Jordan Bertke
Central Piedmont Community College
Cindy Box
Perimeter College at Georgia State University
Kyle Carter
The University of West Georgia
Erin Cooke
Perimeter College at Georgia State University
Wendy Davidson
Perimeter College at Georgia State University

Maria DeLucia
Middlesex County College

Holly Dickin
Ball State University

Julie Gunkelman
Oakland Community College

Debbie Hanus
Brookhaven College

Barbie Hoag
Oakland Community College

Kelly Kohlmetz
University of Wisconsin- Milwaukee

Crystal Lorch
Ball State University

Vikki Maurer
Linn-Benton Community College

Kevin McLeod
University of Wisconsin- Milwaukee

LeAnn Neel-Romine
Ball State University

Eric Neumann
Community College of Philadelphia

Kathy Pinzon
Georgia Gwinnett College

Linda K. Schmidt
Spartanburg Community College

Jennifer Stanley
West Georgia State

Donna Toll
Ball State University

Erin Wilding-Martin
Parkland College

Janet Yi
Ball State University

Prudence York-Hammons
Temple College

We were very lucky to put together a board of advisors, listed below. The board's thoughtful discussions and detailed content reviews helped to shape our somewhat rough-hewn early attempts into the shiny product you're looking at now.

Nikki Armstrong
Southwestern Oregon Community College

Jordan Bertke
Central Piedmont Community College

Cindy Box
Perimeter College at Georgia State University

Erin Cooke
Perimeter College at Georgia State University

Wendy Davidson
Perimeter College at Georgia State University

Debbie Hanus
Brookhaven College

Barbie Hoag
Oakland Community College

Julie Gunkelman
Oakland Community College

Kelly Kohlmetz
University of Wisconsin–Milwaukee

Vikki Maurer
Linn-Benton Community College

Kevin McLeod
University of Wisconsin–Milwaukee

Lori Murphy
Rockingham Community College

Dawn Peterson
Itawamba Community College

Erin Wilding-Martin
Parkland College

We also had the rare opportunity to have the following courageous instructors class testing the book as we wrote it: Deb Hanus, Nikki Armstrong, Barbie Hoag, and Marcus Szwankowski (Front Range Community College).

At McGraw-Hill Higher Education, we thank: Managing Director Ryan Blankenship for supporting us as we sailed off into uncharted waters, and Brand Manager Adam Rooke for piloting the ship; Product developers extraordinaire, Vincent Bradshaw and Elizabeth O'Brien, who provided so much more than what their job description calls for; Tara McDermott and her design team, who continue to amaze with their fresh ideas for visual appeal; Director of Digital Content Rob Brieler (we ask for new features, and he says "I think we can do that"); Marketing Director Sally Yagan, who will NEVER settle for "good enough"; accuracy checker Dawn Peterson—yet another member of the team that did far more than what was expected; copy editors Pat Steele and Julie Kennedy, who never make fun of our silly mistakes; and project manager Peggy Selle who kept it all together.

Supplements

Connect Math® Hosted by ALEKS

Connect Math Hosted by ALEKS Corp. is an exciting assignment and assessment ehomework platform. Starting with an easily viewable, intuitive interface, students will be able to access key information, complete homework assignments, and utilize an integrated, media rich eBook.

SmartBook

SmartBook is the first and only adaptive reading experience available for the higher education market. Powered by the intelligent and adaptive LearnSmart engine, SmartBook facilitates the reading process by identifying what content a student knows and doesn't know. As a student reads, the material continuously adapts to ensure the student is focused on the content he or she needs the most to close specific knowledge gaps.

CREATE

With **McGraw-Hill Create™**, you can easily rearrange chapters, combine material from other content sources, and quickly upload content you have written such as your course syllabus or teaching notes. Find the content you need in Create by searching through thousands of leading McGraw-Hill textbooks. Arrange your book to fit your teaching style. Create even allows you to personalize your book's appearance by selecting the cover and adding your name, school, and course information. Assemble a Create book, and you'll receive a complimentary print review copy in 3—5 business days or a complimentary electronic review copy (eComp) via email in minutes. Go to www.mcgrawhillcreate.com today and experience how McGraw-Hill Create™ empowers you to teach your students your way.

Instructor's Testing and Resource Online

This computerized test bank, available online to adopting instructors, utilizes TestGen® cross-platform test generation software to quickly and easily create customized exams. Using hundreds of test items taken directly from the text, TestGen allows rapid test creation and flexibility for instructors to create their own questions from scratch with ability to randomize number values. Powerful search and sort functions help quickly locate questions and arrange them in any order, and built-in mathematical templates let instructors insert stylized text, symbols, graphics, and equations directly into questions without need for a separate equation editor.

Student and Instructor Solution Manuals

provide comprehensive solutions to all exercises in the text.

Lecture and Exercise Videos

introduce concepts, definitions, theorems, formulas, and problem-solving procedures to help students comprehend topics.

Unit 1
The Importance of Being Numerate

© Creatas/PunchStock RF

Outline

Lesson 1: Giving 110 Percent (Review of Percents)
Lesson 2: Where Does the Time Go? (Using Percentages and Pie Charts)
Lesson 3: It's All About Style (Interpreting and Drawing Bar Graphs)
Lesson 4: Take a Guess! (Estimation and Number Sense)
Lesson 5: Of Planes, Boats, Doll Houses, and Dr. Evil (Dimensional Analysis)
Lesson 6: Relatively Speaking (Interpreting Relative Difference/Relative Error)
Lesson 7: Attraction and Melted Chocolate (Scientific Notation)
Lesson 8: Portion Control (Apportionment)

Lesson 1-1 Prep Skills

This is a short review of skills that will come in handy in the next lesson. In each case, answer the question, then rate your confidence level by checking one of the boxes. If you feel like you're struggling with these skills, consult the online resources provided by your instructor for extra practice.

I KNOW IT | THINK SO | UNSURE | NO IDEA

SKILL 1: REDUCING FRACTIONS

1. Reduce each fraction to lowest terms.

   $\frac{6}{10}$ $\frac{24}{30}$ $\frac{51}{21}$ $\frac{11}{24}$

2. Reduce each fraction to lowest terms.

   $\frac{60}{100}$ $\frac{24}{100}$ $\frac{98}{100}$ $\frac{33}{100}$

SKILL 2: WRITE FRACTIONS IN DECIMAL FORM

3. Write the decimal form of each fraction.

   $\frac{12}{100}$ $\frac{9}{30}$ $\frac{11}{8}$ $\frac{11}{3}$

SKILL 3: SOLVE EQUATIONS USING DIVISION

4. Solve: $12x = 84$

5. Solve: $0.3y = 210$

6. Solve: $0.125x = 47.99$

Lesson 1-1 Giving 110 Percent (Review of Percents)

LEARNING OBJECTIVES

- ☐ 1. Perform conversions and calculations involving percents.
- ☐ 2. Find percent increase or decrease.
- ☐ 3. Solve problems using percents.
- ☐ 4. Evaluate the legitimacy of claims based on percents.

I think I'm a kind of a person who works hard at whatever I do, literally from being a waitress to being on television. I always try to give 110 percent to whatever it is I'm doing.
—Teri Hatcher

© RubberBall Productions RF

Coaches are fond of insisting that their players give 110% effort. Players that are good at math will sometimes counter with the fact that it's impossible to give more than you have, which by the way sounds like a good idea if you want to run laps for about a week. Can you really have 110% of something? Since percents play an important role in many, many different areas where studying numbers is useful, everyone that wants to be successful in our society should be armed with a solid understanding of percents: what they mean, how to compute and use them, and how to interpret information that they're providing. In this lesson, we'll review a lot of useful information about percents, information that will come in handy throughout the rest of the course. Pay attention, or you'll be doing pushups until you drop.

0. After reading the opening paragraph, what do you think the main topic of this lesson will be?

1-1 Class

Surely you're familiar with the word "percent" but have you ever thought about what it literally means? "Percent" can be translated literally as "per hundred." So 1% = 1/100. For example, the International Mass Retail Association reported that 20% of adults plan to buy Valentine's Day cards next year. Of those who plan to buy them, 36% of women and 26% of men plan to buy romantic cards. This tells us that 20 out of every 100 adults plan to buy a card. Also, 36 out of 100 women who buy a card will choose a romantic one, and 26 out of 100 men will do so.

Percent Conversions

To work with percents in calculations, we will need to convert them to either decimal or fractional form. But to interpret the answers to calculations, we will want to convert them back into percents.

1. Using the fact that "percent" means "per hundred," write 84% and 9% as fractions with denominator 100.

$84\% = \frac{84}{100}$ $9\% = \frac{9}{100}$

This simple little question is a great opportunity to set the tone for the course. Rather than just memorizing formulas and procedures, like working with percents, we're going to really understand what it all means.

Here's a simple rule that can be learned from Question 1:

Converting a Percent to a Fraction

To convert a percent to a fraction, drop the percent sign and use the percent number as the numerator of a fraction whose denominator is 100.

If you noticed that the fraction in the first part of your answer to Question 1 could be reduced, good job! When we're working with percents in fraction form, many times it will be easier if the fraction is reduced to lowest terms.

2. Convert each percent to a fraction in lowest terms.

a. 42%

$42\% = \frac{42}{100} = \frac{21}{50}$

b. $37\frac{1}{2}\%$ (Hint: After initially writing as a fraction, multiply by $\frac{2}{2}$ so that the numerator and denominator of your fraction are both whole numbers.)

$37\frac{1}{2}\% = \frac{37.5}{100} \cdot \frac{2}{2} = \frac{75}{200} = \frac{3}{8}$

Often, it's especially useful to write a percent in decimal form when doing calculations, so let's see how that works, using what we already know about fractions.

3. We know that we can write 12% as a fraction in the form $\frac{12}{100}$, and 72% in the form $\frac{72}{100}$. Use a calculator to perform the division in each case, writing the original percent in decimal form.

$12\% = 0.12$

$72\% = 0.72$

4. Use the result of Question 3 to fill in the blank in our next rule for converting percents:

Converting a Percent to a Decimal

To convert a percent to a decimal, drop the percent sign and move the decimal point __2__ places to the __left__.

5. Write each percent in decimal form.

a. 41% 0.41 **b.** 6.7% 0.067 **c.** 120% 1.20

Sometimes, interpreting the story told by a percentage will require a conversion in the other direction: starting in decimal form, and converting to percent form. Not surprisingly, we accomplish this by doing exactly the opposite of the steps we did for converting from percent to decimal form.

6. The opposite of moving the decimal two places to the left is moving the decimal __2 places to the right__. The opposite of dropping the percent sign is __inserting the % sign__.

7. Write a procedure for converting a decimal to percent form, then use it to convert 0.34, 0.07, and 1.658 into percent form.

Move the decimal point two places to the right, then insert a percent sign.

0.34 = 34%

0.07 = 7%

1.658 = 165.8%

This is a key idea in the course: letting students discover and write rules on their own (with some leading) rather than just telling them. Let it happen!

Math Note

Part c of Question 5 shows that our conversion rules work just fine for percentages greater than 100. We'll study the meaning of such a percentage in the Group portion of this lesson.

Did You Get It

Try this problem to see if you understand the concepts we just studied. The answer can be found at the bottom of the portfolio page.

1. Perform each conversion.

   a. 47.3% to decimal b. 0.24 to percent c. 80% to fraction in lowest terms

In the most common calculations involving percents, the goal is to find a percentage of some quantity. To understand how to do so, consider the following example. You probably know that 50% of 10 is 5. Let's rewrite that statement, then turn it into a calculation:

50% of 10 is 5

$0.5 \times 10 = 5$

When writing a percentage statement in symbols, the word "of " becomes multiplication, and the word "is" becomes an equal sign. Also, we need to change the percent into decimal or fractional form. Let's look at some sample calculations using this idea.

A restaurant was running a fundraiser tonight for the local food bank in which 30% of every purchase was donated for food relief. If my wife and I spent \$24.59 at the restaurant, how much was donated?

8. Fill in the blank to describe the calculation described above: ___30___% of ___\$24.59___ is what amount?

9. What was the amount donated?

$0.30 \cdot \$24.59 = \$7.38$

The same idea can be used for other calculations involving percentages. Like this one: A closeout store is advertising leather jackets at 70% off the original price, which means the sale price is 30% of the original price. If one jacket has a sale price of \$59, how much did it cost originally?

10. Fill in the blanks to describe the calculation described above: ___30___% of what amount is ___\$59___?

11. Based on your answer to Question 10, set up and solve an equation to find the original price of the jacket.

$0.30x = \$59 \Rightarrow x = \$59/0.30 = \$196.67$

Guess what? The same idea can be used to find a percentage, but it's more straightforward to use division. For example, of the 32 teams in the National Football League, 12 make the playoffs each year. What percentage of teams make the playoffs?

12. Divide the number of teams that make the playoffs by the total number of teams, and write your answer in decimal form.

$\frac{12}{32} = 0.375$

13. Based on your answer to Question 12, what percentage of teams make the playoffs?

37.5%

Did You Get It

Try this problem to see if you understand the concepts we just studied. The answer can be found at the bottom of the portfolio page.

2. Calculate each amount involving percents.
   a. Find 38% of 400.
   b. $17.49 is 20% of what amount?
   c. What percentage of 42 is 15?

In this book, youll notice that we dont usually specify a number of decimal places to round to, like in part c here. We want students to think about what might be a reasonable amount to round to.

The next page marks the first time students will formally introduce themselves to their groups and work on the Group portion of a lesson. Set the tone for the class by walking around, asking the groups questions about their work, and encouraging groups that are not working together. Make sure theyre not rushing through the problems without really thinking about them.

1-1 Group

1. Numerous studies have shown that one of the best ways to do better in college classes is to study in pairs or groups. To help you get started, if you feel comfortable sharing contact information, exchange the information in the table below. The group you're in now will be your small group for the first unit of this course. When you get used to meeting in class, you'll likely find that meeting outside of class to study and work on homework is a good idea as well, so include some study times that would be convenient for you to meet.

Name	Phone Number	Email	Available times

2. Let's look further into the idea of "giving 110% effort." First, explain why a lot of people feel like saying that doesn't really make sense.

110% is more than the total that you have to give. If giving 100% is giving all of the effort you possibly can, then giving 110% effort is impossible.

3. Convert 110% into decimal form, then multiply the result by 200 to find what 110% of 200 is.

$1.10 \cdot 200 = 220$

4. Suppose that in a normal half hour of practice, a tennis player expends 280 calories of energy. How many calories would she need to expend if she were to "give 110%" effort?

$1.10 \cdot 280 = 308$ *calories*

Did You Get It

Try this problem to see if you understand the concepts we just studied. The answer can be found at the bottom of the portfolio page.

3. A real estate developer is looking for investors to help fund a new condo complex. He promises to pay back 145% of the original investment in four years. If you're interested in investing $30,000 and the developer follows through on his promise, how much money will you get back in four years?

Now let's go shopping. If I had a dollar for every time someone has asked me to calculate a discount price for them when a sign advertises a certain percent off, I'd be able to buy us both a very handsome gift indeed. In fact, shopping is one of the most common ways that people running loose in society are likely to encounter percents.

My wife is an absolute wizard with Macy's coupons: as a high volume shopper there (sigh) she gets quite a few coupons in the mail. On one recent mall outing, she found a lovely red strapless number with a selling price of $79. She had two coupons to choose from: one offers $15 off any purchase of $50 or more, the other offers 20% off any item. Which was the better choice?

Courtesy of David Sobecki

5. Find 20% of $79.

$0.2 \cdot \$79 = \$15.80$

6. Keeping in mind that you just found the amount the price would be discounted, which coupon should Cat have used? Explain.

The 20% off would save an extra $0.80.

7. The sales tax in our county is 6.5%. Find the amount of tax on the dress purchase (after applying the coupon that you chose), and the total cost including tax.

Sale price $= \$79 - \$15.80 = \$63.20$

Sales tax $= 0.065 \cdot \$63.20 = \$4.11$

Cost including tax $= \$63.20 + 4.11 = \$67.31$

Did You Get It

Try this problem to see if you understand the concepts we just studied. The answer can be found at the bottom of the portfolio page.

4. A refrigerator has a regular price of $1,149 at Home Depot, and $1,219 at Lowe's. Home Depot is offering 15% off, but Lowe's offers $200 off if you put the purchase on store credit. Which is the better option?

8. It's common for stores to advertise a certain percentage off, then say "discount is reflected in the marked price," leaving you to trust that you're getting the percent off that they claimed. Suppose that a refrigerator is listed as 30% off, with a sale price of $749. What would that make the original selling price? (Look back at Question 11 in the Class section if you need help.)

$0.70x = 749 \Rightarrow x = 749/0.70 = \$1{,}070$

9. A large latte at Shaky's Caffeine Shack usually sells for the low low price of just $3.50, but today, Shaky has gone goofy and is offering it for just $3.00. Wow! Why not just give them away? Is this a better offer than the 20% off the coffee shop across the street is offering?

$0.50/\$3.50 \approx 0.143 = 14.3\%$

Nope, the 20% off is better, if by "better" you mean "greater percent off". Students might say that you'd need to know the original price across the street, which would be great!

The percentage you found in Question 9 is called the **percent decrease.** It's the amount that some quantity is decreased as a percentage of the original amount. Similarly, **percent increase** is the amount of increase as a percentage of the original amount. That sounds like a good opportunity for a colored formula box.

Computing Percent Increase and Decrease

When some quantity gets larger, we calculate the percent increase using the formula

$$\text{Percent Increase} = \frac{\text{Amount of Increase}}{\text{Original amount}} = \frac{\text{New amount} - \text{Original amount}}{\text{Original amount}}$$

If the quantity gets smaller, we calculate the percent decrease using the formula

$$\text{Percent Decrease} = \frac{\text{Amount of decrease}}{\text{Original amount}} = \frac{\text{New amount} - \text{Original amount}}{\text{Original amount}}$$

In each case, the result is a percentage in decimal form.

10. On July 24, 2009, the federal minimum wage was raised from \$6.55 to \$7.25. Find the percent increase.

$$\frac{7.25 - 6.55}{6.55} \approx 0.107 = 10.7\%$$

Did You Get It

Try this problem to see if you understand the concepts we just studied. The answer can be found at the bottom of the portfolio page.

5. In an effort to show solidarity with his workers, a CEO offers to take a \$300,000 per year pay cut. Last year, his salary was \$11 million dollars. Find the percent decrease in pay.

Using formulas in Excel, we can set up a spreadsheet that will calculate the percent increase or decrease for any change in a quantity, like the sample below.

D3 fx =C3/A3

	A	B	C	D	E
1	Original Amount	New Amount	Amount of Change	% Change (decimal form)	% Change (Percent form)
2	3.5	3	-0.5	-0.142857143	-14.3%
3	6.55	7.25	0.7	0.106870229	10.7%

Source: Microsoft

The entry shown in cell D3 is a sample of an **Excel formula.** This is how Excel is programmed to perform calculations, like percent change, automatically. The = sign tells Excel that you're typing a formula. The formula in this case is C3/A3, which tells Excel to divide the number in cell C3 (the amount of change) by the number in cell A3 (the original amount).

11. Write the formulas that appear in cells C3 ad D2.

C3: = B3 − A3

D2: = C2/A2

Math Note

In Lesson 1-6, we'll study relative change, which is a useful topic closely related to percent increase/decrease.

Using Technology: Entering a Formula in Excel

To enter a formula in Excel:

1. Set up a separate column for each quantity involved in the calculation.
2. Identify the column where you want to store the result of the calculation.
3. In the first cell of that column, type = to indicate that you're entering a formula.
4. Type the calculation that you want completed after the = sign. When you want to include the number in a particular cell as part of the calculation, you can either type the column and row of that cell (A3, for example), or you can use the cursor to click on the cell and it will be entered into your formula automatically.
5. When you've completed the formula, press ENTER.

See the Lesson 1-1 video in class resources for further instruction.

We would never, ever accuse any advertisers of being deceptive (wink wink), but let's just say that it's not uncommon for percents to be used incorrectly in marketing materials. Let's look at one such case to close our review of percents.

12. A department store advertised that certain merchandise was reduced 25%. Also, an additional 10% discount card would be given to the first 200 people who entered the store on a specific day. The advertisement then stated that this amounted to a 35% reduction in the price of an item. Explain why the store marketing manager is either a liar, or bad at math. Show calculations to back up your explanation.

The additional 10% would be taken off the discounted price, not the original price. So it would not be an additional 10% off.

13. In promotional materials, a company that increased their work force from 1,000 to 1,500 workers claimed that they hired 50% more workers. A rival of theirs bragged about decreasing emissions from 1,500 to 1,000 cubic feet per day, claiming a 33% reduction. Which company is lying? Explain.

Neither. Let's calculate the percent change for each:

$$\frac{1{,}500 - 1{,}000}{1{,}000} = 0.5 = 50\% \qquad \frac{1{,}000 - 1{,}500}{1{,}500} \approx -0.33 = -33.3\%$$

Talk about how numbers can sometimes be spun to tell a certain story depending on the context.

1-1 Portfolio

Name ______________________________

Check each box when you've completed the task. Remember that your instructor will want you to turn in the portfolio pages you create.

If groups finish a lesson early, check over their work. Then think about encouraging them to start on their Portfolio assignments together in class.

Technology

1. ☐ Build a spreadsheet that calculates percent increase or decrease, like the one in the Group portion of this lesson. Your spreadsheet should look similar to the one in this lesson, and should use formulas with cell references to complete the calculations for amount of change and percent change. A template to help you get started is included in the online resources for this lesson.

On many of the Excel assignments, students will find it helpful if you project an example, to show them what their finished product should look like.

Online Practice

1. ☐ Include any written work from the online assignment along with any notes or questions about this lesson's content.

Applications

1. ☐ Complete the applications problems.

Responses to the Reflections questions can be used to determine which concepts students really understood, and whether they saw the main point of the lesson. This can help you decide which concepts you may need to clarify in a later class.

Reflections

Type a short answer to each question.

1. ☐ What is the literal translation of the word "percent"?
2. ☐ How can an understanding of percents be valuable to you as a shopper?
3. ☐ Do you think you can give 110% effort to doing well in this class? Explain.
4. ☐ Take another look at your answer to Question 0 at the beginning of this lesson. Would you change your answer now that you've completed the lesson? How would you summarize the topic of this lesson now?
5. ☐ What questions do you have about this lesson?

Looking Ahead

1. ☐ Complete the Lesson 1-2 Prep Skills and read the opening paragraph in Lesson 1–2 carefully, then answer Question 0 in preparation for that lesson.

Answers to "Did You Get It?"

1. a. 0.473 **b.** 24% **c.** $\frac{4}{5}$ **2. a.** 152 **b.** \$87.45 **c.** 35.7% **3.** \$43,500
4. \$976.65 at Home Depot vs. \$1,019 at Lowe's **5.** 2.7%

1-1 Applications

Name ______________________________

Many people probably think that those who are good at math always do exact calculations, but estimation is actually a really useful tool, especially when it comes to things like percent discount. Suppose that you're shopping on a rack that offers 40% off the lowest ticketed price, and you're interested in a sweater with a ticket price of $47.99.

1. Given that 40% is being subtracted from the ticket price, explain why that makes the sale price 60% of the ticket price.

   If you subtract 40% from 100%, what's left over is 60%.

2. The ticket price can be rounded easily to $50, so you can estimate the sale price by multiplying 50 by 0.6. The easy way to do this is to multiply 50 by 6, which you can do in your head, then move the decimal one place to the left. Use this idea to estimate the sale price. (For what it's worth, I used this procedure about 8 times yesterday while clothes shopping.)

   $50 \cdot 6 = 300$; sale price is about $30.

3. Now use the procedures we reviewed in this lesson to find the exact discount price.

   $0.6 \cdot \$47.99 = \$28.79$

4. Would your estimate from Question 2 have been good enough for you to decide whether or not you wanted to buy the item? Discuss.

   Answers will vary.

1-1 Applications

Name ____________________

5. After a round of layoffs, a company that used to have 1,200 employees now has 900. They then issue a press release saying "In spite of trying economic times, we were able to limit job cuts to just 25% of our job force." A couple of years later, they raise the number of employees back to 1,200. This time, they claim "We were able to increase our job force by more than 33%!" In which press release were they lying? Explain.

Check both percent changes:

$$\frac{900 - 1{,}200}{1{,}200} = \frac{-300}{1{,}200} = -0.25 = -25\% \qquad \frac{1{,}200 - 900}{900} = \frac{300}{900} \approx 0.33 = 33\%$$

They were telling the truth on both.

6. An employer offers a retirement program in which 9.8% of a worker's salary is deducted from their pay and deposited into an investment fund. The company then adds an additional 12% of the worker's salary to the fund. How much would an employee have deposited into their fund in the first year if their starting salary is $43,000?

The total is 21.8% of their salary: $0.218 \cdot \$43{,}000 = \$9{,}374$

7. While shopping at Nordstrom, you pick out two pairs of jeans at $49 each and a pair of shoes on sale for $39. Then you splurge on a $95 sweater. You have two coupons but can only use one. The first offers 40% off any single item. The second offers 15% off your entire purchase. Which is the better choice?

40% off of the sweater: $0.60 \cdot \$95 = \$57$. Total price: $\$49 + \$49 + \$39 + \$57 = \$194$

15% off total purchase: $0.85 \cdot (\$49 + \$49 + \$39 + \$95) = \$197.20$

40% off the most expensive item is the better choice.

Lesson 1-2 Prep Skills

This is a short review of skills that will come in handy in the next lesson. In each case, answer the question, then rate your confidence level by checking one of the boxes. If you feel like you're struggling with these skills, consult the online resources provided by your instructor for extra practice.

I KNOW IT | THINK SO | UNSURE | NO IDEA

SKILL 1: CONVERTING DECIMALS TO PERCENTS

1. Write each decimal in percent form.

   0.37 0.125 0.091

SKILL 2: CONVERTING FRACTIONS TO PERCENTS

2. Write each fraction in percent form.

   $\frac{23}{100}$ $\frac{9}{20}$ $\frac{17}{8}$ $\frac{17}{3}$

SKILL 3: FIND PERCENTAGE OF A WHOLE

3. What is 27% of 360?

4. What is 11.2% of 1,000?

5. What is 9.42% of 360?

SKILL 4: FIND PERCENTAGE OF A SHADED FIGURE

6. Find the percentage of each figure that is shaded.

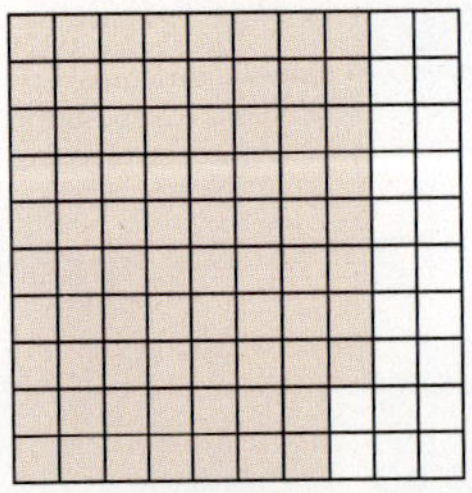

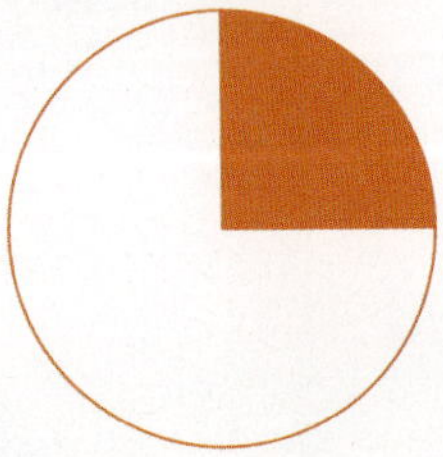

Lesson 1-2 Where Does the Time Go? (Using Percentages and Pie Charts)

LEARNING OBJECTIVES

- ☐ 1. Complete and analyze a weekly time chart.
- ☐ 2. Review a variety of calculations that involve percentages.
- ☐ 3. Create and interpret pie charts by hand.
- ☐ 4. Use a spreadsheet to create a pie chart.

Time is what we want most, but what we use worst.
—William Penn

One of the most important aspects of success in college is very underrated: learning how to manage time. This is a skill that you're unlikely to acquire by chance: in order to understand how to use your time most effectively, you have to first become aware of how you're using your time. Then you'll have to develop a plan that will help you best take advantage of your valuable time. This can help you to do better in your classes, and can also help you to have more time to do the things you enjoy.

0. After reading the opening paragraph, what do you think the main topic of this lesson will be?

1-2 Class

Where does the time go? I'm not asking in the philosophical way, as in "Holy smokes, it's almost Halloween! Where does the time go?" I'm thinking more literally: where exactly does your time go? Have you ever kept track of what exactly you accomplish with the 24 hours you're given every day? If you have very limited funds, you have to be very careful about how you spend what you have. But we ALL have limited time, and not many people think about whether or not they're using that time wisely.

1. What does the quote at the beginning of this lesson mean to you?

Answers will vary.

2. Write down the three activities that you think you spend the most time on during an average week. Don't think about it too much: just come up with three off the top of your head.

Answers will vary.

Here's a sample time chart put together by a student who was interested in identifying the amount of time she spent for one week. Each hour is marked with C (time in class), H (time spent on homework), S (sleep), W (work), or O (other time commitments). In the Portfolio portion of this lesson, you'll be asked to fill out a similar time chart of your own. For now, we'll analyze this chart.

	Sunday	Monday	Tuesday	Wednesday	Thursday	Friday	Saturday
12am–1:00	0	S	S	S	S	S	0
1:00–2:00	0	S	S	S	S	S	0
2:00–3:00	S	S	S	S	S	S	S
3:00–4:00	S	S	S	S	S	S	S
4:00–5:00	S	S	S	S	S	S	S
5:00–6:00	S	S	S	S	S	S	S
6:00–7:00	S	S	0	S	0	S	S
7:00–8:00	S	S	W	S	W	S	S
8:00–9:00	S	0	W	0	W	0	S
9:00–10:00	0	C	W	C	W	C	S
10:00–11:00	0	C	W	C	W	C	S
11:00–Noon	H	H	H	H	H	H	H
Noon–1:00	H	0	C	0	C	0	H
1:00–2:00	W	C	C	C	C	C	0
2:00–3:00	W	H	C	H	C	H	0
3:00–4:00	W	0	H	0	H	0	0
4:00–5:00	W	0	H	0	H	0	H
5:00–6:00	W	0	0	0	0	0	H
6:00–7:00	W	0	0	0	0	0	H
7:00–8:00	W	H	H	H	H	0	0
8:00–9:00	H	H	H	H	H	0	0
9:00–10:00	H	0	H	0	H	0	0
10:00–11:00	H	0	0	0	0	0	0
11:00–12am	0	S	0	S	0	0	0

3. Count the number of spaces containing each letter in the time chart (C, H, S, W, O). Use the results to fill in the following chart, writing the total number of hours devoted to each activity during one week.

	Hours
Class	15
Homework	32
Sleep	54
Work	15
Other	52

Many of the Class portions in this text will also work well in groups. The one in this lesson lends itself especially well to group work. This can give students a chance to get to know their groups right away, and is a good way to set the tone for how this class is different from a traditional lecture course.

4. Without adding the hours in the table, how can you decide what the total number of hours should be?

There are seven days in a week and 24 hours in a day, so there must be 7 × 24, or 168 hours in a week.

This is a good opportunity to emphasize the value of using key information to double check your work.

5. Most college advisors will tell you that a good rule of thumb is to allow 2 hours of study time outside of class for each hour spent in class. Is the student that filled out the table in Question 3 following that advice?

Yes. Two hours of homework for every hour of class would result in 30 hours of homework for every 15 hours of class, and this student did 32 hours.

6. Use your results from Questions 3 and 4 to fill in the next chart with the percentage of hours in a week devoted to each activity. (Round to the nearest percent.)

	Percentage
Class	9%
Homework	19%
Sleep	32%
Work	9%
Other	31%

Math Note

Recall that to find a percentage of a total, you divide the given number by the total, then multiply the result by 100.

7. Studies have shown that college students get about 6 ½ hours of sleep per night on average, but that academic achievement improves when students average 8 hours of sleep. What percentage of the hours in a week would be devoted to sleep if you get 6 ½ hours of sleep per night? What if you get 8 hours? How is our friend from Question 3 doing when it comes to sleep?

6.5/24 = 27%

8/24 = 33%

He or she is close to averaging 8 hours, but not quite there.

Throughout the course, there is a focus on success skills. Sometimes these are discussed explicitly, and sometimes they're snuck in as math problems. As you talk to the groups, point out success tips such as how much time should be spent studying outside of class and the importance of getting enough sleep. Don't be afraid to "get off topic" by chatting about these kinds of things.

8. A circle can be divided into 360 equal units of measure, which we call **degrees.** In other words, an entire circle is made up of 360° (the ° symbol represents degrees). Multiply the decimal form of each percentage in Question 6 by 360°, and put the results into the next chart. Round to the nearest whole degree.

	Degrees
Class	32°
Homework	68°
Sleep	115°
Work	32°
Other	112°

Did You Get It

Try this problem to see if you understand the concepts we just studied. The answer can be found at the bottom of the portfolio page.

1. Find 20%, 75%, and 41.3% of 360°.

A **pie chart** is a diagram used to compare the relative sizes of different parts of a whole. For example, the pie chart below, adapted from a *USA Today* survey, describes people's hand-washing tendencies when using a public restroom. In this case, the whole is the number of people surveyed, and each of the parts is the percentage of folks that gave a certain response.

Since 70% of respondents said "Always," the "Always" category fills 70% of the circle. Remember, the full circle measures 360°, so this portion represents $0.7 \times 360° = 252°$. (See the special box later in this lesson for a quick review of percents.)

9. The "Sometimes" category fills 29%, and the "Never" category just 1%. How many degrees do each of those sections cover in the pie chart?

$0.29 \cdot 360° = 104.4°$

$0.01 \cdot 360° = 3.6°$

How Often Do You Wash Your Hands After Using a Public Restroom?

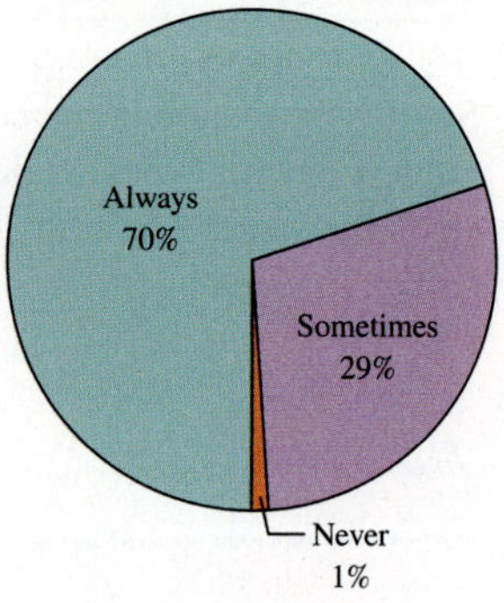

10. What do you think the percentages would be if the question was "How often do you wash your hands after using a public restroom when nobody else is in there with you?"

This is just for fun, but I bet the Always category would be a whole lot less.

Now let's see if we can build a pie chart that illustrates the amount of time spent on various activities for the student survey that began this lesson.

11. Build a pie chart for the information in Question 8 by marking off angles on the circle below, starting at the 0° mark, that correspond to the numbers of degrees in Question 8. In this case, the "whole" refers to the total number of hours in a week, and each part is the portion of that time spent on one of the activities. Each of the light gray lines on the graph represents 10°.

12. Does the pie chart make it easy to analyze the amount of time this student spends on each activity? Explain.

Answers will vary.

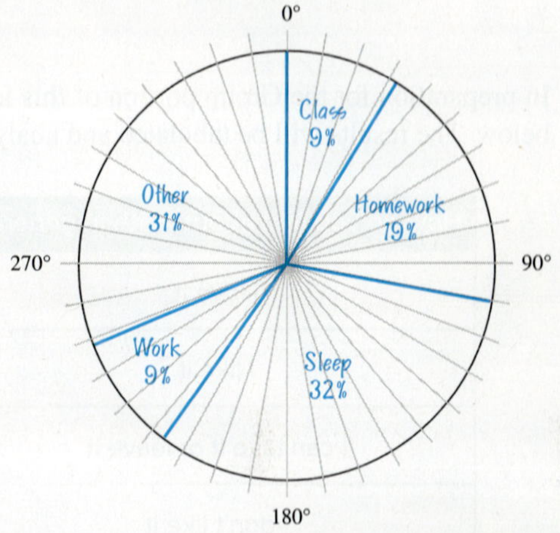

Some students need the clarification that for each pie slice, you start counting the number of degrees from where the last one left off, and not from 0.

Review of Percents

1. The word "percent" literally means "per hundred." When we read that 70 percent of respondents to a survey always wash their hands in a public restroom, it means that 70 people per hundred do so.
2. Translating a percent as "per hundred" makes it easy to convert percents to fraction form for using them in calculations: 70% means 70 per hundred, or $\frac{70}{100}$. And there's your fractional form for 70%!
3. This also shows us how to convert percents to decimal form: dividing any number by 100 moves the decimal two places to the left. That makes 70% equal to 0.70, 5% equal to 0.05, and 23.5% equal to 0.235.
4. If you forget to convert a percent into fractional or decimal form when doing a calculation, you can usually catch your mistake with a bit of thought. In the pie chart on the previous page, if we used 70 × 360° instead of 0.7 × 360°, we'd have found an angle of 25,200°, which is pretty silly given that a full circle is 360°.

Did You Get It

Try this problem to see if you understand the concepts we just studied. The answer can be found at the bottom of the portfolio page.

2. In 2015, there were 54 Republican senators, 44 Democratic senators, and 2 who were Independents. Find the percentage for each category, then the number of degrees that each category would cover in a pie chart.

In preparation for the Group portion of this lesson, everyone in your class should supply their answer to the one-question poll below. The results will be tabulated and analyzed in the group work.

How do you feel about math in general?	
I love it	☐
I like it	☐
I can take it or leave it	☐
I don't like it	☐
I hate it	☐

You can have students raise their hands to vote for a response, and write the totals on the board. Assure students that they won't hurt your feelings if they say they don't like math. One of the best things about teaching this class is the students whose attitude toward math changes during the course.

1-2 Group

1. Use data from the class poll to complete the table. Then use your results to draw a pie chart comparing the responses.

Feelings toward math	Number	Percent	Degrees
Love it	2	8%	29°
Like it	5	20%	72°
Take or leave	4	16%	58°
Don't like	10	40%	144°
Hate it	4	16%	58°

Sample answers shown. (Degrees add to 361 due to rounding.)

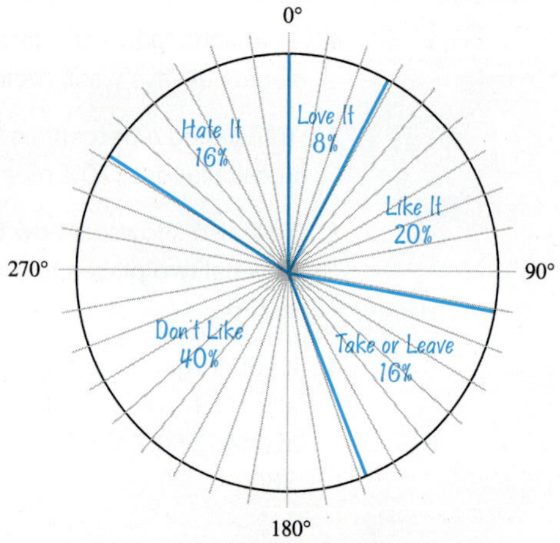

Using Technology: Creating a Pie Chart in Excel

To create a pie chart in Excel:

1. Type the category names in one column (or row).
2. Type the category values in the next column (or row).
3. Use the mouse to drag and select all the data in those two columns (or rows).
4. With the appropriate cells selected, click the **Insert** tab, then choose **Chart** and Pie chart. There are a few different styles you can experiment with, but starting with the simplest is a good idea.

You can add titles, change colors and other formatting elements by right-clicking on certain elements, or using the options on the **Charts** menu. Try some options and see what you can learn!

See the Lesson 1-2 video in class resources for further instruction.

Throughout the course, it's helpful to students if you use a projector to show them how to use new Excel functions. In this case, you can do this before or after the Group portion.

Interpreting Pie Charts

A national survey on college students and work was conducted during the 2003–2004 academic year and reported on the website www.acenet.edu. The results are summarized in the pie chart.

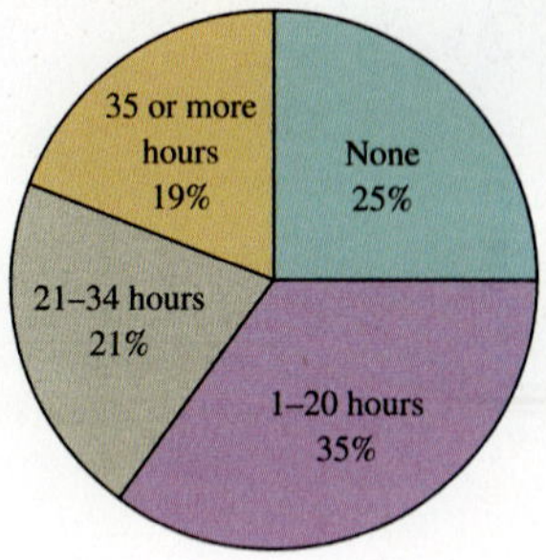

2. What range of hours worked was most common?

1–20

3. What percentage of students worked between 21 and 34 hours per week?

21%

4. How many degrees make up the angle for the sector representing students who worked from 1 to 20 hours per week?

$0.35 \times 360 = 126^\circ$

In Questions 5 and 6, be careful—the questions may refer to more than one sector of the pie chart.

5. On a campus with 7,500 students, on average how many would you expect to work at least 1 hour per week?

75% of 7,500 is 5,625 students.

6. On the campus in Question 5, how many students would you expect to work less than 21 hours?

60% of 7,500 is 4,500 students.

7. How do you think these results would compare with a similar survey conducted for this school year? Why do you feel that way? Explain.

Open to interpretation, but it seems like it's more common for students to work every year, so the None category might be smaller, and the 1–20 might be bigger.

Did You Get It ?

Try this problem to see if you understand the concepts we just studied. The answer can be found at the bottom of the portfolio page.

3. Refer to the pie chart on the previous page.
   a. What percentage of students worked at least part time?
   b. If one campus has 4,705 students, how many would you expect to work more than 20 hours?

1-2 Portfolio

Name ____

Check each box when you've completed the task. Remember that your instructor will want you to turn in the portfolio pages you create.

Technology

1. ☐ Fill in a time chart like the one at the beginning of this lesson based on your schedule. Then compile the results and use Excel to create a pie chart showing where your time is spent. There's a sample template in the online resources for this lesson to help you get started. Copy and paste the pie chart into a Word document, and write a couple of sentences below the chart describing any thoughts you have about what you learned.

Online Practice

1. ☐ Include any written work from the online assignment along with any notes or questions about this lesson's content.

Applications

1. ☐ Complete the applications problems.

Reflections

Type a short answer to each question.

1. ☐ Do you spend your time efficiently? Explain.
2. ☐ Why are you in college? If you've never thought about a good answer to this question, you certainly should have!
3. ☐ What changes do you plan to make in how you budget your time?
4. ☐ What resources do you plan to use to help you do well in this course?
5. ☐ List some benefits of working/studying in groups rather than alone.
6. ☐ Take another look at your answer to Question 0 at the beginning of this lesson. Would you change your answer now that you've completed the lesson? How would you summarize the topic of the lesson now?
7. ☐ What questions do you have about this lesson?

Looking Ahead

1. ☐ In preparation for the next lesson, visit the learning styles website linked under Lesson 1-3 resources, and take the learning style assessment. Your results will be recorded in a table in the Class portion of the next lesson.
2. ☐ Complete the Lesson 1-3 Prep Skills and read the opening paragraph in Lesson 1-3 carefully, then answer Question 0 in preparation for that lesson.

Answers to "Did You Get It?"

1. 72°, 270°, 148.68° **2.** Republican: 54%, 194.4°; Democrat: 44%, 158.4°; Independent: 2%, 7.2° **3. a.** 75% **b.** 1,882 students

1-2 Applications

Name ______________________________

Pizza is easily one of America's favorite foods, and it's pretty interesting to study some of the numbers involved in just how much we like pizza. Each year, a pizza industry organization releases a "State of the Industry Report," which is like the State of the Union address, but a whole lot tastier. According to one recent report, Americans eat approximately 100 acres of pizza each day (an acre is pretty close to the size of a football field). On average, 350 slices of pizza are eaten in the United States every *second.*

Overall, there are 67,554 pizza stores in the United States. Pizza Hut started in Wichita, Kansas, in 1958 and now has over 7,500 stores in the U.S. Little Caesars emerged in 1959 in Michigan and now has over 2,500 stores. Domino's started in 1960, also in Michigan, and now has over 5,000 stores in the United States. Papa John's opened its doors in 1989 in Indiana, and now boasts nearly 3,000 stores in the United States. These are the largest pizza chains in the country, but certainly not the only ones. And chain restaurants actually make up less than half of all pizza places nationally. The number of stores and percentage of sales are displayed in the two pie charts below.

U.S. Pizza Stores

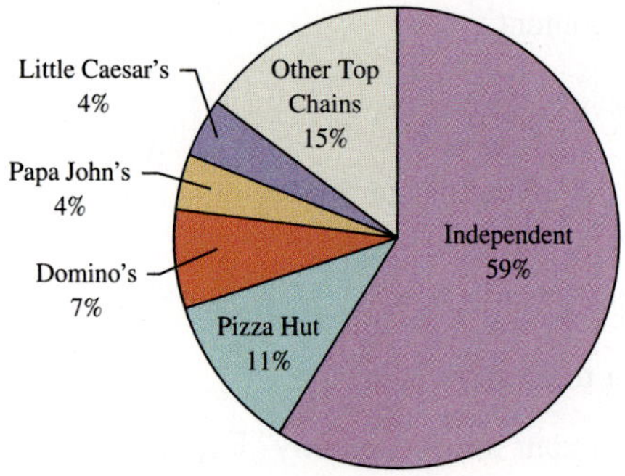

U.S. Pizza Sales

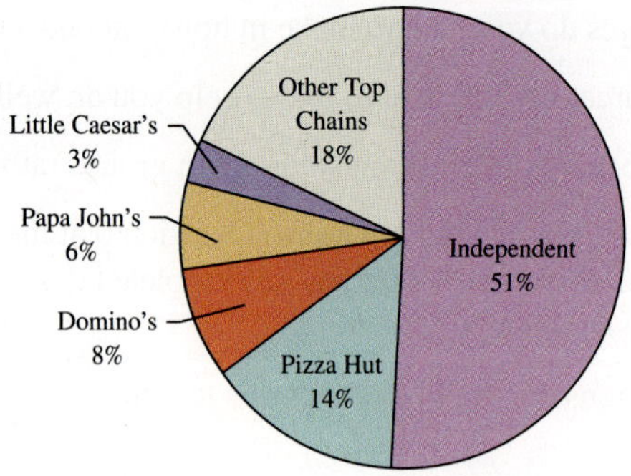

1-2 Applications

Name ____

1. How many slices of pizza are eaten per second in the United States? What about per hour? Per day? Write the last number verbally as well as numerically.

350 per second, 1,260,000 per hour, 30,240,000 per day, which is thirty million, two hundred forty thousand.

2. In the year that report was published, there were about 311 million people in the United States. How many pieces of pizza did the average American eat that year? (Start with your last answer to Question 1.)

$$\frac{30.24 \text{ million} \cdot 365 \text{ pieces of pizza}}{311 \text{ million people}} \approx 35.5 \text{ pieces per person}$$

3. According to the report, how many independent pizza stores are there in the United States? Show work to justify your answer.

59% of 67,554 is 39,857 stores.

4. How many degrees make up the angle for the sector representing sales by other top chains? How many degrees make up the angle for the number of stores for other top chains?

Sales: $0.18 \times 360 = 64.8°$

Stores: $0.15 \times 360 = 54°$

5. According to the report, Pizza Hut has over 7,500 stores in the United States. Based on the pie chart representing number of stores, how many stores does Pizza Hut have? Show work to justify your answer. Compare your result to the number given in the article, and discuss possible reasons for discrepancies.

11% of 67,554 is 7,431 stores, which is 69 less than 7,500. This could be due to rounding error in the 11% figure. Or the total number of pizza locations could have come from different sources.

1-2 Applications

Name ______________________________

6. For every \$20 spent on average on pizza in the United States, how much goes to chain pizzerias?

49% of sales go to chains: $0.49 \times \$20 = \$9.80$

7. Compare the percentage of independent pizza stores to the percentage of sales. What do you think the difference indicates?

59% of the stores are independent, but that accounts for just 51% of sales. This indicates that on average a chain pizza store has more sales than an independent one.

8. This table, based on information provided by *Business Insider,* lists American's favorite pizza toppings. Use the information to draw a pie chart. A pizza pie chart, if you will.

Favorite Topping	Percent	Degrees
Pepperoni	36%	130°
Sausage	14%	50°
Mushrooms	11%	40°
Cheese Only	8%	29°
Other	31%	112°

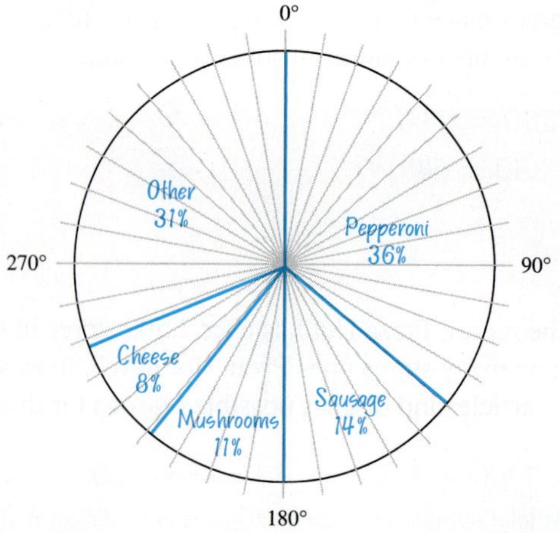

Lesson 1-3 Prep Skills

This is a short review of skills that will come in handy in the next lesson. In each case, answer the question, then rate your confidence level by checking one of the boxes. If you feel like you're struggling with these skills, consult the online resources provided by your instructor for extra practice.

I KNOW IT | THINK SO | UNSURE | NO IDEA

SKILL 1: FIND A TOTAL GIVEN A CERTAIN PERCENTAGE

1. Fifty-five is 32% of what number?

2. $7.99 is 15% of what amount?

SKILL 2: READING INFORMATION FROM A TABLE

3. The table shows the percentage of Americans that accessed the Internet at home and work between 2000 and 2010. How much higher was the percentage accessing at home in the year when the discrepancy was largest?

Year	Home	Work
2000	47%	42%
2002	59%	40%
2005	66%	55%
2008	72%	59%
2010	76%	65%

4. What was the largest year-to-year increase in Internet access, and which category was it in?

SKILL 3: ARITHMETIC WITH SIGNED NUMBERS

5. Perform the computation.

$-12 - (-17) + 9 - 22 + 12$

6. Perform the computation.

$-4.3 + 11.9 - (-32.8) + 6.1 + 9.5$

Lesson 1-3 It's All About Style (Interpreting and Drawing Bar Graphs)

LEARNING OBJECTIVES

- ☐ 1. Identify and understand your learning style.
- ☐ 2. Create and interpret bar graphs using a spreadsheet and by hand.
- ☐ 3. Decide whether a bar graph or pie chart is more appropriate for a given situation.

Alexander the Great so valued learning, that he used to say he was more indebted to Aristotle for giving him knowledge than his father Philip for giving him life.
—Thomas Babington Macaulay

Yes friends, this lesson is about style—learning style, that is. Learning is an active process, not a spectator sport. Think of some other active processes in your life: do you walk the same as everyone else? What about talk, sing, dance, drive, eat, smile, laugh . . . well, you get the picture. The answer, of course, is no. Then why on earth would you think that you learn like everyone else? Part of getting an education is learning how you learn. In this lesson, we'll study some of the different ways that people learn.

0. After reading the opening paragraph, what do you think the main topic of this lesson will be?

1-3 Class

Every student can benefit from finding out a little bit about his or her learning style—in almost any area, understanding your strengths and weaknesses is an important part of improving.

1. Record the scores you found when taking the online learning style assessment from the online resources for Lesson 1-3.

Style	Score	Style	Score
Bodily-Kinesthetic	42	Musical	19
Interpersonal	56	Naturalistic	56
Intrapersonal	62	Verbal-Linguistic	42
Logical-Mathematical	56	Visual-Spatial	50

Sample answers shown.

2. Which style did you expect to get your highest score in? Why?

Answers will vary.

A **bar graph** is a visual way to compare the sizes of different values. Pie charts are great for comparing parts to a whole: bar graphs are effective for comparing parts to other parts. In a bar graph, the length of each bar gives an indication of how the size of a quantity compares to other related quantities.

The bar graph below shows the percentage of global Internet users that visit the top five most-visited sites in the world. Just by looking at the heights of the bars, we can see how the percentages for these sites compare. By looking more closely at the scale on the vertical axis, we can get a good estimate of the actual percentage for each of the top five sites.

3. Which site has the highest daily reach? What percentage of Internet users go to that site on an average day?

Google: about 49%.

Average Daily Reach for Top Five Websites in the World

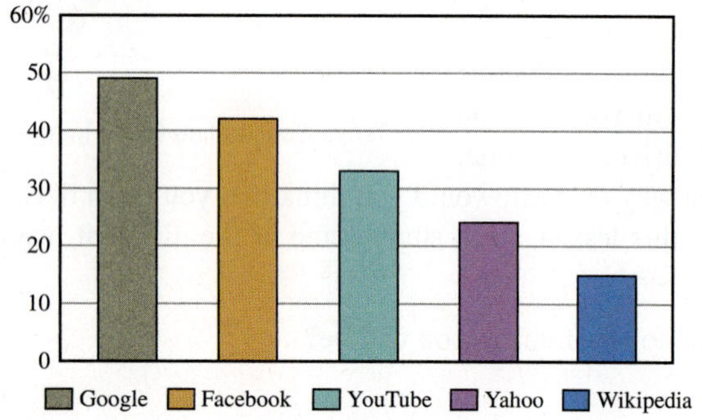

4. Which is higher: Google's daily reach, or the combined daily reach for YouTube and Yahoo? Explain your answer.

The combined reach is at least more than 50%, since the green bar is higher than 30% and the orange one is higher than 20%. So those two combined are more than Google's 49%.

5. I did some poking around on the Internet, and found that in an average day, about 690 million people visit Facebook. Based on Facebook's daily reach, how many people access the Internet each day?

The question: 41% of what number is 690 million?

$0.41x = 690$ million $\Rightarrow$ $x = 1.682$ billion daily users

6. I also found out that an estimated 3.3 billion people are Internet users. What do you think accounts for the big discrepancy between this number and your answer to Question 5?

Answers can vary, but the most likely explanation is that not every Internet user uses the Internet every day.

Did You Get It ?

Try this problem to see if you understand the concepts we just studied. The answer can be found at the bottom of the portfolio page.

1. The bar graph below is based on responses to a survey of 1,000 adults conducted by crucial.com.
   a. What is the thing that the most people would be willing to do rather than give up their phone or laptop for a month?
   b. About how many of those thousand would rather get a migraine or the flu than give up their phone or laptop?
   c. Why did the last question begin with "About"?

Top Things People Would Rather Do Than Give Up Their Phone or Laptop For A Month

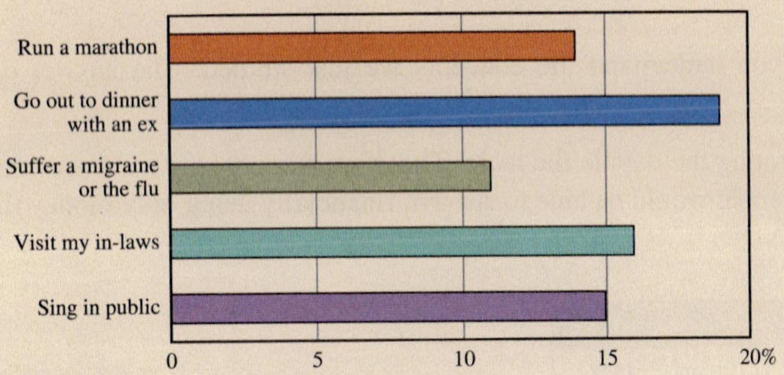

7. Use your learning style scores from Question 1 to draw a bar graph by creating a rectangle for each style with height corresponding to your score for that style.

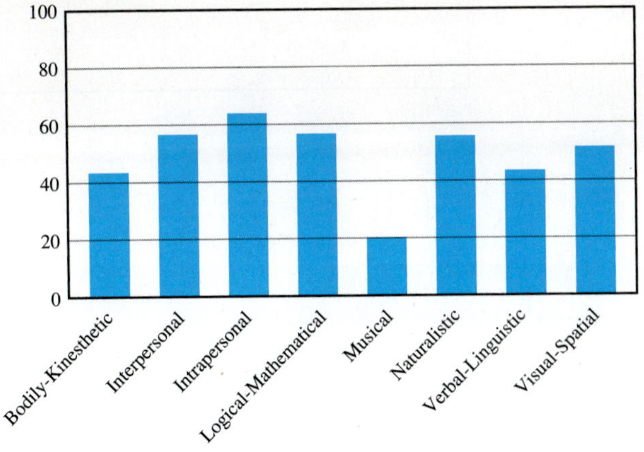

For Questions 7–9, it's important to give students time to work on their own. Then discuss their answers to Question 9, which is the key question of the activity.

8. Is there one style that stands out as having a higher score than the others for you? How does its bar compare in size to the next highest?

Intrapersonal is the highest, and is six units higher than any others.

9. Describe how the bar graph makes it easier to compare the relative scores than the table from Question 1 did.

You can compare the sizes of all the different scores easily without having to compare the sizes of the numbers one pair at a time. It's also very easy to see which styles have the highest and lowest scores.

Did You Get It ?

Try this problem to see if you understand the concepts we just studied. The answer can be found at the bottom of the portfolio page.

2. Draw a bar graph illustrating the data in the table. These are responses to a survey done by State Farm Insurance on how many months people would be able to survive financially using only money that they have currently saved.

Response	%
Less than 1	18%
1	13%
2-6	32%
7-12	12%
More than 1 year	11%
Don't know/refused to answer	13%

Using Technology: Creating a Bar Graph in Excel

To create a bar graph in Excel:

1. Type the category names in one column (or row).
2. Type the category values in the next column (or row).
3. Use the mouse to drag and select all the data in those two columns (or rows).
4. With the appropriate cells selected, click the **Insert** tab, then choose **Chart** and Column or Bar graph. ("Column" gives you vertical bars, "Bar" gives you horizontal.) Again, there are a few different styles you can experiment with, but starting with the simplest is a good idea.

You can add titles, change colors and other formatting elements by right-clicking on certain elements, or using the options on the **Charts** menu. Try some options and see what you can learn!

See the Lesson 1-3 video in class resources for further instruction.

1-3 Group

The Harris Interactive polling agency surveyed 2,094 consumers in October of 2012, asking which former president they would most enjoy having a burger with. The top four vote-getters are summarized in the bar graph.

Which Former President Would You Most Enjoy Having a Burger With?

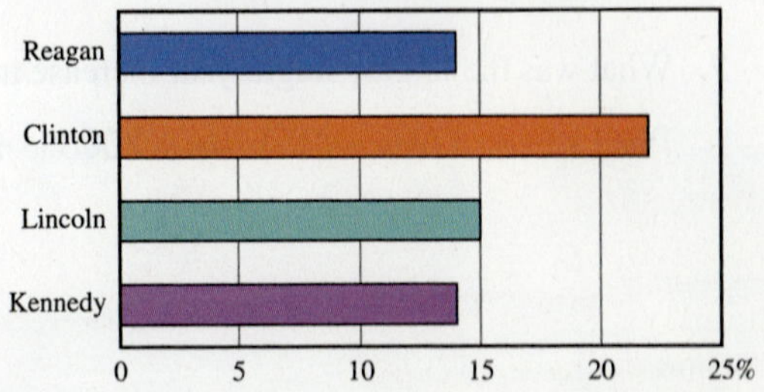

1. Which former president had the highest percentage, and what was the percentage?

   President Clinton: about 22%

2. How many of the respondents would most enjoy having a burger with the most popular choice?

   22% of 2,094 people surveyed is 461 people.

Math Note

Sometimes a bar graph will include the actual value at the end of the bars, making it easier to read accurately.

3. Which two presidents tied in the poll?

   Kennedy and Reagan tied at 14%.

4. How many people chose a president not in the top four?

   The four in the table add up to 65%, so 35% chose a president out of the top four; this is 0.35 x 2,094, or 733 people.

> When talking to groups, emphasize the need to check to see if all respondents are included. Did they consider an "other" category that isn't mentioned?

The yearly net income or loss for the Ford Motor Company from 2007 to 2011 is shown in the next graph.

5. Describe how Ford performed in 2008 and in 2010 in terms of profit or loss.

   In 2008, the company lost 15 billion dollars. In 2010, it made 7 billion dollars.

Ford Net Profit/Loss

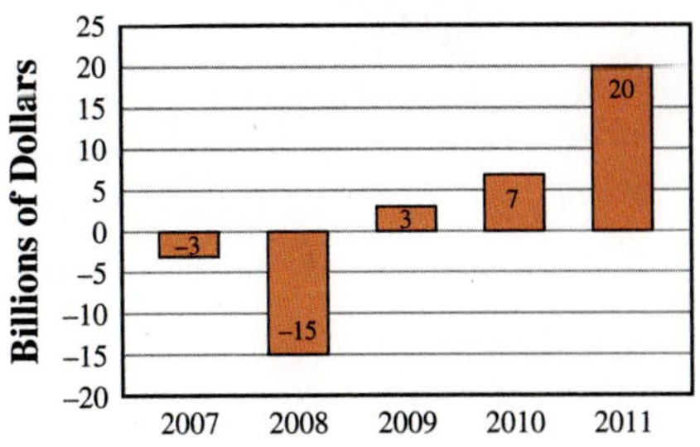

6. What were the best and worst years for Ford from 2007 to 2011? Explain how you decided.

2008 was the worst, with a loss of \$15 billion. 2011 was the best; the company made \$20 billion.

7. What was the largest single-year increase in profit?

Profit increased from −\$15 billion in 2008 to +\$3 billion in 2009, an increase of \$18 billion.

The concept of loss as negative profit will be revisited in the context of studying linear functions.

8. A mathematical way of expressing profit is to use a positive number. A negative number is used to express a loss.

Did You Get It

Try this problem to see if you understand the concepts we just studied. The answer can be found at the bottom of the portfolio page.

3. Did Ford make or lose money for the overall period from 2007 to 2011? How much did they make or lose?

It's really important to pay attention to all aspects of graphs, not just what the pretty picture looks like. Sometimes graphs can be misleading, as seen in the next example. These graphs might be poorly drawn, or they may be intentionally misleading to try to influence your opinions, or get you to buy products that you otherwise wouldn't.

The first bar graph below compares the number of calories in a regular cheeseburger at the three top burger franchises in the United States.

9. Look only at the HEIGHTS of the bars on the graph. Describe how the number of calories appears to compare based on those heights.

Since the bar for the first chain is so much taller than the others, it makes it look like that burger has 2-3 times as many calories.

Calories in a Regular Cheeseburger at the Three Top Burger Chains

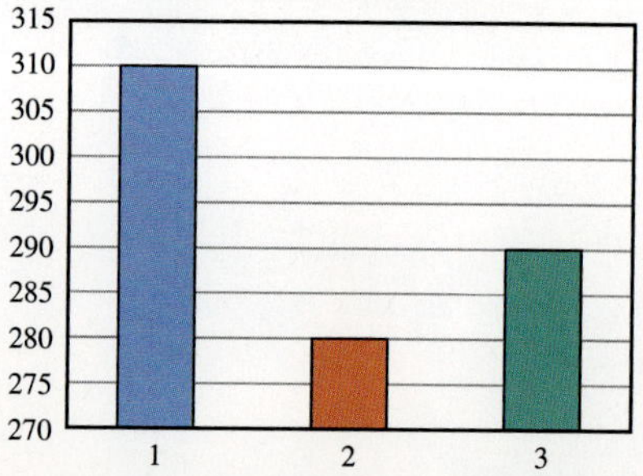

10. Assuming that a consumer enjoys the taste of all three cheeseburgers equally, how likely do you think this graph would be to influence which burger they choose to eat most often?

The huge difference in heights would make many people most likely to choose chain 2.

11. Carefully compare this second graph to the first. Does the second graph provide the same information as the first?

Yes.

12. Again focusing strictly on the heights of the bars, how do the three burgers compare in terms of calorie content?

They appear to be very similar.

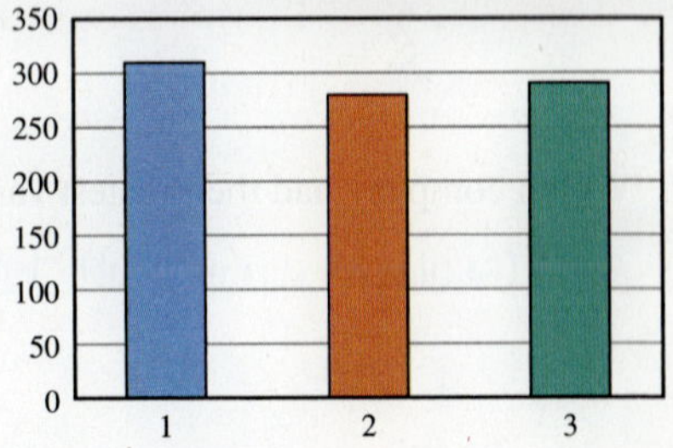

13. What scale is used on the vertical axis for each graph? (**Scale** refers to the spacing of the labels on an axis.)

Each line corresponds to 5 calories in the first graph, and 50 calories in the second.

14. Which of the three franchises would be most likely to use the first graph in advertising? Which would be most likely to use the second? Needless to say, you should explain your answers.

Chain 2 would prefer the first graph: it makes it look like their burger has WAY fewer calories than the competitors. Chain 1 would like the second graph best: even though they have the highest calorie count, that graph shows that the difference isn't very significant.

Emphasize how scale and starting point of a graph can be manipulated to make differences seem more dramatic. Discuss how students should pay attention to this when looking at graphs they see in the media, and that when creating their own graphs they should either start at 0 or indicate that they haven't with a symbol on that axis.

For comparisons involving more complicated data, a **clustered bar graph** can be used. This allows us to compare data values for information in multiple categories on one graph. The sample below illustrates the daily trade volume for three large tech companies on the NASDAQ market for a five-month period in 2015.

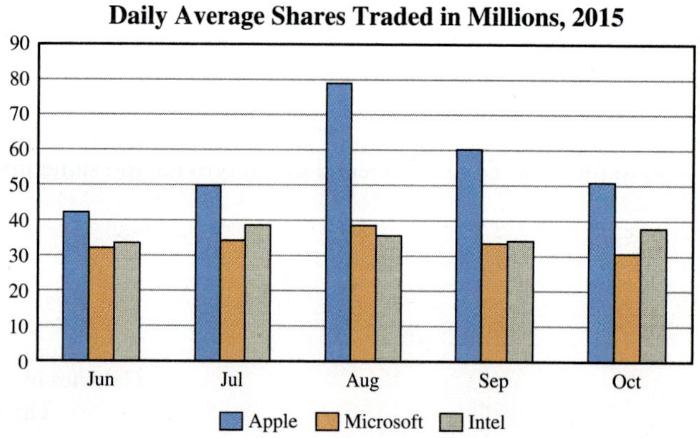

15. Which company had the greatest variance in trade volume over that period? Write a sentence.

Apple. The other two were very stable, but Apple's volume went up a lot, then came back down.

16. Which company had the most consistent trade volume? Write another sentence.

That can actually be debated. It's certainly either Microsoft or Intel. The smallest difference between highest and lowest volume appears to be Intel.

17. Which company had the lowest number of overall shares traded during that period? How can you tell?

You'd have to add the heights of all bars for each company to tell. For Microsoft, it looks like about $31 + 34 + 39 + 32 + 30 = 166$ million shares. For Intel, it looks like $33 + 39 + 36 + 34 + 38 = 180$ million shares, so I'll go with Microsoft as lowest.

1-3 Portfolio

Name ______________________________

Check each box when you've completed the task. Remember that your instructor will want you to turn in the portfolio pages you create.

Technology

1. ☐ Find an example of a bar graph in a book, magazine, newspaper, or online article on a subject that you find interesting. Recreate the bar graph in a spreadsheet, including labeling. Then copy and paste your bar graph into a Word document and type a sentence or two explaining why you find the topic interesting, and what the bar graph tells you. Be sure to make note of your source!

Online Practice

1. ☐ Include any written work from the online assignment along with any notes or questions about this lesson's content.

Applications

1. ☐ Complete the applications problems.

Reflections

Type a short answer to each question.

1. ☐ Write a sentence or two about each of the eight learning styles. Describe each in your own words.
2. ☐ Were you surprised by what your primary learning style is? Why or why not?
3. ☐ How do you plan to take advantage of knowing your primary learning style?
4. ☐ Take another look at your answer to Question 0 at the beginning of this lesson. Would you change your answer now that you've completed the lesson? How would you summarize the topic of this lesson now?
5. ☐ What questions do you have about this lesson?

Looking Ahead

1. ☐ To prepare for the next lesson, you'll need to take a personality test. (You do have a personality, right?) Links to several such tests can be found under Lesson 1-4 resources. Your personality type will be a four-letter code. Make note of it.

2. ☐ Complete the Lesson 1-4 Prep Skills and read the opening paragraph in Lesson 1-4 carefully, then answer Question 0 in preparation for that lesson.

Answers to "Did You Get It?"

1. **a.** Go out to dinner with an ex
   b. 120
   c. Because you can't tell the exact value from the length of the bars.

2. 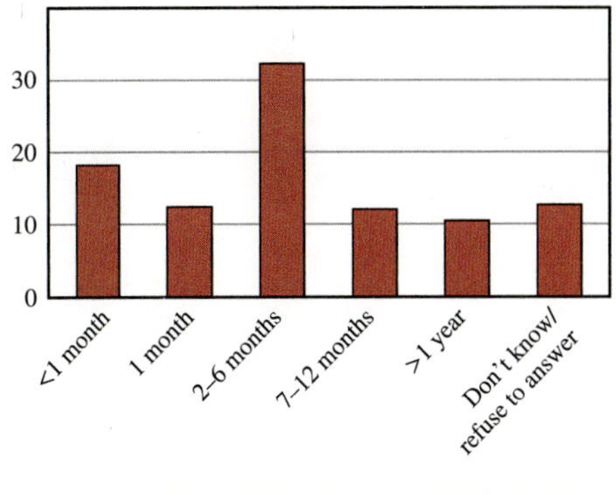

3. They made $12 billion.

1-3 Applications

Name ______________________________

In Questions 1 and 2, two different situations are described that you could illustrate with a graph. For one of the two, it would be most appropriate to use a bar graph, and for the other a pie chart would be a better choice.

Social Network Use: A survey of 100 college freshman found that 82 use Facebook, 10 use MySpace, 48 use Twitter, 60 use Instagram, and 33 use Google+.

Test scores: In a Geology class with 32 students, 6 earned an A on the first exam; 12 earned a B; 7 a C; and 4 earned a D.

1. Which data is best illustrated with a bar graph? Explain why you feel that way, then draw a bar graph representing the data.

A bar graph is a better choice for the social network data. Since many students fall into multiple categories, the data doesn't compare parts of a whole like the test scores do.

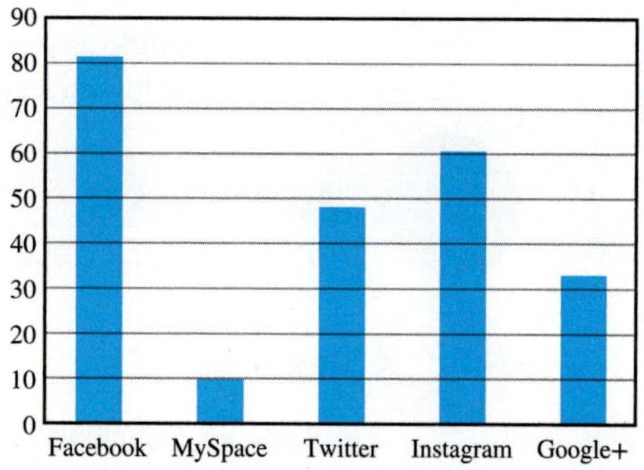

2. Which data is best illustrated with a pie chart? Explain why you feel that way, then draw a pie chart representing the data.

The test score data compares the parts of a whole: the number of people that got each grade compared to the total number of students. Note that percentages add to 100.1 due to rounding error.

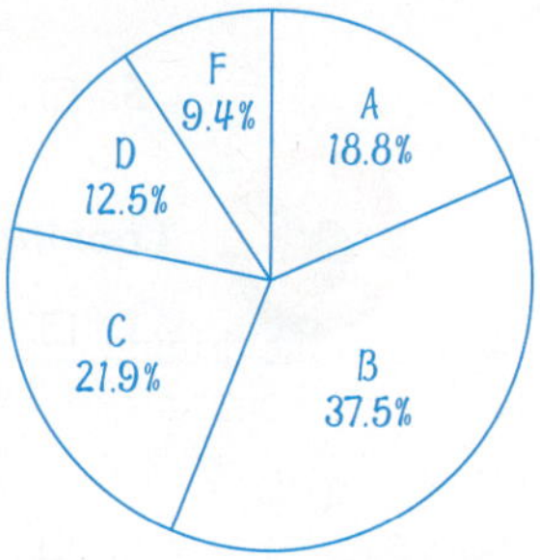

1-3 Applications

Name ______________________________

The graph below was reproduced from the website of the Substance Abuse and Mental Health Services Administration, a division of the U.S. Department of Health and Human Services. A study was done on alcohol use among people in the 12–20 age bracket in 2010, with some results summarized by the graph. The SAMHSA website defines alcohol use as follows:

Current (past month) use—At least one drink in the past 30 days.

Binge use—Five or more drinks on the same occasion (i.e., at the same time or within a couple of hours of each other) on at least 1 day in the past 30 days.

Heavy use—Five or more drinks on the same occasion on each of 5 or more days in the past 30 days.

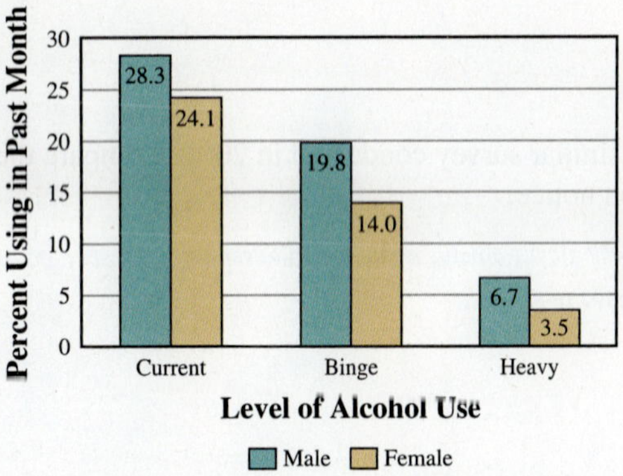

3. What does the graph indicate about the difference in alcohol usage between boys and girls?

At all three levels, boys are more likely to use alcohol than girls.

4. Which of the drinking categories shows the greatest discrepancy between boys and girls?

Binge drinking: the discrepancy is 5.8%, compared to 4.2% in the "current" category, and 3.2% in the "heavy" category.

5. What percentage of boys reporting current alcohol use were in the heavy drinking category? What about girls?

6.7 of the 28.3% are in the "heavy" category for boys: this represents 23.7% of the current drinkers. For girls, it's 3.5 out of 24.1%, which is 14.5%.

1-3 Applications

Name ____

6. Based on percentages, if a person aged 12–20 is a current drinker, are they more likely to be a binge drinker if they are a boy or a girl? Explain how you got your answer.

19.8 of the 28.3% of boys are binge drinkers: this is 70%. 14.0 of the 24.1% of girls are binge drinkers: this is 58%. So a boy who is currently using alcohol is more likely to be a binge drinker than a girl who's also currently using alcohol.

7. The next graph shows the results of a similar survey conducted in 2007. Compare the two graphs and write at least two trends in underage alcohol use that you notice.

Overall, alcohol use is down from 2007, particularly among girls. The discrepancy between boys and girls got larger in binge drinking and current use, but smaller in heavy use.

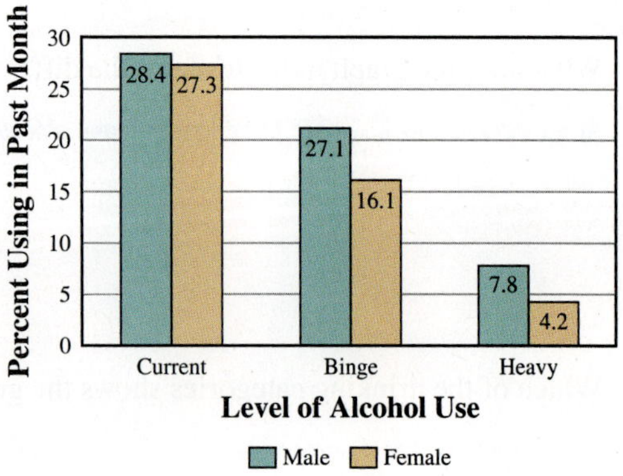

Lesson 1-4 Prep Skills

This is a short review of skills that will come in handy in the next lesson. In each case, answer the question, then rate your confidence level by checking one of the boxes. If you feel like you're struggling with these skills, consult the online resources provided by your instructor for extra practice.

I KNOW IT | THINK SO | UNSURE | NO IDEA

SKILL 1: ORDERING SIGNED NUMBERS

1. Insert <, >, or = in the blank to make each statement true.

   -11 ______ -15

   1.6 ______ 0.93

   -1.6 ______ -0.93

   $-\frac{23}{8}$ ______ -3

SKILL 2: ROUNDING NUMBERS

2. Round each number to the nearest ten.

   147 12,356

3. Round each number to the nearest whole number.

   11.519 146.32

4. Round each number to the nearest tenth.

   0.2345 146.39

5. Round each number to the nearest hundredth.

   0.049 1,893.7445

Lesson 1-4 Take a Guess! (Estimation and Number Sense)

© Rich Legg/Getty Images RF

LEARNING OBJECTIVES

- ☐ 1. Make educated guesses in a variety of settings.
- ☐ 2. Estimate locations on a number line.
- ☐ 3. Compare numbers using inequality symbols.
- ☐ 4. Approximate square roots.

Don't be afraid to fail. Be afraid not to try.
—Michael Jordan

Consider this scenario: You're a contestant on a game show, and you're given a trivia question worth $50,000. Pressure! The question is "Which state was first to secede from the union in 1860?" Do you know the answer? If so, terrific—drinks are on you. But if you don't know the answer, you'd have to guess. Would your list of guesses include the Pacific Ocean, Justin Bieber, Twinkies, and a goat? Of course not. Those would be completely silly. The point is that "guessing" doesn't usually mean just throwing out some random answer and praying for a miracle. Instead, we use information we know to be true and some reasoning ability to make an **educated guess.** In this lesson we'll try to gain some confidence with educated guessing, which sometimes goes by a different name in math: **estimation.**

From an instructor's point of view, it seems like in math classes, a lot of students are afraid to make educated guesses. But sometimes that's a completely reasonable step in problem solving. Nobody likes to be wrong, but if you're not willing to at least answer a question, you have no chance at ever being right. Not being afraid of failure is one of the characteristics shared by almost everyone that's very successful in their chosen field. So let's overcome the fear of being wrong, and build the confidence to take a guess.

0. After reading the opening paragraphs, what do you think the main topic of this lesson will be?

> Show students how to do Questions 1 and 2, then give them time to try the rest on their own. Have them check their answers with their group.

1-4 Class

An **isotherm** is a curve on a map that connects locations where the temperature is the same. By looking at a large map with isotherms, you can get a rough idea of what the temperature is like in many different locations. Using the map below, estimate the temperature in each city.

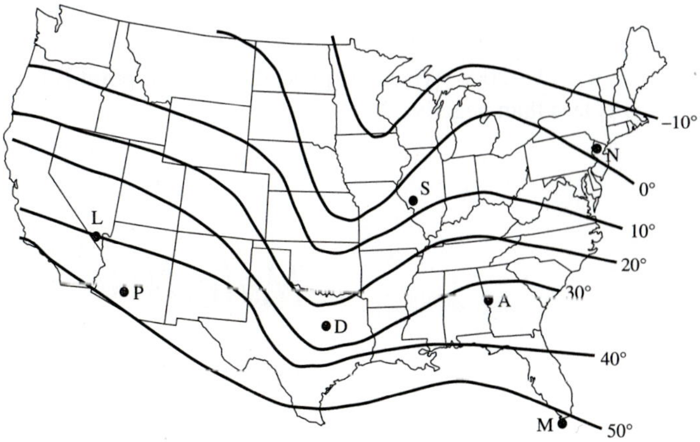

1. Las Vegas, NV (L) __40°__

2. Phoenix, AZ (P) __47°__

3. Dallas, TX (D) __24°__

4. Springfield, IL (S) __6°__

5. Atlanta, GA (A) __32°__

6. Miami, FL (M) __52°__

7. New York, NY (N) __−1°__

> Number 6 is a good question to discuss: since Miami is not between two isotherms, we should be less confident of our estimate.

Estimation is a hugely valuable skill when it comes to reading information from charts and graphs. In fact, we used that skill (without even realizing it!) in Lesson 1-3 when getting information from bar graphs. When the length of a bar falls between two lines, like in the graph on page 32, we have to estimate its height. We can practice this useful skill by looking at number lines.

8. Use the number line to estimate the value represented by each point. Write a short explanation of how you decided on your answer.

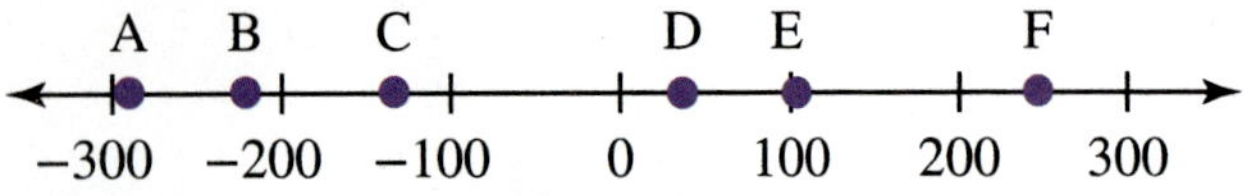

Point	Value	Explanation
A	−290	Just a bit more than −300
B	−220	Less than −200; almost a quarter of the way to −300
C	−130	Less than −100; a little more than a quarter of the way to −200
D	40	A bit less than halfway between zero and 100
E	102	Really close to 100; just a tiny bit larger
F	245	Very close to halfway between 200 and 300

Did You Get It

Try this problem to see if you understand the concepts we just studied. The answer can be found at the bottom of the portfolio page.

1. Estimate the value represented by each point.

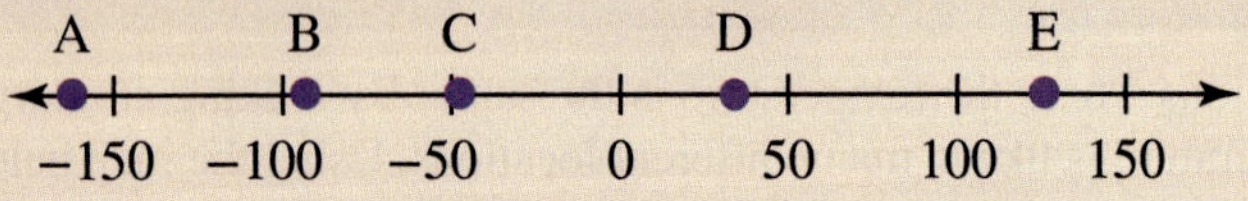

You might have started Question 8 by thinking "Point B is a little bit left of −200." We can express the fact that point B is to the left of −200 in symbols by writing B < −200. This is read as "B is less than negative 200."

Use the next number line to answer Questions 9 and 10.

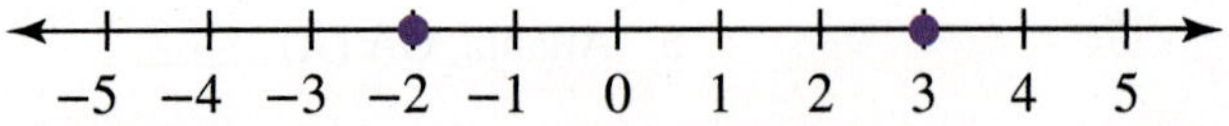

9. We write −2 < 3 because −2 is to the ___left___ of 3.

10. We write 3 > −2 because 3 is to the ___right___ of −2.

Math Note

The expression A > −300, read as "A is greater than negative 300," expresses the fact that A represents a number to the right of −300 on a number line.

Insert <, =, or > in the blank to make each statement true. Then write a sentence describing how each statement would be read aloud.

11. -10 ___$>$___ -15 **12.** -100 ___$<$___ 50 **13.** $\sqrt{25}$ ___$=$___ 5

11. Negative ten is greater than negative 15.

12. Negative one hundred is less than 50.

13. The square root of twenty five is equal to 5.

Did You Get It

Try this problem to see if you understand the concepts we just studied. The answer can be found at the bottom of the portfolio page.

2. Insert <, =, or > in the blank to make each statement true.

-23 ______ -27 $\quad -\frac{11}{3}$ ______ -3 $\quad \sqrt{64}$ ______ 7.8

For Questions 14–19, plot a point on the number line that meets the given condition.

14. A number greater than 3

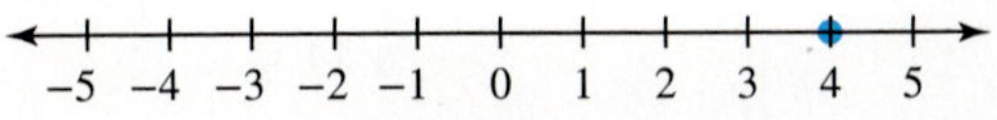

15. A number less than -3

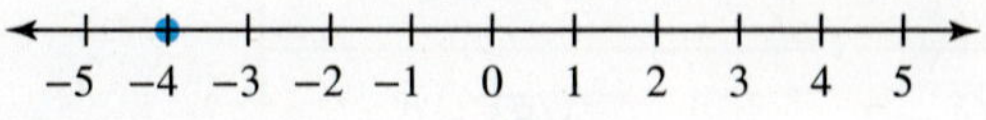

16. A number equal to 3

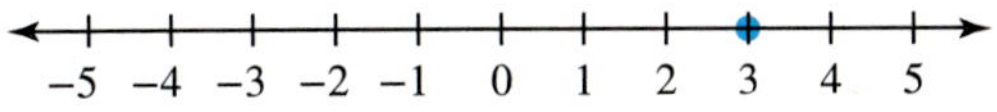

17. A number greater than -2 and less than -1

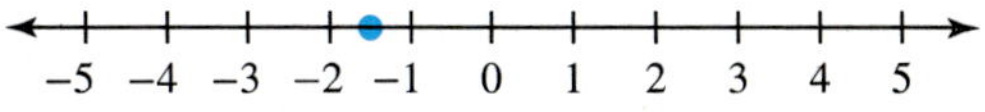

18. A number less than -2 and greater than -4

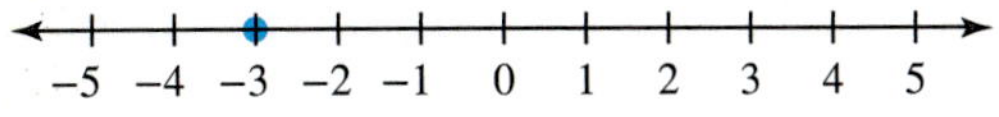

19. A number equal to -2

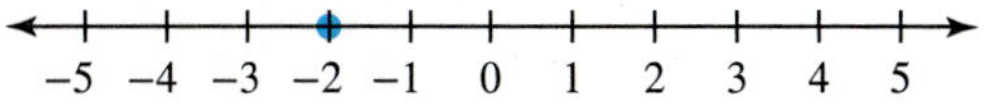

Sample answers shown for 14, 15, 17, 18.

Estimation is commonly used when dealing with square roots. We know, for example, that $\sqrt{49} = 7$. But what is $\sqrt{47}$? Since 47 is a little smaller than 49, it seems reasonable to estimate that $\sqrt{47}$ is a little less than 7. To get more accurate (but still approximate) values for square roots, we can use a calculator or a spreadsheet.

Using Technology: Approximating Square Roots

TI-84 Plus

1. Press 2nd x^2 to access the square root function.
2. Enter 47.
3. Press) ENTER to complete and evaluate this expression.

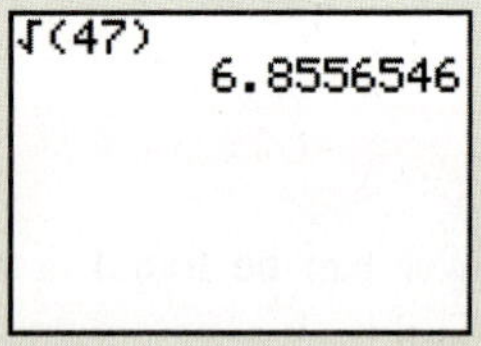

Excel

Enter "=SQRT(47)" in cell A1.
The formula bar, seen here in the upper right corner, displays this entry.
This entry produces a decimal approximation of $\sqrt{47}$ that is shown in cell A1.

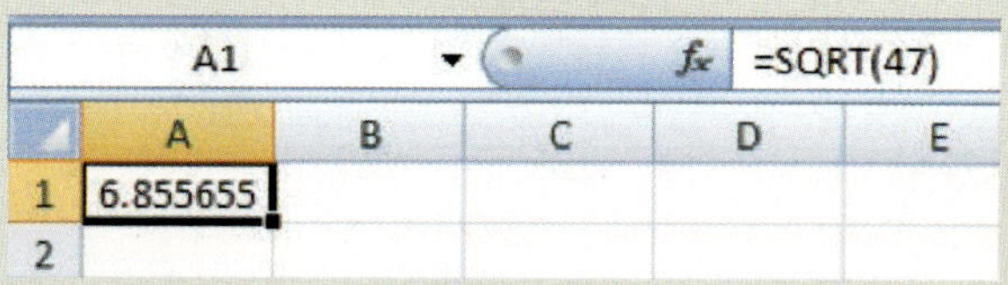

See the Lesson 1-4 video in class resources for further instruction.

20. Complete the table by following the instructions in each column header.

Fill in the correct value to make each equation true. No calculators!		Estimate each square root to the nearest whole number then write a statement using > or < to indicate whether your estimate is more or less than the square root.	Use a calculator or a spreadsheet to approximate the following square roots to the nearest hundredth.
$\sqrt{1} = 1$	$\sqrt{64} = 8$	$\sqrt{11} > 3$	$\sqrt{11} \approx 3.32$
$\sqrt{4} = 2$	$\sqrt{81} = 9$		
$\sqrt{9} = 3$	$\sqrt{100} = 10$	$\sqrt{78} < 9$	$\sqrt{78} \approx 8.83$
$\sqrt{16} = 4$	$\sqrt{121} = 11$		
$\sqrt{25} = 5$	$\sqrt{144} = 12$	$\sqrt{99} < 10$	$\sqrt{99} \approx 9.95$
$\sqrt{36} = 6$	$\sqrt{169} = 13$		
$\sqrt{49} = 7$	$\sqrt{196} = 14$	$\sqrt{124} > 11$	$\sqrt{124} \approx 11.14$
	$\sqrt{225} = 15$		

Students may need help understanding the middle column of this chart. Point out that it makes perfect sense that $\sqrt{11} > \sqrt{9}$, which is why we know that $\sqrt{11} > 3$. You might also find it useful to discuss that each falls between a nearest lower whole number and a nearest higher whole number, so $3 < \sqrt{11} < 4$, but $\sqrt{11}$ is closer to 3 than 4. Do one or two examples, then let your class try the rest.

Did You Get It

Try this problem to see if you understand the concepts we just studied. The answer can be found at the bottom of the portfolio page.

3. For each given square root, find the nearest whole number, then approximate the value to the nearest hundredth using a calculator or spreadsheet.

$\sqrt{32}$ $\sqrt{51}$ $\sqrt{90}$

1-4 Group

Spend some time thinking about, and discussing, each question below. After talking with your group, make your best guess **and explain your reasoning.** Remember, these are supposed to be EDUCATED guesses, not shots in the dark, so think about using some basic calculations to help you. No using phones or computers to look anything up yet! Finally, after making your guess, write down some information you would need to know to find a more accurate answer.

1. How many text messages do you send in a year? 4,000

Reasoning:

Sample answers shown. These are just one guy's opinion, and of course the explanation parts will vary.

Information needed:

2. How much would you spend on gasoline to drive from here to Daytona Beach, FL? $150

Reasoning:

Information needed:

To say the least, these answers can vary by quite a lot. Listen to the groups discuss these, and emphasize that while different approaches may be used, good problem solving requires us to make reasonable estimates based on our own experience. Now is also a good time to point out that when solving problems, you should always think about whether your final answer seems reasonable (no 200 pound rabbits or 3-ft tall buildings, for example).

3. How much more money will you make in your lifetime by going to college instead of entering the workforce without a degree? $600,000

Reasoning:

Information needed:

4. What would your car payment be on a $10,000 car if you were going to pay it off in 3 years? $300

Reasoning:

Information needed:

5. How much time do you spend on Facebook and other social media sites each week? 3 hours

Reasoning:

Information needed:

6. How much time do you spend watching TV each week? 12 hours

Reasoning:

These questions can spark a lot of good discussion. Be flexible. Allow students to be creative, as long as they are making calculations based on estimated initial values that make sense.

Information needed:

7. How many people in the United States do not smoke cigarettes? 250 million

Reasoning:

Information needed:

8. How much money do you spend on soda or coffee each week? $0

Reasoning:

Information needed:

9. If you put $100 into an Individual Retirement Account (IRA) today, how much would it be worth when you retire? $500

Reasoning:

Information needed:

10. How much money do you waste when you skip one class? $20

Reasoning:

Information needed:

11. How many people in your state are on their cell phone right now? 100,000

Reasoning:

Information needed:

12. How tall is a stack of 1,000,000 dollar bills? *(See a cool illustration in Lesson 1-4 resources online.)*

Reasoning: 350 feet

Information needed:

1-4 Portfolio

Name ____________________

Check each box when you've completed the task. Remember that your instructor will want you to turn in the portfolio pages you create.

Point out to students that from now on, all Excel work should be submitted as an Excel file, and not pasted into Word. (This will allow you to check that they used Excel functions to do the calculations rather than typing in the answers manually.)

Technology

1. ☐ Use Excel to make a table of square roots for all of the whole numbers from 1 to 25. Format so the results are rounded to two decimal places. A template to help you get started can be found in the online resources for this lesson; also see the technology video under the resources for Lesson 1-4 for help.

When grading this assignment, check to see that students used the SQRT command.

Online Practice

1. ☐ Include any written work from the online assignment along with any notes or questions about this lesson's content.

Applications

1. ☐ Complete the applications problems.

Reflections

Type a short answer to each question.

1. ☐ What kinds of things would you think about when making an educated guess in a math problem?

2. ☐ Describe at least one situation where estimating is good enough, and one where it isn't.

3. ☐ When using a calculator or spreadsheet to find square roots, why did we say "approximate" rather than compute? An Internet search will help.

4. ☐ Take another look at your answer to Question 0 at the beginning of this lesson. Would you change your answer now that you've completed the lesson? How would you summarize the topic of this lesson now?

5. ☐ What questions do you have about this lesson?

Looking Ahead

1. ☐ Complete the Lesson 1-5 Prep Skills and read the opening paragraph in Lesson 1-5 carefully, then answer Question 0 in preparation for that lesson.

Answers to "Did You Get It?"

1. A: −170, B: −90, C: −47, D: 40, E: 125 **2.** >, <, >

3. 6, 5.66; 7, 7.14; 9, 9.49

1-4 Applications

Name ______________________________

1. Pie charts can be used to describe many things—including pies! A small bakery shop in Hamilton, Ohio, sells an average of 120 pies per month, with the types broken down in the given pie chart. Use the chart to make an educated guess as to the number of each type the shop sold. Explain your reasoning for each estimate.

Type	Number
Apple	58
Other	8
Pumpkin	18
Strawberry	8
Cherry	28

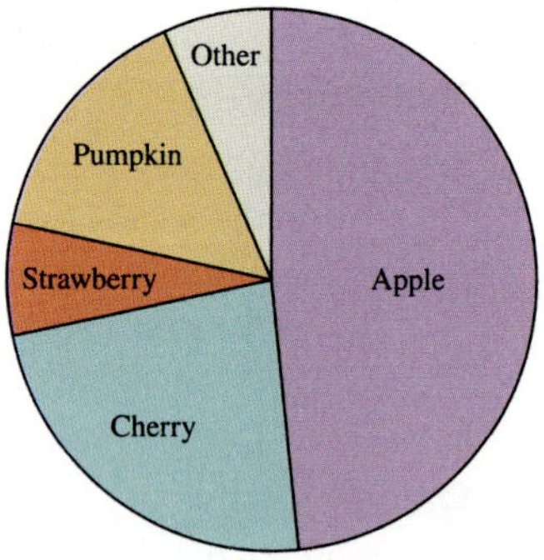

Reasoning:

The apple sector takes up just a bit less than half of the circle; half of the circle would be 60 pies. Cherry is just a bit less than a quarter of the circle, which would be 30. Pumpkin is more than half of cherry; other and strawberry are about the same size. And of course, all the numbers have to add up to 120.

2. Use the map to estimate the distance from St. Louis to New York, and explain your reasoning.

Distance estimate:

A bit less than 800 miles

Reasoning:

I measured the distance at about 8.5 cm, and scale shows that 200 miles corresponds to about 2.2 cm. So the distance is a bit less than 4 times as much as 200 miles.

1-4 Applications

Name ____________________

Shopping provides a really good example of the value of estimation. Most people have a rough idea of how much they plan to spend in a given shopping trip. While you could pull out the old cell phone calculator and add up all of your purchases to the cent, who has time for that? Instead, it's helpful to round prices to easy-to-use numbers and keep rough track of the total. Use this idea to estimate the total cost of each group of items in Questions 3–5. Do NOT use a calculator to get the exact cost! The point is to estimate and do mental arithmetic.

3. Four boxes of mac and cheese (89 cents per box)
A gallon of orange juice ($3.79)
A loaf of bread ($1.79)
Two pounds of ground turkey ($2.97 per pound)

4($1) + $4 + $2 + 2($3) = $16

4. Two pairs of jeans ($24.95 and $32.95)
A three-pack of socks ($7.99)
Three workout shirts ($11.99 each)

$25 + $33 + $8 + 3($12) = $102

5. Two Xbox games (regular price $29.50 each, on sale for 50% off)
iPhone case ($18.95)
Headphones ($22.49)

$30 + $19 + $22 = $71

6. Estimation is useful for tipping when you're dining out, too. If the service is good, 20% of the overall bill is a nice tip to leave. Here's a clever way to find that: move the decimal of the total amount one place to the left: that gives you 10%. Round this value to the nearest dollar and double the amount, and voila! Twenty percent. Use this to find a reasonable tip for good service on each bill.

a. $31.45

$6

b. $129.80

$26

Lesson 1-5 Prep Skills

This is a short review of skills that will come in handy in the next lesson. In each case, answer the question, then rate your confidence level by checking one of the boxes. If you feel like you're struggling with these skills, consult the online resources provided by your instructor for extra practice.

I KNOW IT | THINK SO | UNSURE | NO IDEA

SKILL 1: REVIEW COMMON UNIT EQUIVALENTS

1. Fill in the blanks to make each statement true.

   1 foot = ______ inches

   1 mile = ______ feet

   1 yard = ______ feet

   1 quart = ______ pints

SKILL 2: MULTIPLY INTEGERS BY FRACTIONS

2. Perform each multiplication. Reduce your answer to lowest terms.

   $3 \cdot \frac{7}{4}$ $\quad$ $-120 \cdot \frac{11}{3}$

SKILL 3: MULTIPLY FRACTIONS

3. Perform each multiplication. Reduce your answer to lowest terms.

   $\frac{2}{5} \cdot \frac{7}{4}$

   $-25 \cdot \frac{1}{60} \cdot \frac{12}{1}$

Lesson 1-5 Of Planes, Boats, Doll Houses, and Dr. Evil (Dimensional Analysis)

LEARNING OBJECTIVES

- ☐ 1. Understand the meaning of scale in models and maps.
- ☐ 2. Convert units using dimensional analysis.
- ☐ 3. View percentages in terms of scale.

I shall call him . . . Mini Me.

—Dr. Evil

© New Line/courtesy Everett Collection

Have you ever thought about the number of different ways we have to measure objects? Inches, feet, yards, miles, centimeters, meters, kilometers . . . and these cover only length; there are many more ways to measure length, as well as different units of measure for temperature, time, area, volume, mass, and other quantities. Why so many different measurements? The short answer is efficiency. Things come in so many different sizes that being stuck with just one unit of measurement would be very inconvenient. For example, it's about 5,892,480 inches from Chicago to Milwaukee. Clearly, measuring the distance in miles (93) is far more efficient. On the other hand, our friend Mini Me is 0.000505 miles tall, so measuring his height in inches (32) makes far more sense. In this lesson, we'll consider scale models (like Mini Me) as a way to learn an important skill: converting measurements from one unit to another.

0. After reading the opening paragraph, what do you think the main topic of this lesson will be?

1-5 Group

1. Most people who pay any attention to the news are aware that the U.S. Military uses drones: unmanned aircraft that patrol the skies over unstable regions equipped with video and other surveillance equipment, and even weapons in some cases. These drones come in many different sizes, and the **scale** of a drone indicates its size compared to a similar full-sized plane. For example, a half-scale drone is half the size of a full-sized plane. List some other examples of where you've heard the term "scale" used in this way.

Answers will vary widely.

DoD photo by Petty Officer 3rd Class Jeffrey S. Viano, U.S. Navy

Some students will try to rush through this lesson without thinking about it, while others may get confused and frustrated. Encourage them to take their time and think about how scale and unit conversions work. Let them know that learning how to do dimensional analysis now will come in handy throughout the course.

The best gift I got for my birthday last year was my very own Mini Me; not a living one, which would have been the best gift that *anyone* ever got for their birthday, but a bobblehead doll made to look like me. (In case you're wondering, I don't wear shorts that short. This isn't 1975.) The scale for many objects is given as a ratio of two numbers. Ratios can be written either as fractions, or using a colon. If a scale model is exactly half the size of the original, we could say the ratio of the smaller size to the larger is 1:2, or we could write that ratio as 1/2. My bobblehead is a 1:13 scale model, which means that the handsome little devil is 1/13th of my actual height.

Courtesy of David Sobecki

2. My Mini Me is 5.82 inches tall. How tall am I? Round to the nearest inch.

6'4"

3. If the actual Mini Me (32 inches tall) had a bobblehead made at the same scale, how tall would it be? Round to two decimal places.

2.46 inches

4. Write a verbal description of the process for going from larger to smaller and smaller to larger when given a scale, as in Questions 2 and 3.

In each case we're multiplying: it's just a question of whether we multiply by 1/13, or 13/1. To go from the full-sized version to the smaller scale model, we'd multiply by 1/13. To go from the smaller scale model to the full-sized version, we'd multiply by 13/1.

Did You Get It ?

Try this problem to see if you understand the concepts we just studied. The answer can be found at the bottom of the portfolio page.

1. A model of the space shuttle Challenger has a scale of 1:64.
   a. The actual shuttle was 122 feet in length. How long is the model? Rather than writing a decimal answer in feet, write in number of feet with left over inches, like 5 feet 8 inches.
   b. The wingspan of the model is 15″. What was the wingspan of Challenger in feet?

Here's a look at a map of the east side of Dubuque, Iowa. It shows a driving path from a casino on the river to a golf course on the other side. According to the maps app on my phone, the driving distance is 3.0 miles. We're most interested in the graphic at the bottom left, which indicates that 1 inch on the map (measure if you don't believe me) corresponds to 2,000 feet in real life.

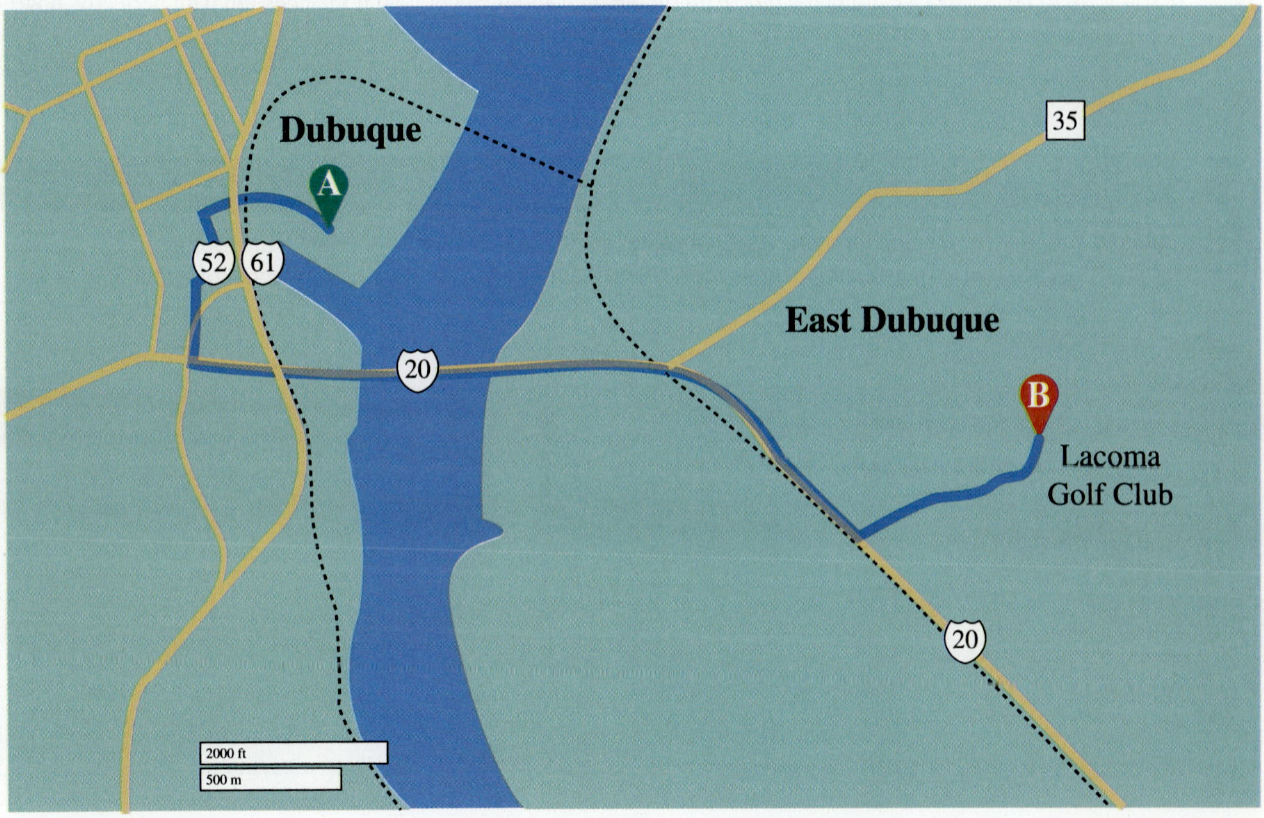

5. Points A and B are 4.06 inches apart on the map. How far is the casino from the golf course in feet and miles? (The straight-line distance, not the driving distance, which we already know.) Round miles to two decimal places.

8,120 feet, or 1.54 miles

6. If we looked at a satellite photo at the same scale, could we see any boats on the river? If a 50-foot boat were on the river, how big would it appear on the photo? Round to three decimal places. Do you think you'd be able to see it?

0.025 inch; you might see a tiny speck, but you sure couldn't recognize it as a boat.

7. Write a verbal description of the process of going from larger to smaller or smaller to larger in Questions 5 and 6.

To go from larger (real life) to smaller (map), divide the length in feet by 2,000. The result will be in inches. To go from smaller (map) to larger (real life), multiply the length in inches by 2,000. The result will be in feet.

Did You Get It

Try this problem to see if you understand the concepts we just studied. The answer can be found at the bottom of the portfolio page.

2. If you were to lay a shoelace along the blue path on the map and then cut it so it matches the path perfectly, how long would the shoelace be if you stretched it back out straight and measured it?

1-5 Class

Our next goal is to simplify the method for converting a measurement from one unit to another. This useful process, known as **dimensional analysis,** is based on a brilliantly simple idea. So brilliant in fact, that it deserves a cool colored box.

The Key Idea in Dimensional Analysis

If measurements in two different units represent the same actual length, like 12 inches and one foot, then a fraction formed from dividing those units is just a fancy way to say 1.

Let's look at an example. Since 1 yard and 3 feet are exactly the same distance,

$$\frac{3\text{ ft}}{1\text{ yd}} = 1 \text{ and } \frac{1\text{ yd}}{3\text{ ft}} = 1$$

The Class portion of this lesson is one that benefits from a mini-lecture. If you usually have students work on the Class sections in groups, consider more structure here.

We call a fraction of this type a **conversion factor,** and *multiplying any measurement by a conversion factor won't change the size of the measurement* because we're just multiplying by one; but it will change the units used to measure, which is exactly what we're trying to accomplish.

Each unit conversion in Questions 1–4 can have two conversion factors associated with it. Write the two conversion factors. And while you're at it, think about the fact that each is just another way to say 1.

1. 1 m = 100 cm Conversion factor #1: $\frac{1\text{ m}}{100\text{ cm}}$ Conversion factor #2: $\frac{100\text{ cm}}{1\text{ m}}$

2. 1 mi = 5,280 ft Conversion factor #1: $\frac{1\text{ mi}}{5{,}280\text{ ft}}$ Conversion factor #2: $\frac{5{,}280\text{ ft}}{1\text{ mi}}$

3. 12 in. = 1 ft Conversion factor #1: $\frac{1\text{ ft}}{12\text{ in.}}$ Conversion factor #2: $\frac{12\text{ in.}}{1\text{ ft}}$

4. 1 in. = 2.54 cm Conversion factor #1: $\frac{1\text{ in.}}{2.54\text{ cm}}$ Conversion factor #2: $\frac{2.54\text{ cm}}{1\text{ in.}}$

Emphasize that we're building an approach that will be used to convert units without having to figure out whether to multiply or divide, and that it is used in many sciences so students might see it in other classes.

Did You Get It

Try this problem to see if you understand the concepts we just studied. The answer can be found at the bottom of the portfolio page.

3. Write two forms of a conversion factor based on the relationship between minutes and hours.

The biggest challenge in converting units like we did in the group portion of this lesson, using multiplication or division, is that it's easy to get confused over whether you should multiply or divide. Now it's time to unveil our clever procedure for converting units:

Converting Units Using Dimensional Analysis

To convert units using dimensional analysis, the goal is to multiply by a conversion factor in a way that makes the units we don't want divide out, leaving behind the units we DO want.

Example time! The goal will be to convert 22 feet to inches.

5. Earlier, you wrote two conversion factors relating feet and inches. Below, multiply each of them by our measurement of 22 feet.

$22 \text{ ft} \times \frac{1 \text{ ft}}{12 \text{ in.}}$ $\qquad$ $22 \text{ ft} \times \frac{12 \text{ in.}}{1 \text{ ft}}$

6. For each multiplication above, what would be the units of the result? Ignore the numbers and just look at the units.

For the first it would be square feet over inches. For the second, inches.

In one of the multiplications above, the units that we did not want (feet) divided out, and the units that we do want (inches) were left behind. That's great! Because we multiplied by a conversion factor, all we really did is multiply by 1, which didn't change the SIZE of the measurement. It just changed the units that it's measured in. And THAT, friends, is the point of dimensional analysis.

7. Back to our calculation: how many inches is 22 feet?

$22 \text{ ft} \times \frac{12 \text{ in.}}{1 \text{ ft}} = 264 \text{ in.}$

Here's the great thing about dimensional analysis: To do that conversion, you didn't have to figure out whether to multiply or divide to get from feet to inches. *The units made that choice for you.* And when conversions get more complicated, it can be REALLY hard to just "figure out" what to do. That's when dimensional analysis is almost essential.

Let's say we want to convert the 3 miles of driving distance from the map in the group activity to yards. Why would anyone want to do that? Because they can.

8. First use your conversion factor from Question 2 to convert 3 miles to feet.

$$3 \text{ mi} \times \frac{5{,}280 \text{ ft}}{1 \text{ mi}} = 15{,}840 \text{ ft}$$

9. Now use another conversion factor to convert the answer from Question 8 to yards.

$$15{,}840 \text{ ft} \times \frac{1 \text{ yd}}{3 \text{ ft}} = 5{,}280 \text{ yd}$$

A lot of students are going to be resistant to dimensional analysis at first. (Trust me, I'm a doctor.) There's a strong tendency to keep trying to just figure out whether to multiply or divide. Doing a little bit of a sales job in front of the class is always a good idea.

10. Now here's a better idea: Convert 3 miles directly to yards by doing one two-part calculation. Again, focus on eliminating the units you don't want by putting that unit on the opposite side of the fraction when you multiply by your conversion factors. (This is why we had you write conversion factors two ways—so you get accustomed to the idea that YOU get to pick which unit goes on top and which goes on the bottom.)

$$3 \text{ mi} \times \frac{5{,}280 \text{ ft}}{1 \text{ mi}} \times \frac{1 \text{ yd}}{3 \text{ ft}} = 15{,}840 \text{ yd}$$

11. We've been focusing on units of length, but dimensional analysis can be used to convert ANYTHING that can be measured in more than one unit. There's a list of conversion factors on page 64. Use factors in that table to find how many cups are in a 2-liter bottle of Dr. Pepper. Round to two decimal places.

$$2 \text{ L} \times \frac{1.06 \text{ qt}}{1 \text{ L}} \times \frac{2 \text{ pt}}{1 \text{ qt}} \times \frac{2 \text{ c}}{1 \text{ pt}} = 8.48 \text{ c}$$

Did You Get It

Try this problem to see if you understand the concepts we just studied. The answer can be found at the bottom of the portfolio page.

4. Convert 138 ounces to pounds, and 36,800 grams to tons.

Sometimes scale is described in terms of percentage. For example, for the filming of the Academy Award-winning film *Titanic,* a model of the stern of the ship was created at 12.5% of the actual size.

© Pictorial Press Ltd/Alamy Stock Photo

12. A section of rail on the stern of the real Titanic was 24 feet long. How long was the corresponding section on the model used for the movie?

$0.125 \times 24 \text{ ft} = 3 \text{ ft}$

Students will probably need some help understanding the connection between percentage scale and scale using ratios and fractions.

13. If an I-beam on the model spanned 30 feet of deck space, how big was the corresponding I-beam on the actual ship?

$30 \text{ ft}/0.125 = 240 \text{ ft}$

14. Now let's turn this into a problem we already know how to solve: given that the model's size was 12.5% of the real ship's size, write a reduced fraction that describes the scale in fractional form, like we used earlier. (Hint: Recall that "percent" means "per hundred.")

$$\frac{12.5}{100} = \frac{25}{200} = \frac{1}{8}$$

Quick quiz: Is it illegal to make color copies of United States currency? Most people think it is, but that's not entirely true. According to the Counterfeit Detection Act of 1992, you can make color copies as long as they're less than 75% or more than 150% of the original size.

15. Write a reduced fraction (like you did in Question 14) to describe each of those scales in fractional form.

$$\frac{75}{100} = \frac{3}{4} \qquad \frac{150}{100} = \frac{3}{2}$$

Have students try Questions 15 and 16 on their own and compare answers.

16. An actual bill measures 2.61 in. by 6.14 in. Are we in trouble with the Feds for the reproduction below? Describe how you decided. Bonus points if you use the scale fractions you wrote in Question 15.

Courtesy of Brian Mercer

Students will need a ruler here, so make sure some are available.

$\frac{3}{4} \times 6.14 \text{ in.} = 4.605 \text{ in.}$

We are not. A 75% scale would be 4.605 inches in length, so the 4.3 inch reproduction is fine. This is fortunate, because we would look terrible in prison garb.

English Measure and Equivalents

Length	Weight
12 inches (in.) = 1 foot (ft) 3 ft = 1 yard (yd) 5,280 ft = 1 mile (mi)	16 ounces (oz) = 1 pound (lb) 2,000 lb = 1 ton (T)
Liquid Volume	**Time**
3 teaspoons (tsp) = 1 tablespoon (tbsp) 8 fluid ounces (oz) = 1 cup (c) 2 c = 1 pint (pt) 2 pt = 1 quart (qt) 4 qt = 1 gallon (gal)	60 seconds (sec) = 1 minute (min) 60 min = 1 hour (hr) 24 hr = 1 day 7 days = 1 week 365 days = 1 year

The Metric System

Kilo (1,000)	Hecto (100)	Deka (10)	Base Unit	Deci (1/10)	Centi (1/100)	Milli (1/1,000)
km kg kL	hm hg hL	dam dag daL	**Length** Meter (m) **Weight** Gram (g) **Volume** Liter (L)	dm dg dL	cm cg cL	mm mg mL

Conversions Between Systems

Length	Weight	Volume
2.54 cm = 1 in. 1 m = 3.28 ft 1.61 km = 1 mi	28.3 g = 1 oz 2.2 lb = 1 kg	1.06 qt = 1 L 3.79 L = 1 gal

1-5 Portfolio

Name ____________________

Check each box when you've completed the task. Remember that your instructor will want you to turn in the portfolio pages you create.

Technology

1. ☐ Complete the following spreadsheet to calculate the lengths of models of different modes of transportation. Each model will be 1:16 scale. Use a formula that calculates the length of each model in feet, then another formula to convert those lengths to inches. A template to help you get started can be found in the online resources for this lesson.

	A	B	C	D
1		True Length (ft)	Model Length (ft)	Model Length (in)
2	Boat	17		
3	Car	20		
4	Truck	24		
5	Tractor	34		
6	Train	330		

Online Practice

1. ☐ Include any written work from the online assignment along with any notes or questions about this lesson's content.

Applications

1. ☐ Complete the applications problems.

Reflections

Type a short answer to each question.

1. ☐ Describe the process we developed for converting units; focus on the fact that we're multiplying by one.
2. ☐ Why do maps (like the one on page 59) have a scale in one corner?
3. ☐ Take another look at your answer to Question 0 at the beginning of this lesson. Would you revise your answer now that you've completed the lesson? How would you summarize the topic of this lesson now?
4. ☐ What questions do you have about this lesson?

Looking Ahead

1. ☐ Complete the Lesson 1-6 Prep Skills and read the opening paragraph in Lesson 1-6 carefully, then answer Question 0 in preparation for that lesson.

Answers to "Did You Get It?"

1. a. 1 foot, 11 inches **b.** 80 feet **2.** 7.92 in.

3. $\frac{60\text{ min}}{1\text{ hr}}, \frac{1\text{ hr}}{60\text{ min}}$ **4.** 8.6 lb, 0.041 T

1-5 Applications

Name ______________________________

1. Forensic scientists often examine scaled-up crime scene photos to search for evidence that might be hard to see at regular magnification. Studying bite marks on a photo of a victim's leg at 500% magnification, a technician measures the distance between two punctures from the bite to be 165 mm. What is the actual distance between the punctures?

1/5 of 165 mm is 33 mm.

© Mikael Karlsson

© Columbia Pictures/Newscom

2. In the 1984 movie *Ghostbusters,* a 112.5-foot-tall Stay Puft marshmallow man goes on a murderous rampage in New York City. After the film's successful box office run, a 20-inch replica of the big guy was sold in toy stores. What scale was used to make the replica? Give your answer both as a reduced fraction or ratio and as a percentage.

$112.5\text{ ft} \times \frac{12\text{ in.}}{1\text{ ft}} = 1{,}350\text{ in.}$

$\frac{20\text{ in.}}{1{,}350\text{ in.}} = \frac{2}{135}$ This is about 1.48%.

Doll houses provide an excellent example of scale modeling. The 1/2-inch scale is very common for doll houses; this means that 1/2 inch on the doll house corresponds to 1 foot for life-sized objects. Make sure you clearly show all calculations and units as you answer Questions 3–5.

© Andy Crawford/Getty Images RF

3. What is an equivalent ratio form for the 1/2-inch scale as described above?

$\frac{1/2\text{ in.}}{1\text{ ft}} = \frac{1/2\text{ in.}}{12\text{ in.}} = \frac{1}{24}$ The ratio form is 1:24.

4. The width of one doll house is 34 in. Assuming that it uses the 1/2-inch scale, what's the width of the full-sized house that it's modeled after?

$34\text{ in.} \times 24 = 816\text{ in.}$, or 68 feet

5. If you're building a model stove for the doll house in Question 4, how tall should it be if the standard height for a full-sized stove is 36 inches?

$36\text{ in.}/24 = 1.5$ inches

Emphasize to students that they should think about whether their answers are physically reasonable!

1-5 Applications

Name ____________________

For Questions 6–9, use dimensional analysis with conversion factors to convert each measurement to the given units. You'll need unit equivalences from this lesson to build your conversion factors. Round to two decimal places if necessary.

6. How many miles is a 10-kilometer race?

$$10 \text{ km} \times \frac{1 \text{ mi}}{1.61 \text{ km}} = 6.21 \text{ mi}$$

7. How many feet is a 10-kilometer race?

$$6.21 \text{ mi} \times \frac{5{,}280 \text{ ft}}{1 \text{ mi}} = 32{,}788.80 \text{ ft}$$

8. The average offensive lineman in the National Football League weighs 141.3 kilograms. How many pounds is that?

$$141.3 \text{ kg} \times \frac{2.2 \text{ lb}}{1 \text{ kg}} = 310.86 \text{ lb}$$

9. (This one requires more than one conversion factor.) If one liter of freshwater weighs 2.2 pounds, how much does the water in this 22,000 gallon swimming pool weigh?

$$22{,}000 \text{ gal} \times \frac{3.79 \text{ L}}{1 \text{ gal}} \times \frac{2.2 \text{ lb}}{1 \text{ L}} = 183{,}436 \text{ lb}$$

Courtesy of David Sobecki

Lesson 1-6 Prep Skills

This is a short review of skills that will come in handy in the next lesson. In each case, answer the question, then rate your confidence level by checking one of the boxes. If you feel like you're struggling with these skills, consult the online resources provided by your instructor for extra practice.

SKILL 1: CONVERT DECIMALS TO PERCENT FORM

1. Write each decimal in percent form.

   0.23

   0.794

   0.035

SKILL 2: COMPUTE AREA

2. What is the area of a rectangle that is 22 ft long and 13 ft wide?

3. Find the area of the figure below.

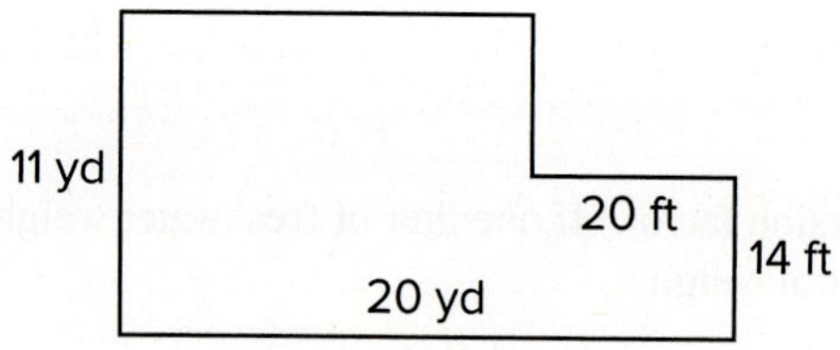

SKILL 3: COMPUTE VOLUME

4. What is the volume of a rectangular box that is 14 in. long, 9 in. wide, and 2 ft high?

Lesson 1-6 Relatively Speaking (Interpreting Relative Difference/Relative Error)

LEARNING OBJECTIVES

- ☐ 1. Compare difference and relative difference, and understand why relative difference is often a more meaningful comparison.
- ☐ 2. Apply relative error.
- ☐ 3. Find conversion factors for square and cubic units.

Library of Congress, Prints and Photographs Division [LC-USZ62-60242]

Put your hand on a hot stove for a minute, and it seems like an hour. Sit with a pretty girl for an hour and it seems like a minute. That's relativity.

—Albert Einstein

In a nutshell, Einstein's theory of relativity says that even the passage of time is not an absolute, but rather is relative to your frame of reference. In this lesson, we'll study appropriate ways to compare numbers, focusing on the importance of relativity. Consider these two scenarios: (1) I borrow \$20 from you, and when it's time to pay up, I give you \$10 instead. (2) You win \$5 million in a lottery, but you only get paid \$4,999,990. In which scenario was the shortage significant? You're out ten bucks in each, but clearly that's a much bigger deal when the amount owed is \$20. If that makes sense to you, then you have an instinctive understanding of the concept of relative difference.

0. After reading the opening paragraph, what do you think the main topic of this lesson will be?

1-6 Class

When asked to compare the sizes of two numbers, most people think of subtraction, but the example above shows that this isn't necessarily the best choice. The **difference** between two values is found by simply subtracting the old value from the new:

$$\textbf{Difference} = \textbf{New value} - \textbf{Old value}$$

In both scenarios above, the difference in what was owed and what was paid is \$10, which leads us to conclude the significance of that \$10 is the same in each case. But doesn't it seem clear that the \$10 shortfall is more significant when the total amount owed is \$20? That's where relative difference comes in. The **relative difference** between two values measures the difference as a fraction of the original value:

$$\textbf{Relative difference} = \frac{\textbf{Difference}}{\textbf{Original value}} = \frac{\textbf{New value} - \textbf{Original value}}{\textbf{Original value}}$$

In Scenario 1 above:

$$\text{Relative difference} = \frac{\$10 - \$20}{\$20} = -\frac{1}{2} \text{ or } -50\%. \text{ (The negative indicates that you got shorted. Bummer.)}$$

This shows that the difference between the amount owed and the amount received is half of the amount owed. That's a big deal!

1. Find the relative difference if you win \$5,000,000 and get shorted by \$10. Write as a fraction, not a decimal.

$$\frac{4{,}999{,}990 - 5{,}000{,}000}{5{,}000{,}000} = -\frac{10}{5{,}000{,}000} = -\frac{1}{500{,}000}$$

Math Note

Relative differences are often written as percentages, but they don't have to be. Fraction form is fine, too.

2. Convert your answer to Question 1 into a percent.

$-1/500{,}000 = -0.0002\%$

3. Compare the relative differences for the two scenarios in the opening paragraph from this lesson. Does this match your intuition about the scenarios?

If you get shorted by \$10 out of \$20, that's half of what you were owed, which is significant. If you get shorted \$10 out of \$5,000,000, that's just 1/500,000th of what you were owed, which is almost meaningless. Who gets bent out of shape over 0.0002 percent of anything?

Did You Get It

Try this problem to see if you understand the concepts we just studied. The answer can be found at the bottom of the portfolio page.

1. A major corporation's profit projection for the recent fiscal year was \$3.7 billion. The actual profit was \$3.45 billion. Find the actual difference and relative difference between the company's projection and the real profit. Write each in percent form.

4. Both Shawna and John found out that they're getting a \$1,000 annual raise this year. John went out to celebrate, while Shawna yawned and said "Yeah, whatever." Use the topic of this lesson to describe what you think might be likely to account for this discrepancy in their reactions to the news.

Most likely, the relative change (difference) in salary would account for these reactions. Shawna probably makes a LOT more than John, and the \$1,000 raise represents a small relative difference in salary compared to John.

Have students discuss Questions 4 and 5 with a partner, then discuss as a class.

5. Your instructor tells you that you're going to get an additional 2 points on a recent test or quiz. Describe when this would be great news, and when it wouldn't help very much.

If it's a quiz worth a small amount of points, like maybe 5, an extra 2 points would be a significant help. But if it's a big test worth 200 points, an extra 2 points wouldn't be a very big relative change. It may also depend on how many points are possible for the entire course.

6. The U.S. federal budget deficit for 2011 was about 1.298 trillion dollars. (This means that the government spent almost $1.3 trillion dollars more than it took in.) If you round that amount up to 1.3 trillion dollars, which seems perfectly harmless, you cost the taxpayers of this country 2 BILLION dollars. Thanks. What's the relative difference between the actual amount and the higher rounded amount? (Hint: If you're clever, you won't need to work with huge numbers. . . .) Discuss whether or not you think this difference is significant. It's up to you to decide (and defend) what you think "significant" means in this context.

$$\frac{1.3 \text{ trillion} - 1.298 \text{ trillion}}{1.298 \text{ trillion}} = \frac{0.002 \text{ trillion}}{1.298 \text{ trillion}} \approx 0.0015 \text{ or } 0.15\%$$

Answers will vary in terms of how significant this is. Try to get students to focus on the relative difference, not the fact that we're talking about 2 billion dollars.

7. A college football team won 5 games and lost 7 in 2014, then turned that around, winning 7 and losing 5 in 2015. At the same school the basketball team won 10 of 34 games in 2014, then won 15 of 34 the following year. Which team had the more significant improvement? Discuss and back up with evidence!

$$\text{Football: } \frac{7-5}{5} = \frac{2}{5} = 40\% \text{ increase in wins}$$

$$\text{Basketball: } \frac{15-10}{10} = \frac{5}{10} = .50\% \text{ increase in wins}$$

In terms of relative difference, the basketball team's increase was more significant, at 50% compared to 40%. On the other hand, the football team went from a losing to a winning record, while the basketball team still had a losing record, so from that standpoint you could certainly argue that their improvement was more significant.

8. In my Calc 2 class yesterday, we were doing approximations of various outputs using advanced calculus techniques. Fancy, right? Anyhow, in one case, the approximation was 1, while the exact value was 0.406. In another case, the approximation was 21, while the exact value was 22.176. Which was the worse approximation? Use the techniques we've used in this lesson, show all work, and write a sentence or two explaining your choice.

Actual differences: 0.594 for the first approximation, 1.176 for the second

This is a good opportunity to discuss the value of considering relative difference, as the result is nonintuitive for most students.

Relative differences:

$\frac{0.594}{0.406} \approx 1.46 = 146\%$ *for the first approximation,*

$\frac{-1.176}{22.176} \approx -0.053 = -5.3\%$ *for the second*

Even though the actual difference is quite a bit more for the second approximation, the relative differences show that the second approximation is WAY better. In the first case, the approximation is too big by 146% of the exact value. The second is too small by just 5.3% of the exact value.

Did You Get It

Try this problem to see if you understand the concepts we just studied. The answer can be found at the bottom of the portfolio page.

2. On one fall day, the expected high temperature was 62° and actual high was 68°. A month later, the expected high was 48° and the actual high was 43°. On which day was the departure from the expected temperature most significant?

1-6 Group

One interesting application of relative difference is relative error. When some quantity is estimated, the difference between the actual amount and the estimated amount is the **actual error.** The **relative error** is the relative difference between those amounts, or the actual error divided by the actual amount.

If you're ever in the market for some new carpet, you might consider searching the Web for the Carpet Buyer's Handbook. The site has many useful resources on buying and installing carpet. It turns out that figuring out how much you're going to spend is more complicated then you might think. Carpet is sold by the square foot, and of course, you can measure the square footage of a room. But you can't buy exactly a certain number of square feet because carpet comes on rolls of a certain width (usually 12 feet). If you need a 2 ft × 3 ft piece for a closet, you can't buy just 6 square feet: you'd have to buy 2 feet off of the roll, which would be 24 square feet. Ouch.

In addition, there are certain guidelines that installers will adhere to in order to get the best result. These govern how the pieces off the roll can be situated, leading to more potential waste. The bottom line is that when a professional installer measures a room to estimate the amount of carpet needed, the result will usually be quite a bit more than the actual square footage that will be covered.

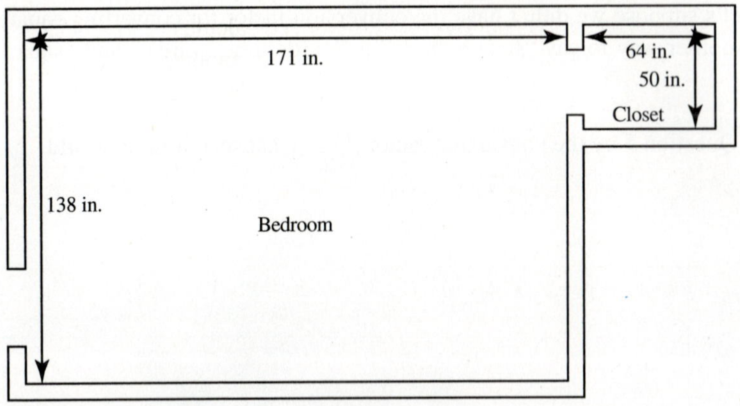

Using professional guidelines, an installer estimated that $24\frac{2}{3}$ square yards of carpet would be needed for the bedroom and closet shown in the diagram.

1. Find the area of the bedroom only in square inches. (Recall that the area of a rectangle is length times width.)

   171 in. × 138 in. = 23,598 sq. in.

2. Find the area of the closet in square inches.

   64 in. × 50 in. = 3,200 sq. in.

3. Each of the doorways shown in 32 inches wide, and the walls are 4 inches thick. Carpet is needed for the entire doorway between the bedroom and closet, and in half of the doorway into the bedroom. How many square inches does this add to the total carpeted area?

   32 in. × 4 in. + 32 in. × 2 in. = 192 sq. in.

4. We know that there are 36 inches in one yard. How many square inches are there in a square yard? The diagrams below should help. (If you don't get the point here, try looking ahead to Question 6.)

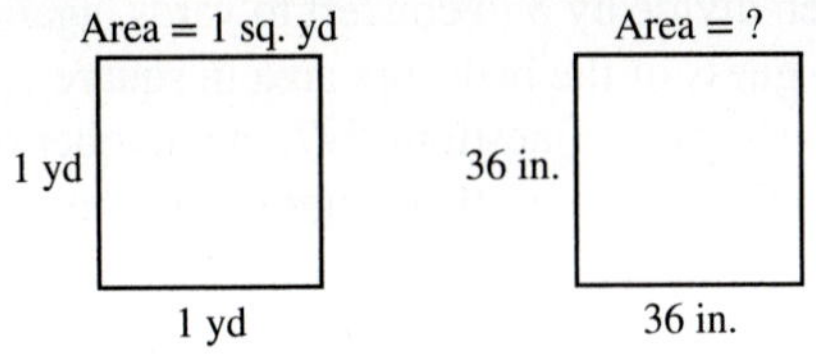

   1,296 sq. in.

5. Use the conversion factor you developed in Question 4 to convert the total carpeted area to square yards. Round to one decimal place.

   23,598 sq. in. + 3,200 sq. in. + 192 sq. in. = 26,990 sq. in.

   $26{,}990 \text{ sq. in.} \times \frac{1 \text{ sq. yd}}{1{,}296 \text{ sq. in.}} \approx 20.83 \text{ sq. yd}$

To understand the significance of Questions 4 and 5, let's suppose we didn't have the conversion factor for converting square inches to square yards we developed in Question 4. Another way to say "square yards" is "yards times yards," and "square inches" is "inches times inches." With that in mind . . .

6. If we multiply the total square inches you used in Question 5 by the conversion factor $\frac{1\text{ yd}}{36\text{ in.}}$, what sort of units would remain? Try it, don't just guess.

$$26{,}990\text{ sq. in.} \times \frac{1\text{ yd}}{36\text{ in.}} \approx 749.72\text{ in.} \cdot \text{yd}$$

7. What if we multiplied that result by $\frac{1\text{ yd}}{36\text{ in.}}$ again? What are the units now?

$$749.72\text{ in.} \cdot \text{yd} \times \frac{1\text{ yd}}{36\text{ in.}} \approx 20.83\text{ sq. yd}$$

What's the point? When dealing with area units (where units are squared) or volume units (where units are cubed), you have two options: You can either create a new conversion factor that will divide away area or volume units (like we did in Questions 4 and 5), or you can use an existing conversion factor as many times as you need to divide away area or volume units (like we did in Questions 6 and 7).

8. Now let's use some estimation skills to see if your answer to Questions 5 and 7 is reasonable, because if it's way off, that'll totally mess up the rest of this activity. The bulk of the area comes from the bedroom, right? Find the length and width of the bedroom in feet, ROUNDING TO THE NEAREST FOOT. Then divide by 3 to convert to yards, again rounding to the nearest yard. Finally, use these measurements to get a rough guess of the bedroom area in square yards. If you have the wrong conversion factor in Question 4, or did the conversion wrong in Questions 5–7, the number of square yards you got will probably be radically different from the estimate in this question. If so, figure out what you did wrong in 4 and/or 5.

Bedroom is roughly 14 ft by 12 ft, which is about 5 yards by 4 yards. This makes the area 20 square yards. That's pretty darn close to the amount we got in Question 5, so our calculation is very reasonable. Cool.

9. Write a sentence or two describing how the professional estimate compares to the actual carpeted area. Use your answer from Question 7.

The estimate is almost 4 square yards more than the actual carpeted area.

10. Find the relative difference between the estimate and the actual area of the carpeted portion of the room. Round to two decimal places. Write a sentence explaining what this tells us about how the estimate compares to the actual carpeted area. Writing the relative difference as a percentage is a good idea.

$$\frac{24.67 - 20.83}{20.83} \approx 0.18$$

The estimate is about 18% more than the actual carpeted area.

11. Not everyone realizes this, but when you buy carpet, you pay for the estimated amount, not the actual amount covering your floors. If the family that owns the house in the diagram picked carpet that costs $28 per square yard installed, how much would they have paid?

$$\$28 \times 24\frac{2}{3} \approx \$690.67$$

12. Use the actual difference between the estimate and the carpeted area to find the dollar value of wasted carpet.

$$\$28 \times (24\frac{2}{3} - 20.83) = \$107.43$$

Did You Get It

Try this problem to see if you understand the concepts we just studied. The answer can be found at the bottom of the portfolio page.

3. Lawn services typically charge for their fertilizer treatments based on the estimated size of your lawn in square feet. The first company I used when I moved into my current house estimated our lawn at 10,000 square feet. My lot is 16,800 square feet, and by my calculation, about 8,340 square feet are covered by concrete, house, pool, and mulch. Find the relative error in the lawn company's estimate.

Percent error plays a really important role in the sciences and engineering. In electrical engineering, for example, much of the equipment used in transmitting power is rated for a certain amount of voltage, amperage, or resistance. Systems are designed so that fluctuations in these values can be absorbed without damage, as long as those fluctuations are within a certain range. These ranges are measured using relative error.

13. Suppose that one transformer is designed to handle a maximum of 2.4 megavolts (MV), and it has the ability to handle a 15% overage. If the voltage spikes at 2.91 MV during a power surge, can the transformer handle this? Show your work.

The relative difference is $\frac{2.91 - 2.4}{2.4} \approx 0.2125$. This is an overage of 21.25%, which would have damaged the transformer.

14. Based on Question 13, it would seem to be a natural question to ask exactly how many megavolts the transformer can handle without sustaining damage. We can use some algebra and the calculation above to figure that out. Using the variable V for the voltage, and 0.15 for the relative error, set up and solve an equation that tells us the voltage that corresponds to a 15% overage.

$$\frac{V - 2.4}{2.4} = 0.15 \Rightarrow V - 2.4 = 0.15(2.4) \Rightarrow V = 0.15(2.4) + 2.4 = 2.76$$

The transformer can handle up to 2.76 MV.

15. List some situations where relative difference is a much more useful number than actual difference. Be creative!

Answers vary widely. Share some of your own with the students!

1-6 Portfolio

Name ______________________________

Check each box when you've completed the task. Remember that your instructor will want you to turn in the portfolio pages you create.

Technology

1. ☐ Complete the following spreadsheet. Use formulas in all the cells in row 3, then copy them to fill out the remaining rows. Then write any observations you have based on the results. A template to help you get started can be found in the online resources for this lesson.

Point out to students that the technology exercise is asking for the difference and relative difference from one amount to the next, not between each amount and the starting value. Mention that relative difference can sometimes refer to relative change.

	A	B	C	D	E	F	G
1	**Account #1**	**Difference**	**Relative Difference**		**Account #2**	**Difference**	**Relative Difference**
2	$10,000.00				$10,000.00		
3	$10,700.00				$10,500.00		
4	$11,400.00				$11,025.00		
5	$12,100.00				$11,576.25		
6	$12,800.00				$12,155.06		
7	$13,500.00				$12,762.82		
8	$14,200.00				$13,400.96		
9	$14,900.00				$14,071.00		

Online Practice

1. ☐ Include any written work from the online assignment along with any notes or questions about this lesson's content.

Applications

1. ☐ Complete the applications problems.

Reflections

Type a short answer to each question.

1. ☐ How would you explain to someone you know that relative difference is often much more meaningful than actual difference?

2. ☐ Think of a situation in which relative difference would be more meaningful than actual difference. Using one of the situations in the book would be totally cheating.

3. ☐ Take another look at your answer to Question 0 at the beginning of this lesson. Would you change your answer now that you've completed the lesson? How would you summarize the topic of this lesson now?

4. ☐ What questions do you have about this lesson?

Looking Ahead

1. ☐ Complete the Lesson 1-7 Prep Skills and read the opening paragraph in Lesson 1-7 carefully, then answer Question 0 in preparation for that lesson.

Answers to "Did You Get It?"

1. 0.25 billion; 7.2% **2.** Second day: –10.42% compared to 9.68%

3. 0.182, or 18.2%

1-6 Applications

Name ______________________________

After moving into a new house, a couple wants to have a concrete patio poured to support a hot tub, because . . . well, because it's a hot tub. The plans call for a 14-ft by 16-ft slab of concrete 4 inches thick. Round any calculations to two decimal places if you need to.

1. How many cubic feet of concrete will be needed?

$14 \times 16 \times (1/3) = 74.67$ cubic feet

2. Convert the volume of concrete to cubic yards. (Hint: A cube that is one yard on each side is also 3 feet on each side. You can use that to find a conversion factor from cubic feet to cubic yards.)

$$74.67 \text{ cu ft} \times \frac{1 \text{ cu yd}}{27 \text{ cu ft}} = 2.77 \text{ cu yd}$$

3. The coolest thing about ordering ready-mix concrete is that a concrete mixer will show up at your house. Most companies require orders by the cubic yard. The patio above was estimated at 3 cubic yards. How much extra concrete was ordered?

0.23 cu yd

If this baby shows up in your driveway, you're super-cool.

4. Should the relative error in the amount of concrete be positive or negative? Don't use a formula! Just think about the amount needed and the amount ordered. Explain.

It should be positive, because the amount ordered is more than the amount needed.

5. Find the relative error in the estimate compared to the actual amount of concrete needed.

$$\frac{3 - 2.77}{2.77} = \frac{0.23}{2.77} \approx 0.083 \text{ or about } 8.3\%$$

1-6 Applications

Name ______________________________

6. In 2015, an average cost for concrete was $140 per cubic yard. How much would be spent on concrete for the patio at that price?

$3 \times \$140 = \$420$

7. Find the dollar value of the wasted concrete.

$0.23 \times \$140 = \$32.20$

Everett Dirksen, a senator from Illinois, is widely reported to have said "A billion here, a billion there, pretty soon you're talking about real money" when speaking about the federal budget deficit. On the surface, that statement sounds pretty ridiculous—if you had a billion dollars, you'd be one of the 2,000 richest people in the world. But let's look at some numbers to see what the good senator was getting at.

8. In 1961, around the time that this quote is typically attributed to Dirksen, total federal spending was $554 billion, and the deficit was +$18.9 billion. (This means that the government spent $18.9 billion more than it took in.) Suppose that an extra $1 billion dollars got lost in the shuffle. What would be the relative difference in the amount spent? What about in the deficit? Find these relative differences, then discuss how significant you think that $1 billion is.

Spending: $\frac{555 - 554}{554} = 0.0018$ or 0.18% difference

Deficit: $\frac{19.9 - 18.9}{18.9} = 0.0529$ or 5.29% difference

Significance is an opinion, so answers will vary.

9. In 2014, total federal spending was $3.506 trillion, and the deficit was +$484.6 billion. If an extra $1 billion was lost in the shuffle, find the relative difference in both spending and the deficit. Again, discuss how significant you think that $1 billion is. You should probably contrast your results from Questions 7 and 8. (By the way, a trillion is a thousand billion.)

Spending: $\frac{3{,}507 - 3{,}506}{3{,}506} = 0.000285$ or 0.028% difference

Deficit: $\frac{485.6 - 484.6}{484.6} = 0.0021$ or 0.21% difference

Significance is an opinion, so answers will vary. But obviously a good answer will point out that the significance is much less than it was in 1961. Maybe Dirksen was on to something.

Lesson 1-7 Prep Skills

This is a short review of skills that will come in handy in the next lesson. In each case, answer the question, then rate your confidence level by checking one of the boxes. If you feel like you're struggling with these skills, consult the online resources provided by your instructor for extra practice.

I KNOW IT | THINK SO | UNSURE | NO IDEA

SKILL 1: EVALUATE POWERS OF TEN

1. Evaluate 10^5
2. Evaluate 10^{-3}

SKILL 2: WRITE REPEATED MULTIPLICATION AS AN EXPONENT

3. Write the expression in exponent form: $8 \cdot 8 \cdot 8 \cdot 8 \cdot 8 \cdot 8 \cdot 8$
4. Write the expression in exponent form, with no fraction:

$$\frac{1}{10 \cdot 10 \cdot 10 \cdot 10 \cdot 10}$$

SKILL 3: MULTIPLY EXPONENTIAL EXPRESSIONS

5. Perform each multiplication, writing your answer with a single exponent:

$10^5 \cdot 10^{12}$ $\quad$ $8^{-3} \cdot 8^2$

SKILL 4: DIVIDE EXPONENTIAL EXPRESSIONS

6. Perform each division, writing your answer with a single exponent:

$\dfrac{5^7}{5^3}$ $\quad$ $\dfrac{10^4}{10^{11}}$

SKILL 5: SOLVE A FORMULA FOR ONE VARIABLE

7. Solve the formula for the area of a rectangle for the length, *l*.

$A = lw$

Lesson 1-7 Attraction and Melted Chocolate (Scientific Notation)

LEARNING OBJECTIVES

- ☐ 1. Write large and small numbers in scientific notation.
- ☐ 2. Describe the significance of writing numbers in scientific notation.
- ☐ 3. Study Newton's law of gravitation.
- ☐ 4. Compute the speed of light experimentally.

© Image Source/Getty Images RF

Mysteries of attraction could not always be explained by logic.
—Lisa Kleypas

Attraction: a mysterious topic that is like oxygen to the romance novelist and movie producer. We all feel emotional or physical attractions to certain people. But did you know that there's a real attraction between you and every person you come into contact with? It's called "gravity," and one of the physical laws of our universe is that any two objects have a gravitational attraction that can be quantified. This attraction is so tiny that it's hard to quantify (and you can't feel it) when the objects are small, like people. But when objects are really big, like the Earth and Sun, the gravitational attraction is very strong. Dealing with the force of gravity, by its very nature, requires working with both very small and very large numbers. Using regular decimal notation, this can be pretty cumbersome: do 0.0000004 and 0.000000004 look very different to you? One is a *hundred times as big* as the other! In this lesson, we'll use attraction and melted chocolate to study a special way of writing very small or very large numbers in a more readable, convenient way.

0. After reading the opening paragraph, what do you think the main topic of this lesson will be?

1-7 Class

Newton's law of universal gravitation describes the gravitational attraction between any two objects. The formula is

$$F = G\frac{m_1 m_2}{r^2}$$

where F is the force of the gravitational attraction, m_1 and m_2 are the masses of the two objects, and r is the distance between the centers of the two objects. G is a number known as the **gravitational constant:** We'll see what that number is in a bit.

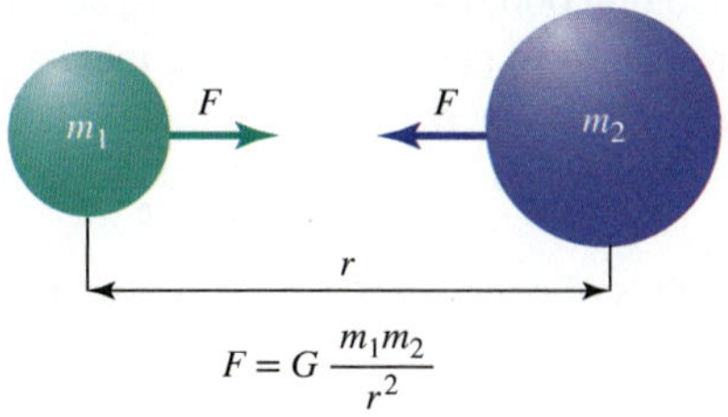

1. In Newton's law of universal gravitation, we were told that G is a constant. Look carefully at the equation that describes gravitational attraction. Describe the relationship between the gravitational force between two objects and the distance separating their centers. If the distance goes up, what happens to the force? What if the distance goes down?

   As the objects get closer (that is, the distance between them gets smaller), the gravitational force between them gets greater, and as the objects move further apart, the gravitational force between them gets smaller.

2. Repeat Question 1, but this time describe the relationship between the gravitational force and the mass of the first object, m_1. (It might help to pretend that everything except F and m_1 are just numbers.)

   If the mass of the first object is larger, the force will be greater. If the object is replaced with something smaller, the force gets smaller.

3. Why can you feel the gravitational attraction between you and the Earth, but not between you and the person sitting next to you? (Notice that I said "gravitational attraction." 😉) Refer to your answer to Question 2.

   Because the Earth has such a large mass. The masses of both you and the person next to you are basically negligible when compared to the mass of the Earth.

4. Why can you feel the gravitational attraction between you and the Earth, but not between you and Jupiter? Refer to your answer to Question 1.

   Because I live on the Earth; the distance between us is small. But Jupiter is millions of miles away, so the gravitational force becomes negligible.

The force of gravity between the Earth and the Sun is strong enough to keep us in orbit because even though we're about 149,597,870,700 meters apart, the mass of the Earth is 5,973,600,000,000,000,000,000,000 kilograms, and the mass of the Sun is 1,989,000,000,000,000,000,000,000,000,000 kilograms. If you're like most people (including me, so don't feel bad) you have NO idea what those numbers actually mean. They're just too big to get any perspective on them other than "wow, those are really big." To make it easier to understand and work with numbers of ludicrous size, we'll use a clever method for writing them using exponents.

Consider the number 300, which can also be written as 3×10^2, which is of course 3×100. If you're thinking "that's silly," hang in there . . . there is a point. Take a look at the first handful of powers of 10:

$10^1 = 10 \qquad 10^2 = 100 \qquad 10^3 = 1{,}000 \qquad 10^4 = 10{,}000 \qquad 10^5 = 100{,}000$

Do you see the pattern? Each extra power of 10 puts another zero on the decimal equivalent. In fact, the power indicates exactly the number of zeros included. Here's why this is helpful: Five billion is a tremendously large number: 5,000,000,000. Notice that there are nine zeros, so we can write five billion as 5×10^9, which is more concise and easier to read. If that makes sense to you, then you understand the idea behind *scientific notation.*

Writing Large Numbers in Scientific Notation

1. Write a decimal point at the end of the number, then move it left so that there is just one digit before it.
2. Multiply the result by 10 to some power. The power will be the number of places you moved the decimal.

Example: 3,968,000

3,968,000. (decimal moved six places)

3.968 (the extra zeros are unnecessary)

3.968×10^6

For Questions 5 and 6, write each number in scientific notation.

Many students are at least somewhat familiar with scientific notation, so the rest of this Class portion could also be done in groups.

5. The number of people in the United States in November 2015: 322,146,000

3.22146×10^8 people

6. The distance in miles from Earth to the Moon: 238,900

2.389×10^5 miles

We can also use scientific notation to write very small numbers, using negative powers of 10:

$$10^{-1} = \frac{1}{10} = 0.1 \qquad 10^{-2} = \frac{1}{10^2} = 0.01 \qquad 10^{-3} = \frac{1}{10^3} = 0.001 \qquad 10^{-4} = \frac{1}{10^4} = 0.0001$$

This time, the pattern is that when the negative exponent gets larger by one (meaning more negative), the decimal point slides one spot to the left. So, for example, 0.03 is the same as 3×10^{-2}.

Writing Small Numbers in Scientific Notation

1. Move the decimal point right so that there is just one nonzero digit before it. Drop all zeros that come before that digit.
2. Multiply the result by 10 to some power. The power will be the negative of the number of places you moved the decimal.

Example: 0.000437

0.000437 (decimal moved four places)

4.37

4.37×10^{-4}

In Questions 7 and 8, write each number in scientific notation.

7. The weight in grams of one grain of spruce pollen: 0.00007

7×10^{-5} grams

8. The average number of hours it took the U.S. government to spend $1 million dollars in 2012: 0.0021605

2.1605×10^{-3} hours

Did You Get It

Try this problem to see if you understand the concepts we just studied. The answer can be found at the bottom of the portfolio page.

1. Convert each number to scientific notation:

a. 0.0098 b. 1,478 c. 54,302,000 d. 0.000 0058

Now that we're good at converting from decimal to scientific notation, it would seem appropriate, if not expected, to work on the reverse: converting back from scientific to decimal notation.

Coverting from Scientific to Decimal Notation

1. If the power of ten is positive, move the decimal point to the right the same number of places as the exponent. You might need to put in zeros as placeholders when you run out of digits. Putting in commas will make it easier to read a large number.

Example: 4.01×10^8

4.01 (decimal moved eight places)

401,000,000 (fill in extra zeros)

2. If the power of ten is negative, move the decimal point to the left the same number of places as the exponent, again putting in zeros as needed.

Example: 3.2×10^{-5}

3.2 (decimal moved five places)

0.000032

9. Write each number in decimal notation.

a. The number of drops in a gallon of water: 9.085×10^4

90,850 drops

b. The mass of an average dust particle: 7.53×10^{-10} grams

0.000 000 000 753 grams

Did You Get It

Try this problem to see if you understand the concepts we just studied. The answer can be found at the bottom of the portfolio page.

2. Convert each number to decimal notation:
   a. 6.3×10^7 b. 2.973×10^{-3} c. 4.999×10^5 d. 9.103×10^{-6}

Now let's see if you really understand scientific notation.

10. When a number written in scientific notation has a positive exponent on 10, what can we say for sure about the number?

That it's bigger than 10.

11. If a number between 0 and 1 is written in scientific notation, what can we say about the exponent on 10?

That it's negative.

12. Fill in the blanks: If a number in scientific notation has a negative exponent on 10, to convert to decimal form, move the decimal place ___left___ because the number is ___less than 1___.

Using Technology: Scientific Notation in Calculators and Excel

Spreadsheets and graphing calculators are both programmed to provide answers in scientific notation only if the size of the result is especially large or small.

TI-84 Plus Calculator

```
99*125
                12375
990000*125000
            1.2375E11
(1.5E12)(3.2E-8)
                48000
```

Notice that whether the numbers are entered in decimal or scientific notation, the result is in scientific notation only if it's very large. The calculator displays 1.2375×10^{11} as 1.2375E11.

To input 1.5×10^{12} in scientific notation:

Press 1.5 [2nd] [,] 12. This is listed as "EE" on the calculator.

If you want to force the results of calculations to be in scientific notation, put the calculator in scientific mode:

Press [MODE] [›] [ENTER] to set the calculator to scientific mode. The result:

```
NORMAL SCI ENG
99*125
             1.2375E4
```

Excel

C1 = =A1*B1

	A	B	C
1	99	125	12375
2	990000	125000	1.24E+11
3	1.50E+12	3.20E-08	48000
4			
5	99	125	1.24E+04

Notice that in rows 1–3, Excel decides whether or not to put the result in scientific notation depending on its size.

To input 1.5×10^{12} in scientific notation, as in cell A3, type 1.5E 12. Excel will interpret this as scientific notation.

If you compare rows 1 and 5, you can see that the product is the same in each case: 99 * 125. The difference in the format of the output is based on the formatting option chosen for the cell. In cell C1, the **General** format was chosen from the **Number** menu, in which case Excel decides on the format as mentioned. In cell C5, **Scientific** was chosen from the **Number** menu, so the result is shown in scientific notation regardless of its size.

See the Lesson 1-7 video in class resources for further instruction.

13. Write the result of the calculation on the calculator screen in scientific and decimal notation.

```
(2.4E-8)*(3.1E11
)
               7.44E3
```

7.44×10^3

7,440

14. A typical MP3 file fills about 4×10^6 bytes of disk space. How many average songs can my iPod classic, which has 80 gigabytes of disk space, hold? Each gigabyte is 1×10^9 bytes.

$80 \times (1 \times 10^9) = 80 \times 10^9 = 8 \times 10^{10}$

$\frac{8 \times 10^{10}}{4 \times 10^6} = 2 \times 10^4$ or 20,000 songs

15. Use Newton's law of universal gravitation to find the gravitational force between the Earth and Sun. The gravitational constant G is 6.674×10^{-11}; the distance from the Earth to the Sun is 1.496×10^{11} meters; the mass of the Earth is 5.974×10^{24} kg; the mass of the Sun is 1.989×10^{30} kilograms. (The units of force are the kilogram meter per second squared, also known as the Newton.)

$F = G\frac{m_1 m_2}{r^2} \Rightarrow F = 6.674 \times 10^{-11} \cdot \frac{5.974 \times 10^{24} \cdot 1.989 \times 10^{30}}{(1.496 \times 10^{11})^2} = 3.543 \times 10^{22}$ Newtons

16. Newton's second law of motion is $F = ma$, where F is force, m is mass, and a is acceleration. For a situation where the force is caused by gravity, say when you drop a bowling ball off of a building, we can use the universal law of gravitation with $a = G\frac{m_1}{r^2}$. (Legal disclaimer: If you drop a bowling ball off of a building, you're a jerk.) Use the value of G, the mass of the Earth, and the radius of the Earth (6.3781×10^6 meters) to find the acceleration due to gravity near the Earth's surface. How does your answer compare to the value used in a typical physics course, $a = 9.8$? (For the record, the units for a are meters per second per second, or meters per second squared if you like.)

$a = G\frac{m_1}{r^2} = 6.674 \times 10^{-11} \cdot \frac{5.974 \times 10^{24}}{(6.378 \times 106)^2} = 9.801$

This matches the value to two decimal places.

Did You Get It

Try this problem to see if you understand the concepts we just studied. The answer can be found at the bottom of the portfolio page.

3. The fastest spacecraft ever made is the New Horizons probe, launched in January 2006, at 5.8×10^7 meters per hour (which is about 36,000 mph). Find how long it would take this craft to travel from New York's JFK airport to LAX in Los Angeles (3.994×10^6 meters), and how long it would take to get to the Sun (distance provided in Question 15).

1-7 Group

A fun fact: you can use a microwave oven and a bar of chocolate to measure the speed of light. Really! Microwaves are electromagnetic waves, just like visible light. The difference is the wavelength of these waves: For visible light, the wavelengths are in the neighborhood of 10^{-6} m, while microwaves are close to 10^{-2} m. Basically, microwaves emitted in an oven make molecules in food vibrate, which causes heat. The first places to heat up are separated by half of a wavelength, so by using something that starts to melt visibly in certain spots, we can calculate the wavelength. We can then use that to calculate the speed of light, which I think we can all agree is pretty darn cool.

If your teacher is unusually adventurous, he or she might bring in a microwave and chocolate bars, in which case you can perform the experiment on your own. Start with around 20 seconds, and make sure you take out the carousel: the chocolate has to stay stationary. (And now you know why a lot of microwaves have a carousel: food heats more thoroughly at the spots corresponding to half of the wavelength.)

In case no microwave and chocolate are available, I did the experiment at home and took a picture for you, which is reproduced below. It's a life-size scale, so you can measure the distance between the melted spots with a ruler.

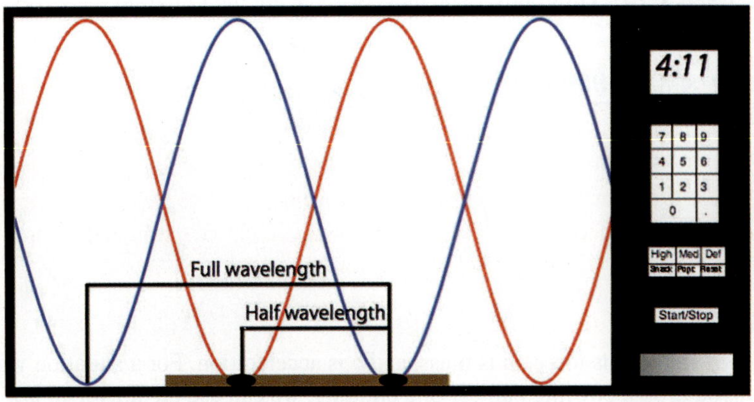

This is fun to do in class. Try it out beforehand—if you're not getting two spots, you may need to use a larger chocolate bar, change its direction in the microwave, or move it to another part of the microwave. Bring plates and do one for each group.

Courtesy of David Sobecki

1. Measure the distance between melted spots on your chocolate, trying to measure from center to center. Record your distance below, remembering that this is half the wavelength of the microwaves.

Distance: 6.1 cm

2. Convert the half-wavelength to meters. (Think about what the prefix "centi" means!)

Distance: 0.061 m

3. Write the full wavelength: 0.122 m

4. If you have any, eat the chocolate.

Mmmmmmmm chocolate.

5. For electromagnetic waves, the equation $c = \lambda f$ relates the speed of light (c), the wavelength (λ), and the frequency (f), which is the number of cycles per second (Hz). The frequency of most microwaves is 2.45 GHz: one GHz represents 10^9 cycles per second. Write the microwave frequency in scientific notation.

2.45×10^9 cycles per second

6. Use your answers to Questions 3 and 5 to calculate the speed of light in meters per second.

$c = (0.122)(2.45 \times 10^9) = 2.989 \times 10^8$ m/sec

7. Use your phone or computer to look up the speed of light and write it here: 299,792,458 m/sec

8. Find the error and relative error in your estimate as a decimal, and as a percentage. If your answer is WAY off, you might want to check the microwave to see if the frequency is different from 2.45 GHz.

Error: $299{,}792{,}458 - 2.989 \times 10^8 = 892{,}458$

Relative error: $\dfrac{892{,}458}{299{,}792{,}458} \approx 0.0030$ or 0.3%

Math Note

There's a link to a really cool video about how microwave ovens work in the online resources for this lesson.

9. This one requires some thought, but it will help you in the homework if you're successful. When a number is written in scientific notation with a negative exponent, it's almost always going to begin with a decimal point and a string of zeros in decimal notation. Looking over the conversion process, develop a rule for the number of zeros to start with.

The number of zeros is always one less than the absolute value of the exponent. For example, if the exponent is −8, the decimal form will start with 7 zeros.

10. Describe a couple of quantities that your group finds interesting that would be efficiently written in scientific notation. Describe why you think scientific notation would be reasonable for the quantities.

Answers vary. This is another good place to encourage creativity.

1-7 Portfolio

Name ______________________________

Check each box when you've completed the task. Remember that your instructor will want you to turn in the portfolio pages you create.

Technology

1. ☐ Watch the video on scale linked in the online resources for Lesson 1-7. Then tell everyone you know how cool it is. Even people you don't like.

2. ☐ After completing the Applications problems, create a spreadsheet that calculates the wavelength for the three FM radio stations you listen to most often. The spreadsheet should contain formulas in columns C and D to compute those values when you enter the information in columns A and B. A template to help you get started can be found in the online resources for this lesson.

	A	B	C	D
1	Radio Station	MHz	Hz	Wavelength
2	WLRW	94.5	94500000	3.172407

Online Practice

1. ☐ Include any written work from the online assignment along with any notes or questions about this lesson's content.

Applications

1. ☐ Complete the applications problems.

Reflections

Type a short answer to each question.

1. ☐ What is the point of scientific notation?

2. ☐ When a number is in scientific notation, how can you tell if it's really large or really small? Explain.

3. ☐ Take another look at your answer to Question 0 at the beginning of this lesson. Would you change your answer now that you've completed the lesson? How would you summarize the topic of this lesson now?

4. ☐ What questions do you have about this lesson?

Looking Ahead

1. ☐ Complete the Lesson 1-8 Prep Skills and read the opening paragraph in Lesson 1-8 carefully, then answer Question 0 in preparation for that lesson.

Answers to "Did You Get It?"

1. a. 9.8×10^{-3} **b.** 1.478×10^{3} **c.** 5.4302×10^{7} **d.** 5.8×10^{-6}

2. a. 63,000,000 **b.** 0.002973 **c.** 499,900 **d.** 0.000009103

3. About 4.13 minutes, about 107.5 days

1-7 Applications

Name ______________________________

Radio waves are also examples of electromagnetic waves. When you listen to a radio station, the frequency that you find on the dial describes the frequency of the waves sent out by that particular station. Since all electromagnetic waves move at the speed of light, you can calculate the wavelength for a station. For FM stations, like 94.5, this frequency is in megahertz, which is millions of cycles per second. AM stations on the other hand, like 1100, are in kilohertz, which is thousands of cycles per second.

1. Explain why the value of c never changes in the formula $c = \lambda f$.

Because c represents the speed of light, which is constant. Depending on the wave, both the frequency and wavelength can vary but the speed of light never changes.

2. Since c doesn't change in the formula $c = \lambda f$, what happens to f if λ increases? What happens to λ if f increases? Describe each situation carefully.

If the wavelength increases, the frequency has to go down in order for the equation to remain true. If the frequency increases, the wavelength has to go down.

3. Write an equation that calculates λ if you know c and f. (Hint: Start with $c = \lambda f$.)

$$\lambda = \frac{c}{f}$$

4. If you're listening to the Bob and Tom show on 94.7 FM, what is the frequency? Write it in scientific notation in terms of hertz.

94.7 million cycles per second, which is 9.47×10^7 hertz.

5. What is the wavelength of the radio waves from this station?

$$\lambda = \frac{299{,}792{,}458}{9.47 \times 10^7} \approx 3.17 \text{ m}$$

1-7 Applications

Name ______________________________

6. How many times larger is your answer to Question 5 than the wavelength for microwaves calculated in the Group portion of this lesson?

$\frac{3.17\ m}{0.122\ m} \approx 26$ The waves are about 26 times longer.

7. Based on your answer to Question 6, do you think it would be possible for you to get interference sounds on an FM radio that's sitting close to a microwave? Why or why not?

This is an opinion question. Look for the student talking about whether the radio would be likely to pick up sound from waves that differ in length. A really good answer might also talk about the two waves combining and interfering, like waves in water can do.

8. Repeat Questions 4-7 for a baseball game on 1120 AM.

1120 thousand cycles per second, which is 1.120×10^6 hertz.

$\lambda = \frac{299{,}792{,}458}{1.12 \times 10^6} \approx 267.7\ m$ $\frac{267.7\ m}{0.122} \approx 2{,}194.3$

The waves are about 2,194 times longer. Answers can again vary about radio interference, but it sure seems less likely here.

Lesson 1-8 Prep Skills

This is a short review of skills that will come in handy in the next lesson. In each case, answer the question, then rate your confidence level by checking one of the boxes. If you feel like you're struggling with these skills, consult the online resources provided by your instructor for extra practice.

I KNOW IT	THINK SO	UNSURE	NO IDEA	
				SKILL 1: DIVIDE FRACTIONS
				1. Perform the division: $\dfrac{\frac{7}{3}}{\frac{35}{2}}$
				2. Simplify: $\dfrac{a}{\frac{b}{c}}$
				SKILL 2: ROUND TO NEAREST WHOLE NUMBER
				3. Round each value to the nearest whole number. 3.257 8.501 57.499
				SKILL 3: DIVIDE TWO NUMBERS, THEN ROUND RESULT
				4. Perform the division and round to the nearest whole number: $247 \div 7$
				SKILL 4: APPROXIMATE THE SQUARE ROOT OF A PRODUCT
				5. Use a calculator or spreadsheet to approximate the value of each calculation to one decimal place. $\sqrt{6 \cdot 7}$ $\sqrt{22 \cdot 23}$

Lesson 1-8 Portion Control (Apportionment)

© Brand X Pictures/PunchStock RF

LEARNING OBJECTIVES

- ☐ 1. Describe what apportionment is and why it's used.
- ☐ 2. Compute standard divisors and quotas.
- ☐ 3. Apportion items using a variety of methods

The principal difficulty lies, and the greatest care should be employed in constituting this representative assembly. It should be in miniature an exact portrait of the people at large.
—John Adams

You arrive at an 8 A.M. final (ugh!) and find that at least your professor was kind enough to bring donuts. Let's say she brought 4 dozen donuts and there are 24 people in class. How can she fairly distribute them? The obvious answer is two donuts per person. This is a very simple example of *apportionment.* It's a simple example because everyone gets the same number of donuts and there are none left over. But what if some people are entitled to more donuts for one reason or another? And what if the number of students doesn't evenly divide the number of donuts? Then the situation gets more complicated.

If you had to, of course, you could cut the donuts into pieces. But what if you're trying to fairly dole out computers to schools, cars to police stations, or sonogram machines to hospitals? Half of a computer won't do anyone any good. These items are called **indivisible** because they can't be divided into smaller pieces. The study of apportionment is all about coming up with fair ways to divide and assign indivisible items.

0. After reading the opening paragraph, what do you think the main topic of this lesson will be?

1-8 Class

The most common example of apportionment (and, as we'll see, where the study came from in the first place) is the United States Congress. There are 100 senators and 50 states: 2 senators per state—easy! But in the House of Representatives, the number of representatives is based on the population of the state. There are 435 seats in the House, and they're divided according to population. For example, Florida has 25 seats, Pennsylvania has 19, and Rhode Island has just 2.

So why can't we just divide 435 by the population to find out how many representatives per person, then multiply by the state's population? Because it's against the law to divide politicians into pieces. (Insert politician joke here.) Almost every state would be entitled to a noninteger number of seats, which won't work. That's where apportionment comes in. **Apportionment** is the process of fairly assigning identical indivisible objects among individuals entitled to shares that may be unequal.

We just pointed out that the most logical way to attack dividing up the seats in Congress would be to divide the entire population by the number of seats: this would tell us how many people on average would be represented by one congressperson. This number is known as the **standard divisor,** and it plays a really important role in apportionment, so we'll highlight the formula next. But don't think about memorizing it: think about what it MEANS, and you'll have a much better chance of understanding this lesson.

Throughout this lesson, we'll refer to the objects being assigned as "seats" (like seats in Congress). The individuals they're allocated to will be called "districts."

The Standard Divisor

The standard divisor for an apportionment process is the average number of people per seat:

$$\text{Standard divisor} = \frac{\text{Total population}}{\text{Number of seats}}$$

Here's the apportionment problem we'll be working on in this lesson:

A large city has 20 seats to be allocated proportionally among 5 districts according to the population of each district. The populations of the districts (in thousands) are shown in the table.

District	1	2	3	4	5	Total
Population (in thousands)	32	80	41	67	20	240

1. Explain why it wouldn't be fair to just assign 4 representatives to each district.

Because the districts have widely different populations. District 2, for example, has four times as many citizens as District 5. It wouldn't be reasonable for them to have the same number of representatives.

2. Find the standard divisor for this apportionment, and briefly describe what it means.

240,000/20 = 12,000. This means that on average, there should be one seat for every 12,000 citizens.

Questions 2 and 3 are where students ideally get a real understanding of why these "fancy" methods of apportionment are necessary. Make sure they focus on these questions.

3. Notice that District 2 has exactly one-third of the population. It would seem perfectly reasonable then for that district to get one-third of the seats. Why won't District 2 get exactly one-third of the seats?

Because the seats are indivisible, and 20 doesn't divide evenly by 3. That district would have to get seven and a third seats, which isn't possible.

Did You Get It ?

Try this problem to see if you understand the concepts we just studied. The answer can be found at the bottom of the portfolio page.

1.

District	1	2	3	4	5	6	7	Total
Population (in thousands)	125	89	235	97	102	128	184	960

The table above gives the population in thousands of seven districts of a particular county. The county assembly has 30 seats that are to be allocated according to the population of each district. Find the standard divisor.

While we've seen that you can't assign seats simply based on what percentage of the total population a given district has, that percentage still plays an important role in apportionment. If we divide the population of a given district by the standard divisor, something interesting happens.

$$\frac{\text{Population of district}}{\text{Standard divisor}} = \frac{\text{Population of district}}{\frac{\text{Total population}}{\text{Number of seats}}}$$

4. Given the fact that dividing by a fraction is the same as multiplying by its reciprocal, fill in the blank to rewrite the division above as a multiplication.

$$\frac{\text{Population of district}}{\text{Standard divisor}} = \text{Population of district} \times \frac{\text{Number of seats}}{\text{Total population}}$$

Now we can rearrange the multiplication a bit using properties of multiplying fractions:

$$\frac{\text{Population of district}}{\text{Total population}} \times \text{Number of seats}$$

Compare this to the line of reasoning in Question 3: the fraction part of the above formula is exactly the percentage of the total population that lives in a given district (in decimal form). With that in mind . . .

5. Perform the calculation in the formula above for District 3 in our example problem. Then write a description of what the result of the calculation tells you.

$$\frac{41}{240} \times 20 \approx 3.42$$

This means that District 3 would be entitled to 3.42 seats if the seats were not indivisible.

The formula we built, which is the percentage of the total population that lives in a given district multiplied by the total number of seats, is the theoretical number of seats that each district would get if it were possible to hand out fractional portions of seats. We call this number the **standard quota.**

The Standard Quota

The standard quota for a district is the population of that district divided by the standard divisor:

$$\text{Standard quota} = \frac{\text{Population of district}}{\text{Standard divisor}}$$

It describes the number of seats that each district would be assigned if it made sense to assign fractional portions of seats.

6. Find the standard quota for each district in the city described on page 96.

Dist. 1: $\frac{32{,}000}{12{,}000} \approx 2.67$

Dist. 2: $\frac{80{,}000}{12{,}000} \approx 6.67$

Dist. 3: $\frac{41{,}000}{12{,}000} \approx 3.42$

Dist. 4: $\frac{67{,}000}{12{,}000} \approx 5.58$

Dist. 5: $\frac{20{,}000}{12{,}000} \approx 1.67$

Keep referring students back to Question 5 to remind them of what exactly the standard quota is, and why it makes sense that EVERY reasonable apportionment method will give each district something very close to its standard quota.

Math Note

For a given apportionment problem, if it just so happens that all of the standard quotas work out to be whole numbers, then you have a perfectly fair apportionment, and life is good. But in most situations, that's spectacularly unlikely, which is why we have more work to do.

Did You Get It

Try this problem to see if you understand the concepts we just studied. The answer can be found at the bottom of the portfolio page.

2. Find the standard quota for each of the districts in Did You Get It 1.

Now that we know how to *almost* find the number of seats that should go to each district (aside from the pesky issue of fractional parts), the rest of the study of apportionment is about deciding if each district's number of seats should come from rounding its standard quota up to the next bigger whole number, or down to the next smaller number. Seriously, that's pretty much it. A logical place to start is by simply using familiar rounding rules, so let's give that a try.

7. Round each of the standard quotas you calculated in Question 6 to the nearest whole number. Then add the results and explain why that won't work as the number of seats assigned to each district.

District 1: 3

District 2: 7

District 3: 3

District 4: 6

District 5: 2

Unfortunately, this adds up to 21 seats, but only 20 are available.

Library of Congress, Prints and Photographs Division [LC-DIG-det-4a26166]

Question 7 shows that simply rounding each standard quotient to the nearest whole number isn't a reliable method of apportionment. This realization dates back to at least 1790, when both Alexander Hamilton (the first Secretary of the Treasury) and Thomas Jefferson (hopefully you know who he was) suggested methods for apportioning congressional seats after our nation's first official census.

The methods of apportionment that we'll study are all based on first calculating the standard quota, then deciding whether to round up or down. For any district, the **lower quota** is the result of rounding the standard quota down to the next lower whole number. The **upper quota** is the result of rounding up to the next larger whole number.

Library of Congress, Prints and Photographs Division [LC-USZ62-117117]

8. Use your results from Question 6 to fill in the table below for each district.

District	1	2	3	4	5
Standard quota	2.67	6.67	3.42	5.58	1.67
Lower quota	2	6	3	5	1
Upper quota	3	7	4	6	2

Did You Get It

Try this problem to see if you understand the concepts we just studied. The answer can be found at the bottom of the portfolio page.

3. Find the upper and lower quotas for each of the districts in Did You Get It 1.

Now we're all set up for the group portion of the lesson, where we'll study several apportionment methods. Make sure you think about the key idea in the remainder of this lesson: *all of our apportionment methods are based on finding the standard quota, then deciding whether to round up or down for a given district.*

1-8 Group

The first method we'll study is Hamilton's method, which was approved by Congress for use in 1791, but was never actually used since President Washington vetoed the bill that introduced it.

Hamilton's Method

1. Assign each district the number of seats according to its lower quota. If this adds to the right number of seats, you're done.
2. If it does not, assign extra seats one by one to the districts in descending order of size of the fractional parts in their standard quota. For example, if three districts have standard quotas 2.93, 4.70, and 3.15, the first extra seat would go to the district with fractional part 0.93, and a second extra seat would go to the district with fractional part 0.70.

1. Use Hamilton's method to apportion the seats for the city described in the Class portion of this lesson. In completing the table on the previous page, you've done most of the work already. Cool. Make sure you clearly write out how many seats each district gets. A table wouldn't be the worst idea ever.

The lower quotas add to 17, so we need to allocate three more seats. The three districts with the largest fractional parts are 1, 2, and 5 (all at 0.67), so they get the extra seats, making the allocation:

District 1: 3, District 2: 7, District 3: 3, District 4: 5, District 5: 2

Math Note

Interesting trivia fact: The bill that would have adopted Hamilton's method for congressional apportionment was the first bill ever vetoed by an American president.

Did You Get It

Try this problem to see if you understand the concepts we just studied. The answer can be found at the bottom of the portfolio page.

4. Use Hamilton's method to apportion the seats for the districts in Did You Get It 1.

The apportionment method that was adopted by congress after Washington's veto in 1792 was designed by Thomas Jefferson, and shockingly is called Jefferson's method. It's a bit more complicated than Hamilton's method: it uses what we call a **modified divisor.** The idea is to make sure that every district gets its lower quota. We know that won't work in general because usually the number of seats will add up to something less than the correct total. In Jefferson's method, we'll make the quotas get bigger by making the standard divisor lower. (Remember, the standard quota comes from dividing a district's population by the standard divisor, so if the divisor gets smaller, the quota gets bigger.)

Jefferson's Method

1. Assign each district the number of seats according to its lower quota. If this adds to the right number of seats, you're done.
2. If the sum of the lower quotas is less than the number of seats, choose a modified divisor that is less than the standard divisor, and recalculate the lower quotas. Use this as the apportionment if the lower quotas now add up to the total number of seats. If not, adjust the modified divisor and try again.

Jefferson's method ends up being a trial and error process; we keep trying different divisors until the lower quotas add up to the right total. Let's give it a shot.

2. Recall that the standard divisor calculated earlier was 12. Fill in the table below, in each case getting the modified quota from dividing a district's population by the modified divisor 11.

District	1	2	3	4	5
Population in thousands	32	80	41	67	20
Modified quota	2.91	7.27	3.73	6.09	1.82
Modified lower quota	2	7	3	6	1

3. Add up the modified lower quotas. Does this apportionment work? Why or why not?

We get 19 seats, which is one too few, so it doesn't work.

This is a GREAT opportunity to show students how they can save work in calculating quotas by setting up a formula in Excel, or using the Table feature on a graphing calculator, with independent variable in Ask mode. Just enter y = x/standard divisor, and put in the population of each district to get the modified quota. After that it's easy to change the divisor and recalculate all quotas instantly.

4. Our next attempt will be to change the divisor we're using. Do you think we should make it bigger or smaller? Why?

Smaller. There are too few seats, so we want at least one quota to go up. That means we want to divide by a smaller number.

5. Try again with modified divisor 10.5. Does the apportionment work now? Is it the same as the one we got from Hamilton's method?

District	1	2	3	4	5
Population in thousands	32	80	41	67	20
Modified quota	3.05	7.62	3.90	6.38	1.90
Modified lower quota	3	7	3	6	1

Now the apportionment works, giving each district the modified lower quota. The seats now add up to 20. Yay. Also, the apportionment is different than Hamilton's method: District 4 got one more seat and District 5 got one less.

Math Note

When using a modified divisor in Jefferson's method, if the sum of the lower quotas ends up GREATER than the number of seats, it means you made the divisor too low.

Jefferson's method was used in Congress until 1842, when it was replaced with a new method devised by Daniel Webster, a distinguished Congressman, Senator, and two-time Secretary of State. Webster's method also uses modified divisors, but is based on standard rules of rounding rather than lower quotas.

Webster's Method

1. Assign each district the number of seats by using standard rounding rules on the standard quota, rounding up if the fractional part is 0.5 or greater and down if the fractional part is less than 0.5.
2. If the sum of the seats is equal to the number of seats available, use this as the apportionment. If not, modify the divisor and try again. If there are too many seats assigned, use a larger divisor. If there are too few seats assigned, use a smaller one.

We already know from Question 7 in the Class portion that using standard rounding rules won't work for the apportionment problem we've been studying: that resulted in 21 seats. So we can skip right to Step 2 in Webster's method. Since we have too many seats, we should try a larger divisor.

6. Using a modified divisor of 12.5, fill in the table below.

District	1	2	3	4	5
Population in thousands	32	80	41	67	20
Modified quota	2.56	6.40	3.28	5.36	1.60
Modified rounded quota	3	6	3	5	2

7. If all went well, you'll find that the number of seats is too low, which means we made the divisor too big. Choose a different modified divisor and try, try again, until you find the apportionment that results from Webster's method. Remember, we know that using divisor 12 makes the number of seats too high. That was a hint. Again, make sure that you clearly write out the number of seats that each district will get.

Using divisor 12.25, the modified quotas are 2.61, 6.53, 3.35, 5.47, 1.63. Using standard rounding rules we get 3, 7, 3, 5, 2. This adds up to 20!

District 1: 3, District 2: 7, District 3: 3, District 4: 5, District 5: 2

Did You Get It

Try this problem to see if you understand the concepts we just studied. The answer can be found at the bottom of the portfolio page.

5. Use Jefferson's and Webster's methods to apportion the seats for the districts in Did You Get It 1.

At this point, you might be wondering "What's the big deal? Why so many different methods? Don't they all result in pretty much the same apportionment?" Keep in mind that we're working with an example that has 5 districts and 20 seats to be divided up. For congressional apportionment, there are 50 "districts" (states) and 435 seats! In that case, different methods can result in significant differences.

Another interesting note: An apportionment problem of that size requires a tremendous amount of calculation, so it's no surprise that computers are used. But think back to, say, 1842. At the time, there were 26 states and 242 seats, and the apportionments were done with NO help on the calculations whatsoever. Those guys were pretty smart.

In the Technology portion of the Portfolio, you'll do a little research on the history of congressional apportionment, filling in some of the gaps that we haven't mentioned within the lesson. In the Applications portion, you'll learn about the method that's been in use since 1941, the Huntington-Hill method.

Some of the examples below ARE apportionment questions, and some aren't. In Questions 8–11, write whether each one is or is not an apportionment problem, and of course justify your answer.

8. A school district raises an extra $1.2 million dollars when a levy passes, and decides to divide it among all of the schools in the system based on enrollment.

Not apportionment. Money isn't indivisible, although if you want to quibble I guess you could worry about the very last penny.

9. A manufacturing company with four locations buys 22 new stamping machines. The plan is to divide them among the locations based on production levels over the last four years.

This is apportionment. Stamping machines are indivisible.

10. Each day, a team of engineers that works for a utility company decides how to allocate power on the grid to various neighborhoods based on past usage statistics.

Not apportionment. The power can be doled out however they like; it's not indivisible.

11. Each day, the manager of a large roofing company decides how to allocate workers to that day's jobs based on the size of the job and the time to completion.

This is apportionment. To say the very least, workers are not divisible.

1-8 Portfolio

Name ____________________

Check each box when you've completed the task. Remember that your instructor will want you to turn in the portfolio pages you create.

Technology

1. ☐ Do an Internet search to build a timeline of the apportionment methods used for Congress since the field was invented in 1791.

Online Practice

1. ☐ Include any written work from the online assignment along with any notes or questions about this lesson's content.

Applications

1. ☐ Complete the applications problems.

Reflections

Type a short answer to each question.

1. ☐ What makes something indivisible?
2. ☐ Describe what the point of apportionment is.
3. ☐ Think of an example aside from seats for representatives where apportionment might be used. (No using the other examples in the lesson!)
4. ☐ Take another look at your answer to Question 0 at the beginning of this lesson. Would you change your answer now that you've completed the lesson? How would you summarize the topic of this lesson now?
5. ☐ What questions do you have about this lesson?

Looking Ahead

1. ☐ Complete the Lesson 2-1 Prep Skills and read the opening paragraph in Lesson 2-1 carefully, then answer Question 0 in preparation for that lesson.

Answers to "Did You Get It?"

1. 32,000
2. Dist. 1: 3.91, Dist. 2: 2.78, Dist. 3: 7.34, Dist. 4: 3.03, Dist. 5: 3.19, Dist. 6: 4.00, Dist. 7: 5.75
3. Dist. 1: upper 4, lower 3, Dist. 2: upper 3, lower 2, Dist. 3: upper 8, lower 7, Dist. 4: upper 4, lower 3, Dist. 5: upper 4, lower 3, Dist. 6: upper 4, lower 4, Dist. 7: upper 6, lower 5
4. 4, 3, 7, 3, 3, 4, and 6 from Districts 1 to 7.
5. Both assign seats 4, 3, 7, 3, 3, 4, and 6 from Districts 1 to 7. (Use divisor 29.5 for Jefferson's method.)

1-8 Applications

Name ______________________________

Here's an apportionment problem you'll be working on in this activity: A large company decided to donate 17 computers to the West Hills school district. The director of computer services decided to allot them to the three middle schools based on number of students, as seen in the table below. Our goal will be to decide how many computers each school gets.

© Creatas/PunchStock RF

School	Jefferson MS	Hamilton MS	Webster MS	Total
Number of Students	1,328	1,008	418	2,754

1. Calculate the standard divisor for this problem.

2,754/17 = 162

2. Fill in the table below using the standard divisor you calculated in Question 1.

School	1	2	3
Number of students	1,328	1,008	418
Standard quota	8.20	6.22	2.58
Lower quota	8	6	2
Upper quota	9	7	3

3. The **geometric mean** of two numbers is defined to be the square root of their product. For example, the geometric mean of 2 and 8 is $\sqrt{2 \cdot 8} = \sqrt{16} = 4$; the geometric mean of 15 and 18 is $\sqrt{15 \cdot 18} = \sqrt{270} \approx 16.43$. Copy the above table, then fill in the next row with the geometric mean of upper and lower quotas for each school.

School	1	2	3
Number of students	1,328	1,008	418
Lower quota	8	6	2
Upper quota	9	7	3
Geometric mean of quotas	8.49	6.48	2.45

1-8 Applications

Name ______________________________

In the Huntington-Hill method of apportionment, either the upper or lower quota is assigned to each district, depending on whether the standard quota is more or less than the geometric mean for that district.

4. Compare the standard quota for each school with the geometric mean that you calculated, and fill in each blank with > or <.

Jefferson: The standard quota is ___<___ the geometric mean.

Hamilton: The standard quota is ___<___ the geometric mean.

Webster: The standard quota is ___>___ the geometric mean.

5. For each school, assign the lower quota if the standard quota is less than the geometric mean, and assign the upper quota if the standard quota is greater than the geometric mean.

Jefferson gets ___8___ computers.

Hamilton gets ___6___ computers.

Webster gets ___3___ computers.

6. Explain why this is a reasonable apportionment.

Because this adds up to 17 computers, which is the number to be apportioned.

7. Use the table from Question 2 to decide if Hamilton's method would result in the same apportionment as Huntington-Hill.

It would. The lower quotas add to 16, so one more computer needs to be allocated. And Webster MS has the largest fractional part on its standard quota, so it would go up to the upper quota.

Final note: When using the Huntington-Hill method, the number of seats doesn't always add up to the correct number initially. If it does not, we would use modified divisors, like we did in Jefferson's and Webster's method, then try again until we get a proper apportionment.

Unit 2
Contending with Change

Outline

Lesson 2-1 Prep Skills

This is a short review of skills that will come in handy in the next lesson. In each case, answer the question, then rate your confidence level by checking one of the boxes. If you feel like you're struggling with these skills, consult the online resources provided by your instructor for extra practice.

I KNOW IT	THINK SO	UNSURE	NO IDEA	
				SKILL 1: ADD LONG STRINGS OF INTEGERS
				1. Perform each addition: $2 + 5 + 1 + 0 + 5 + 7 + 3 + 4 + 4 + 4 + 1$ $8 + 7 + 6 + 9 + 8 + 7 + 1 + 2 + 3 + 7 + 1 + 9$
				SKILL 2: MULTIPLY EXPRESSIONS INVOLVING DECIMALS
				2. Perform the multiplication and round to the nearest hundredth: $21{,}419 \times 1.07$
				3. Perform the multiplication and round to the nearest cent: $\$1{,}436 \times 1.055$
				SKILL 3: COMPUTE PERCENTAGE OF A DOLLAR AMOUNT
				4. If you have $4,308 in a savings account, and that amount increases by 4.5% in the next year, what would the new amount be at the end of next year?
				SKILL 4: RECOGNIZE PATTERNS IN NUMBERS
				5. Find the most likely next three numbers in each pattern. 11, 17, 23, 29, 35, . . . 3, 6, 12, 24, 48, . . .

Lesson 2-1 Follow the Pattern (Comparing Linear and Exponential Growth)

LEARNING OBJECTIVES

☐ 1. Recognize patterns and use them to make predictions.

☐ 2. Distinguish between linear and exponential growth.

What we call chaos is just patterns we haven't recognized. What we call random is just patterns we can't decipher.

—Chuck Palahniuk

The world that we live in is a dynamic place: change is inevitable, and those that don't adjust and adapt tend to get left in the dust. In many walks of life, adapting and adjusting is based on recognizing patterns and trends. Societal trends, financial trends, behavioral trends . . . if you can recognize these patterns, you have a much better chance at anticipating what's likely to happen next. This type of anticipation is often the edge you need against competitors, whether it's in school, a career, or your personal life.

I'm sure you know that people often ask in math classes, "When am I going to use this?" When they do, they're focusing too much on specific formulas and procedures, and not enough on the big picture. Being able to work through complex procedures and being able to carefully follow a series of instructions is important in almost any job, so we'll practice that in this lesson as well.

0. After reading the opening paragraphs, what do you think the main topic of this lesson will be?

2-1 Group

If it doesn't make you uncomfortable, exchange the following information with the classmates in your Unit 2 group. This will be your small group for the second unit. It would be a good idea to schedule a time for the group to meet to go over homework, ask/answer questions, or prepare for exams. You can use this table to help schedule a mutually agreeable time.

Name	Phone Number	Email	Available times
			Assign new groups for Unit 2, and allow them some time to trade information before working on the lesson.

How good are you at recognizing patterns? If you look for them, they're all around you. In some cases, there are patterns in things you see all the time, but never really thought about in terms of patterns. Two examples of visual patterns are shown here. The QR code will direct a smart phone to the website of the St. Louis Cardinals; the UPC is for a 12-pack of 7-Up. (Try them if you don't believe me.) They may look like a bunch of squares and lines at first glance, but there are definite patterns for the discerning eye that knows what to look for.

One good way to practice recognizing patterns and anticipating where they lead is to study strings of numbers and letters, looking for a pattern. In Questions 1-8, find AND DESCRIBE the pattern, then use it to make a reasonable prediction for the next three symbols.

1. 3, 6, 9, 12, 15, 18, 21

Each entry is three more than the previous.

2. 3, 6, 12, 24, 48, 96, 192

Each entry is twice the previous.

3. 320, 160, 80, 40, 20, 10, 5

Each entry is half of the previous.

4. 320, 240, 160, 80, 0, −80, −160

Each entry is 80 less than the previous.

5. A, D, G, J, M, P, S

Starting with A, go through the alphabet, skipping two letters each time.

6. 1, b, 3, d, 5, f, 7, h, 9

Alternate numbers and letters, starting with 1 and b. For the numbers, add 2 each time, and for the letters skip one letter each time.

7. 1, 1, 2, 3, 5, 8, 13, 21, 34, 55

Each entry is obtained from adding the previous two entries.

8. J, F, M, A, M, J, J, A, S

These are the first letters of the months of the year, in order.

Source: St. Louis Cardinals

Source: 7-Up

Did You Get It ?

Try this problem to see if you understand the concepts we just studied. The answer can be found at the bottom of the portfolio page.

1. Find a pattern and use it to make a reasonable prediction for the next three symbols.

   a. 150, 135, 120, 105, ____, ____, ____

   b. Sp, Su, Fa, Wi, Sp, Su, ____, ____, ____

   c. A, 2, C, 4, E, 6, ____, ____, ____

The familiar UPC codes that are used to find the price of an item have something called a **check digit** that's important in avoiding scanning errors. Obviously, it's important both to the store and the buyer that prices are entered correctly. To practice following a procedure, read the instructions below carefully, then use them to answer Questions 9 and 10.

UPCs are 12-digit codes. A 12-digit Universal Product Code (UPC) code has three parts. The UPC for a 12-pack of 7-Up is shown on the previous page: **0 78000 01180 7.**

- The first 6 digits of this code—0 78000—identify the manufacturer.
- The next 5 digits—01180—are the item number; identifying a specific product from this manufacturer.
- The 12th digit is the check digit. When the item is scanned, the reader performs a calculation on the first 11 digits, and if the result doesn't match the check digit, the machine knows something is wrong and the scan is rejected.

There's a five-step process to calculate the check digit in a UPC:

Computing a UPC Check Digit

Step 1. Add all of the digits in the odd numbered positions—the 1st, 3rd, 5th, 7th, 9th, and 11th.

Step 2. Multiply the result from Step 1 by 3.

Step 3. Add the digits in the even numbered positions—the 2nd, 4th, 6th, 8th, and 10th. (Don't add the 12th digit!)

Step 4. Add the results of Steps 2 and 3.

Step 5. The check digit is the number you have to add to the result of Step 4 to get to the next multiple of 10 (that is, the next number that ends with a zero).

9. Check that the UPC for 7-Up provided above has the right check digit. Show all of your calculations!

Step 1. Add digits in odd positions: $0 + 8 + 0 + 0 + 1 + 0 = 9$

Step 2. Multiply by 3: $9 \times 3 = 27$

Step 3. Add digits in even positions: $7 + 0 + 0 + 1 + 8 = 16$

Step 4. Add results of Steps 2 and 3: $27 + 16 = 43$

Step 5. Need to add 7 to get to the next multiple of 10 (which is 50), so check digit is 7.

10. Repeat Question 9 for this UPC: **8 85909 59964 6.** (You'll be asked later to find what product the code corresponds to.)

Step 1. Add digits in odd positions: $8 + 5 + 0 + 5 + 9 + 4 = 31$

Step 2. Multiply by 3: $31 \times 3 = 93$

Step 3. Add digits in even positions: $8 + 9 + 9 + 9 + 6 = 41$

Step 4. Add results of Steps 2 and 3: $93 + 41 = 134$

Step 5. Need to add 6 to get to the next multiple of 10 (which is 140), so check digit is 6.

11. Use the Internet to look up the UPC code for a product that you've bought recently, write the UPC below, and verify the check digit.

Answers vary.

Did You Get It

Try this problem to see if you understand the concepts we just studied. The answer can be found at the bottom of the portfolio page.

2. Find the correct check digit for the following UPC: 7 27361 12182

2-1 Class

1. If I offered you one of the two accounts detailed below, which would you choose? More importantly, WHY would you make that choice?

	Account 1	Account 2
Start	$1,000.00	$1,000.00
After 1 year	$1,060.00	$1,050.00
After 2 years	$1,120.00	$1,102.50
After 3 years	$1,180.00	$1,157.63
After 4 years	$1,240.00	$1,215.51

This is an excellent question for class discussion. Try to involve students that speak up less often.

Answers vary, but Account 2 is the better long-term choice due to compound interest.

Let's take a more in-depth look at some patterns like the ones that began this lesson.

2. Describe the change from one number to the next in this pattern: 2, 5, 8, 11, 14, 17, ... Be *very* specific. That's way more than just specific. Way.

You get each successive number by adding 3 to the previous number.

After discussing Question 1, give students some time to try questions 2 and 3 before discussing them as a class.

3. Repeat Question 2 for this pattern: 2, 6, 18, 54, 162, 486, ... Still expecting way specific.

You get each successive number by multiplying the previous number by 3.

Both of the lists above start at two and get increasingly bigger, but in different ways.

4. In the first list, the values grow by *adding* the same constant number.

This type of growth is called **linear growth.**

5. In the second list, the values grow by *multiplying* by the same constant number.

This type of growth is called **exponential growth.**

Did You Get It ?

Try this problem to see if you understand the concepts we just studied. The answer can be found at the bottom of the portfolio page.

3. Classify each pattern as either linear or exponential growth.
   a. 1, 9, 81, 729, . . .
   b. 4, 18, 32, 46, 60, . . .
   c. 200, 208.5, 217, 225.5, 234, 242.5, . . .
   d. 200, 212, 224.72, 238.20, 252.49, . . .

6. Is exponential growth always going to result in faster growth than linear? Think about it, try some examples, and explain your answer.

Answers can vary widely. The result depends on what you mean by faster, what the growth rate is, and how far out you look for a given example. Look for quality of explanation here, as well as having looked at a variety of different cases. Encourage this, in fact.

The spreadsheet to the right shows an example of linear growth.

	A
1	35
2	45
3	55
4	65
5	
6	

7. What's the next value on the list?

75

8. Write a verbal description of how you would obtain the value in cell A6 from the value in cell A5.

Add 10 to the value in cell A5.

9. What would you type in cell A6 to calculate this value?

= A5 + 10

10. What value would appear in cell A12? This takes either repeated calculations, or some ingenuity. If you have Excel available in class, feel free to use it! Indicate how you got your answer.

145

The spreadsheet to the right shows an example of exponential growth.

	A
1	2
2	20
3	200
4	2,000
5	
6	

11. What's the next value on the list?

20,000

12. Write a verbal description of how you would obtain the value in cell A6 from the value in cell A5.

Multiply the value in cell A5 by 10.

13. What would you type in cell A6 to calculate this value?

$= A5 * 10$

14. What value would appear in cell A12? This takes either repeated calculations, or some ingenuity. If you have Excel available in class, feel free to use it! Indicate how you got your answer.

200,000,000,000

Did You Get It

Try this problem to see if you understand the concepts we just studied. The answer can be found at the bottom of the portfolio page.

4. Find the next entry on the list for each of the patterns in Did You Get It 3.

Now let's study these two types of growth in a little more depth. (By the way, we'll study linear change more in Lessons 2-2 and 2-4, then study exponential change in Lessons 2-5 and 2-6.)

15. Suppose that you're offered two jobs by a very eccentric but lovable billionaire: one pays $40,000 in the first year, with a salary that increases by $10,000 every year. The second starts out at 50 cents a day (which I imagine probably violates minimum wage laws, but whatever), and doubles every month. Which job would you be inclined to choose, and why?

Answers vary, but most non-mathematicians will choose Job 1 with no hesitation. This is a nice opportunity to talk about delayed gratification!

16. How much would you make per month in the first year with the $40,000 per year job?

$40,000/12 = $3,333.33

17. Going with a 30-day month to keep the calculations simpler, how much would you make in your first month on the second job?

$0.50 × 30 = $15

18. Fill in the table below with the monthly salary for each job over the first year. Remember, the first one offers a raise only after the first year, while the second doubles every month.

Monthly Salary For Each Job

Month	Job 1	Job 2
1	$3,333.33	$15
2	$3,333.33	$30
3	$3,333.33	$60
4	$3,333.33	$120
5	$3,333.33	$240
6	$3,333.33	$480
7	$3,333.33	$960
8	$3,333.33	$1,920
9	$3,333.33	$3,840
10	$3,333.33	$7,680
11	$3,333.33	$15,360
12	$3,333.33	$30,720

19. How much more money would you make in the first year if you had the foresight (and math skills) to choose Job 2?

$21,425

2-1 Portfolio

Name ______________________________

Check each box when you've completed the task. Remember that your instructor will want you to turn in the portfolio pages you create.

Technology

1. ☐ Five scenarios are illustrated by a portion of a spreadsheet in the Applications section of this lesson. For the first four, create a spreadsheet that will calculate the next ten entries on the list. A template to help you get started can be found in the online resources for this lesson.

2. ☐ Do an Internet search to find the product that has the UPC provided in Question 10 on page 114.

Online Practice

1. ☐ Include any written work from the online assignment along with any notes or questions about this lesson's content.

Applications

1. ☐ Complete the applications problems.

Reflections

Type a short answer to each question.

1. ☐ Explain the difference between linear growth and exponential growth in your own words.

2. ☐ Why does exponential growth usually result in much larger quantities than linear growth?

3. ☐ Describe one or two situations outside of school where recognizing a pattern would be useful. Be specific!

4. ☐ Take another look at your answer to Question 0 at the beginning of this lesson. Would you change your answer now that you've completed the lesson? How would you summarize the topic of this lesson now?

5. ☐ What questions do you have about this lesson?

Looking Ahead

1. ☐ Complete the Lesson 2-2 Prep Skills and read the opening paragraph in Lesson 2-2 carefully, then answer Question 0 in preparation for that lesson.

Answers to "Did You Get It?"

1. a. 90, 75, 60 **b.** Fa, Wi, Sp **c.** G, 8, I **2.** 2 **3. a.** Exponential **b.** Linear **c.** Linear **d.** Exponential **4. a.** 6,561 **b.** 74 **c.** 251 **d.** 267.64

2-1 Applications

Name ______________________________

There are many different ways for quantities to grow. In this lesson we studied two of them, linearly and exponentially. Identify each scenario below as illustrating either linear growth, exponential growth, or neither. Show enough work to justify your answer, and if you can, find the next three values on the list. If you can't actually compute the next three values, make your best guess on what they would be, and describe how you got your estimates verbally.

1.

	A	B
1		**Taxi Fare**
2	Start	$3.30
3	After 1 mile	$5.70
4	After 2 miles	$8.10
5	After 3 miles	$10.50
6	After 4 miles	$12.90
7	After 5 miles	$15.30

Type of Growth: Linear

Next three values:

$17.70, $20.10, $22.50

Work/explanation space:

Each fare is $2.40 added to the previous.

2.

	A	B
1		**Account value**
2	Start	$12,000.00
3	After 1 year	$13,200.00
4	After 2 years	$14,520.00
5	After 3 years	$15,972.00
6	After 4 years	$17,569.20
7	After 5 years	$19,326.12

Type of Growth: Exponential

Next three values:

$21,258.73, $23,384.61, $25,723.07

Work/explanation space:

Each value is 1.1 times the previous.

3.

	A	B
1		**Population**
2	Start	10,000
3	After 1 year	10,200
4	After 2 years	10,404
5	After 3 years	10,612
6	After 4 years	10,824
7	After 5 years	11,041

Type of Growth: Exponential

Next three values:

11,262, 11,487, 11,717 (Note that only whole number answers make sense here!)

Work/explanation space:

Each population is 1.02 times the previous.

2-1 Applications

Name ______________________

4.

	A	B
1		**Salary**
2	Start	$40,000
3	After 1 year	$41,500
4	After 2 years	$43,000
5	After 3 years	$44,500
6	After 4 years	$46,000
7	After 5 years	$47,500

Type of Growth: Linear

Next three values:

$49,000, $50,500, $52,000

Work/explanation space:

Each salary is $1,500 added to the previous.

5.

	A	B
1		**Stock price**
2	Start	$40.00
3	After 1 month	$42.00
4	After 2 months	$44.50
5	After 3 months	$48.00
6	After 4 months	$52.00
7	After 5 months	$60.00

Type of Growth: Neither

Next three values:

We can't find the next 3 values, but we could estimate them to be around $75, $105, and maybe $165

Work/explanation space:

Neither: The amount of increase keeps going up, but there's no fixed number that each value is multiplied by to get the next. To estimate the next few values, we can see that the amount of increase has a pattern of $2, $2.50, $3.50, $4, $8. That seems to be growing, so I estimated increases of $15, $30, and $60

Lesson 2-2 Prep Skills

This is a short review of skills that will come in handy in the next lesson. In each case, answer the question, then rate your confidence level by checking one of the boxes. If you feel like you're struggling with these skills, consult the online resources provided by your instructor for extra practice.

I KNOW IT | THINK SO | UNSURE | NO IDEA

SKILL 1: WRITE FRACTIONS IN LOWEST TERMS

1. Reduce each fraction to lowest terms.

$\frac{25}{100}$ $\frac{18}{50}$

SKILL 2: WRITE FRACTIONS WITH A DIFFERENT DENOMINATOR

2. Rewrite each fraction with the given denominator.

$\frac{4}{3}$; denominator 12

$\frac{20}{1}$; denominator 8

SKILL 3: FIND UNIT PRICES

3. Remember when Hostess went bankrupt and stopped making Ho Hos? Worst day ever. Fortunately, they're back. A box of 20 Ho Hos goes for $14.99 on Amazon.com. What is the cost per unit?

SKILL 4: USE DIMENSIONAL ANALYSIS

4. Convert 23 days to hours.

5. Convert 37 pounds to ounces.

6. Convert 37 miles per hour to feet per second.

Lesson 2-2 88 Miles Per Hour! (Rates of Change)

LEARNING OBJECTIVES

- ☐ 1. Compute rates of change and compare to ratios and conversion factors.
- ☐ 2. Interpret and use rates of change.
- ☐ 3. Convert units involving rates.

The future is something which everyone reaches at the rate of 60 minutes an hour, whatever he does, whoever he is.
—C.S. Lewis

Anyone familiar with the classic movie *Back to the Future* instantly recognizes the significance of 88 miles per hour: It's the speed the DeLorean time machine needed to reach in order to travel through time. Speeds (like 88 miles per hour) are familiar examples of the topic of this lesson, *rates of change*. Our world is a dynamic, ever-changing place, so studying the rate at which things change is an excellent way to study the world around us. Fortunately, studying rates will tie together some of the skills we've already practiced in this course, meaning we're in a good position to make the most of this useful topic.

0. After reading the opening paragraph, what do you think the main topic of this lesson will be?

2-2 Class

A rate is a ratio that compares two quantities. By **rate of change,** we mean a rate that compares the change in one quantity to the change in another. A speed like 88 miles per hour qualifies, because we can write it this way:

$$\frac{88 \text{ mi}}{1 \text{ hr}}$$

This quite literally compares a change in distance (88 miles) to a change in time (1 hour). We call it a **unit rate** because its denominator is 1.

1. List some unit rates you're familiar with.

Some examples we've seen are 12 in./1 ft, 5,280 ft/1 mi, 4 qt/1 gal, 2.54 cm/1 in., 2.2 lb/1 kg. Unit rates of change we've seen include things like 50 ft/sec, and $7.75/hr.

Emphasize that calculations can be done using rates of change in a very similar way to conversion factors, using dimensional analysis. Refer the students back to the prep skills for this Lesson, and to Lesson 1-5, if they need a refresher on dimensional analysis.

2. List some rates that aren't unit rates.

Some examples: \$100/8 hr, 240 mi/4 hr, \$1/quarter mile

3. Compare rates of change and conversion factors, which we studied in Lesson 1-5. How are they similar? How are they different?

They're both written as ratios, making them comparisons between two different quantities. In conversion factors, those quantities are always equivalent in size, but given in different units. In rates of change, the quantities can be completely different. You might also point out that any conversion factor is a rate, but not every rate is a conversion factor.

Let's explore the idea of rates of change as a comparison between changes in two quantities a little more. According to the U.S. Census Bureau, about 2.7 percent of registered nurses in 1970 were men, compared to 9.6 percent in 2011. In 2011, the average salary for male nurses was \$60,700 per year, while the average annual salary for female nurses was \$51,100.

4. How much did the percentage of male nurses increase from 1970 to 2011?

By 6.9%.

5. How many years passed in that time span?

41

6. Divide your answer from Question 4 by your answer from Question 5. Include units in your answer (which come from the units in the division). What does your answer tell you? Why is it a rate of change?

6.9%/41 years = 0.168% per year. This tells us that the percentage of male nurses went up on average by 0.168% each year. This is a rate of change because it compares change in percentage to change in time.

7. Find the rate of pay for women compared to men in 2011. Write your rate so that it has \$1 in the denominator.

$$\frac{\$51{,}100}{\$60{,}700} \approx \frac{\$0.84}{\$1}$$

8. What does your answer to Question 7 tell you about the gender gap in pay for nurses? What do you think is responsible for the discrepancy in earnings?

That female nurses make, on average, about 84 cents for every dollar a male nurse makes. There is no good answer as to why. That's strictly open to interpretation.

9. Explain why your answer from Question 7 is NOT a rate of change.

It's not comparing change in salary to change in salary: It's just comparing salaries.

Did You Get It

Try this problem to see if you understand the concepts we just studied. The answer can be found at the bottom of the portfolio page.

1. In 1970, 96.1% of licensed practical and licensed vocational nurses were women. In 2011, 91.9% were women. Find the average rate at which this percentage changed over that time period.

2-2 Group

The word **equivalent** is an important word in math, and one that we'll use often in this course. Make sure you clearly understand what that word means. Look it up and discuss it with your group if necessary.

1. If you make $100 a day for doing a certain job, how much money would you make in a 5-day work week?

$5 \times \$100 = \$500$

2. Are $\frac{\$100}{\text{day}}$ and $\frac{\$500}{5 \text{ days}}$ equivalent rates? What does that mean?

Yes. They both describe the same relationship between number of days worked and amount of money earned. Also, the fraction $500/5 days reduces to the fraction $100/1 day.

Question 2 demonstrates a useful fact about rates of change: They can be scaled up or scaled down to fit a given situation. Scaling up a fraction is done by writing an equivalent fraction (that is, one that has the same value) using larger numbers. Scaling down is writing an equivalent fraction with smaller numbers. These processes are done by either multiplying BOTH the numerator and denominator by the same number, or dividing BOTH the numerator and denominator by the same number.

3. Write a fraction equivalent to $\frac{60 \text{ mi}}{4 \text{ hr}}$ that has 2 in the denominator.

30 mi/2 hr

4. Write a fraction equivalent to $\frac{60 \text{ mi}}{4 \text{ hr}}$ that has 1 in the denominator.

15 mi/1 hr

5. Write a fraction equivalent to $\frac{\$320}{4 \text{ days}}$ that has 10 in the denominator. (Hint: First scale down, then up.)

Scale down to $80/1 day, then up to $800/10 days.

6. If one bicyclist is pedaling at a rate of 40 miles in 3 hours, and another at a rate of 50 miles in 4 hours, which is faster? (Scaling either up or down can be used!)

Scale both up so they have denominator 12: 40 mi/3 hr = 160 mi/12 hr; 50 mi/4 hr = 150 mi/12 hr. This shows that 40 miles in 3 hours is faster.

Did You Get It ?

Try this problem to see if you understand the concepts we just studied. The answer can be found at the bottom of the portfolio page.

2. If the median salary for one health care profession increased by $11,249 over 15 years, and the median salary for a different but related field went up by $25,111 over 35 years, which increased at a greater average rate?

Most experienced runners get to a point where they can comfortably jog long distances at a consistent pace. This "pace," of course, is another way to say "rate of change," because speed is the rate at which distance changes compared to time. One particular runner jogs one lap around a 400-meter track in 2 minutes. In Questions 7–15, you can assume that the runner can maintain this pace for a long time.

7. For every __2__ minutes, the runner jogs __400__ meters.

Time (Min)	Distance (m)
2	400
4	800
6	1,200
8	1,600
10	2,000
12	2,400

8. Complete the table, continuing the pattern for times and finding the associated distances for this runner.

9. Notice that the rate at which the runner's distance changes is constant. When the rate of change is constant, a quantity illustrates what type of growth (that we encountered in Lesson 2-1)?

Linear

10. Write the runner's rate as a fraction using meters in the numerator.

400 m/2 min

Math Note

Using rates of change to calculate sizes of quantities is one of the most important applications of dimensional analysis. It keeps us from having to memorize a bunch of formulas: the units tell you what calculation to perform.

11. Write this rate as a unit rate in meters per minute.

200 m/min

12. Convert your rate from Question 11 to meters per second using dimensional analysis. (Round to two decimal places.)

$$\frac{200\text{ m}}{\text{min}} \times \frac{1\text{ min}}{60\text{ sec}} = 3.33\frac{\text{m}}{\text{sec}}$$

13. What distance will the runner cover in 40 minutes?

$\frac{200 \text{ m}}{\text{min}} \times 40 \text{ min} = 8{,}000 \text{ m}$

14. What distance will he cover in 40 seconds?

$\frac{3.33 \text{ m}}{\text{sec}} \times 40 \text{ sec} = 133.2 \text{ m}$

15. What would this runner's total time be for a 10k race? (Ten kilometers, that is. Recall that 1 km = 1,000 m. Your answer to Question 11 will help.)

$10 \text{ km} \times \frac{1{,}000 \text{ m}}{1 \text{ km}} \times \frac{1 \text{ min}}{200 \text{ m}} = 50 \text{ min}$

As Question 15 shows, when dealing with rates, sometimes it's necessary to convert the units in both the numerator and denominator of a fraction. (For example, it's common to convert speeds from feet per second to miles per hour; both feet and seconds need to be converted.) Fortunately, dimensional analysis is a perfect tool for this purpose, because the conversion factors we use equal one regardless of which unit is in the numerator.

16. The Earth moves at about 98,000 feet per second as it revolves around the Sun. How fast is that in miles per hour? (Recall that 1 mile is 5,280 feet.)

$\frac{98{,}000 \text{ ft}}{\text{sec}} \times \frac{1 \text{ mi}}{5{,}280 \text{ ft}} \times \frac{60 \text{ sec}}{1 \text{ min}} \times \frac{60 \text{ min}}{1 \text{ hr}}$

$\approx 66{,}818 \frac{\text{mi}}{\text{hr}}$

© Digital Vision/Getty Images RF

Did You Get It

Try this problem to see if you understand the concepts we just studied. The answer can be found at the bottom of the portfolio page.

3. For the rate of increase you calculated for the first health care profession in Did You Get It 2:
   a. Find how much median salaries would be expected to increase in the next 8 years if the rate of increase stays the same.
   b. Convert that rate to cents per day.

2-2 Portfolio

Name ______________________________

Check each box when you've completed the task. Remember that your instructor will want you to turn in the portfolio pages you create.

Technology

1. ☐ When you know the rate at which a quantity changes, you can build a spreadsheet to calculate the size of that quantity. Build a spreadsheet that calculates pay given hours worked and hourly rate, like the one shown here. You should be able to enter any rate you like in dollars per hour, and have a formula to calculate the pay corresponding to any time you enter. Also enter a formula to calculate total pay for all employees. A template to help you get started can be found in the online resources for this lesson.

	A	B	C	D
1	**Employee**	**Time (hrs)**	**Rate ($/hr)**	**Pay ($)**
2	Napoleon	10	$8.75	$87.50
3	Deb	20		
4	Rico	32		
5	Me	40		
6			**Total Pay**	
7				

Online Practice

1. ☐ Include any written work from the online assignment along with any notes or questions about this lesson's content.

Applications

1. ☐ Complete the applications problems.

Reflections

Type a short answer to each question.

1. ☐ In this lesson, we learned that the speed of an object is a rate of change; specifically, the rate at which distance traveled changes as time changes. Think of at least three other rates of change, and describe specifically what they measure. Extra points if you come up with one that doesn't have a unit of time in it.
2. ☐ Why does dimensional analysis work for calculations involving rates? (Recall that we developed it as a method for converting units.)
3. ☐ Take another look at your answer to Question 0 at the beginning of this lesson. Would you change your answer now that you've completed the lesson? How would you summarize the topic of this lesson now?
4. ☐ What questions do you have about this lesson?

Looking Ahead

1. ☐ Complete the Lesson 2-3 Prep Skills and read the opening paragraph in Lesson 2-3 carefully, then answer Question 0 in preparation for that lesson.

Answers to "Did You Get It?"

1. –0.102% per year (decrease) **2.** The first, $749.93 per year compared to $717.46 per year
3. a. $5,999.44 **b.** 205.5 cents per day

2-2 Applications

Name ________________________________

When advertising special financing deals, car makers often use rates to describe what your monthly payment will look like. (This is because the amount you need to borrow varies widely based on the car you choose and the down payment you make.) In January 2013, Hyundai offered 2.9% financing on the Genesis coupe, which would result in monthly payments of $22.09 per $1,000 borrowed for 48 months.

Courtesy of David Sobecki

1. For every $1,000 borrowed, the monthly payment will be $22.09.

2. Complete the table of loan payments using your answer to Question 1.

Amount Borrowed	Loan Payment
$1,000	$ 22.09
$2,000	$44.18
$3,000	$66.27
$4,000	$88.36
$5,000	$110.45
$6,000	$132.54
$7,000	$154.63
$8,000	$176.72
$9,000	$198.81
$10,000	$220.90

3. Write the rate Hyundai advertised as a fraction.

$$\frac{\$22.09 \text{ payment}}{\$1{,}000 \text{ borrowed}}$$

Use your answer to Question 3 and dimensional analysis for Questions 4 and 5.

4. What would your monthly payment be if you borrowed $18,000?

$$\$18{,}000 \text{ borrowed} \times \frac{\$22.09 \text{ payment}}{\$1{,}000 \text{ borrowed}} = \$397.62 \text{ payment}$$

5. If your monthly payment is $331.35, how much did you borrow?

$$\$331.35 \text{ payment} \times \frac{\$1{,}000 \text{ borrowed}}{\$22.09 \text{ payment}} = \$15{,}000 \text{ borrowed}$$

6. If you agree on a price of $23,900 (including taxes and fees) for a Genesis and the dealership offers you $7,500 in trade for your old car, how much would your monthly payment be?

$\$23{,}900 - \$7{,}500 = \$16{,}400$ borrowed

$$\$16{,}400 \text{ borrowed} \times \frac{\$22.09 \text{ payment}}{\$1{,}000 \text{ borrowed}} = \$362.28 \text{ payment}$$

2-2 Applications

Name ______________________________

7. When shopping for Doritos for your large office party, you find that you can buy several 11.5 oz bags for $4 each or a case of 64 small 1.75 oz bags for a total of $38.99. Always trying to save money, your boss says, "Just get the one with the lower unit price." What should you do? First, if you need to, you should go straight to Google and look up unit price. Then calculate the unit price of each option. Which is the better option? (Rounding the unit price might be a bad idea here.)

Small bags

$64 \times 1.75\text{ oz} = 112\text{ oz total}$

$\frac{\$38.99}{112\text{ oz}} = \frac{\$0.348125}{\text{oz}}$

Large bags

$\frac{\$4}{11.5\text{ oz}} = \frac{\$0.34782609}{\text{oz}}$

Realistically the costs are identical. This is a good opportunity to point out that every problem in life doesn't have a clear-cut solution. Also point out that Doritos are awesome.

Glaciers are large masses of ice that flow like rivers across the ground. Really, really slow rivers—did I mention that they're ice? Most move less than a foot per day. At one point, the San Rafael glacier in Chile was moving 203 millimeters per day.

Source: Bruce F. Molnia, U.S. Geological Survey

8. How fast was it moving in inches per hour? (1 inch is 25.4 mm.)

$$\frac{203\text{ mm}}{\text{day}} \times \frac{1\text{ in.}}{25.4\text{ mm}} \times \frac{1\text{ day}}{24\text{ hr}} = 0.333\frac{\text{in.}}{\text{hr}}$$

9. Find the speed in miles per hour, then explain why that's a silly unit of speed in this case. (Note that your calculator will probably give you an answer in scientific notation.)

$$\frac{0.333\text{ in.}}{\text{hr}} \times \frac{1\text{ ft}}{12\text{ in.}} \times \frac{1\text{ mi}}{5{,}280\text{ ft}} = 0.00000526\frac{\text{mi}}{\text{hr}}$$

This is such a tiny number that it's hard to get any perspective on what that speed means. In fact, based on that size, you'd guess the glacier's movement is negligible. But 1/3 inch per hour means it is in fact moving noticeably.

10. How long would it take the glacier to move the length of a football field (100 yards)? Choose a unit of time that you think is most reasonable for your answer.

$$100\text{ yd} \times \frac{3\text{ ft}}{1\text{ yd}} \times \frac{12\text{ in.}}{1\text{ ft}} \times \frac{1\text{ hr}}{0.333\text{ in.}} \times \frac{1\text{ day}}{24\text{ hr}} \approx 450.5\text{ days}$$

Lesson 2-3 Prep Skills

This is a short review of skills that will come in handy in the next lesson. In each case, answer the question, then rate your confidence level by checking one of the boxes. If you feel like you're struggling with these skills, consult the online resources provided by your instructor for extra practice.

SKILL 1: RECOGNIZE COORDINATES OF PLOTTED POINTS

1. Write the coordinates of the four points graphed on the axes below.

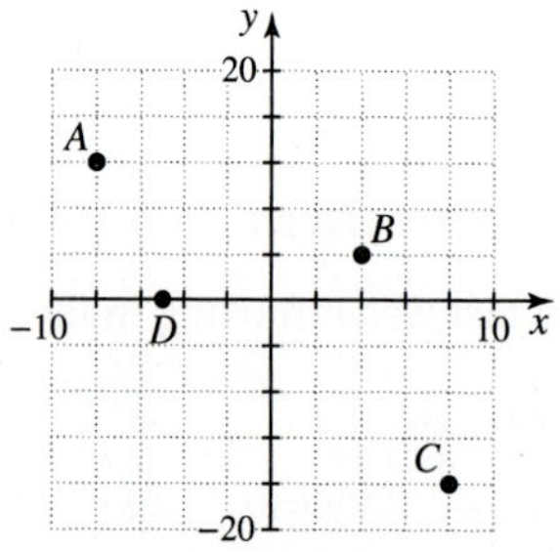

SKILL 2: PLOT POINTS

2. Plot each point on the axes. Make sure you indicate an appropriate scale on each axis!

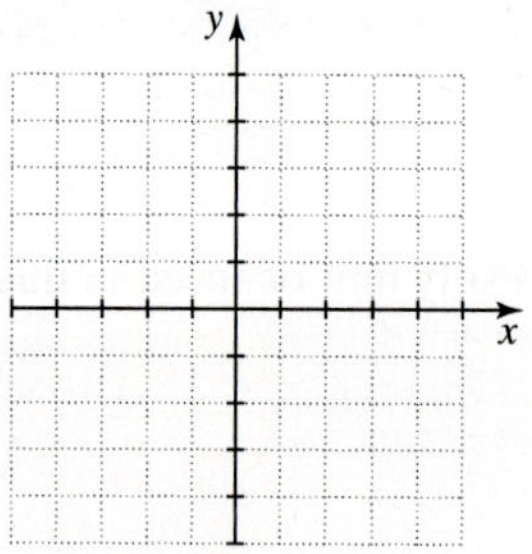

A. (−12, 100)
B. (18, 0)
C. (−6, −60)
D. (4, −10)

Lesson 2-3 Warning: Graphic Content (Graphing, Slope, and Rate of Change)

LEARNING OBJECTIVES

- ☐ 1. Understand and use a rectangular coordinate system.
- ☐ 2. Connect data to graphs and interpret graphs.
- ☐ 3. Define slope as a constant rate of change.
- ☐ 4. Define and interpret the *y* intercept of a line.
- ☐ 5. Write an equation of a line that models data from a description, table, or graph.

© belterz/iStockphoto RF

Believe you can and you're halfway there.
—Theodore Roosevelt

We've already used the word "graph" in this book, when referring to bar graphs. The bar graphs we looked at weren't just pretty pictures, or bars drawn at random heights for no particular reason: They were used to illustrate and understand real data. But the types of graphs we'll study in this unit might be the most misunderstood feature of math: people tend to think of them as just "plotting points and connecting the dots." Nothing could be further from the truth! Just like bar graphs, the graphs that illustrate connections between two variables are all about a visual representation of useful information. So before we talk about the mechanics and terminology involved, we'll use unemployment numbers to vividly illustrate what it's really all about.

0. After reading the opening paragraph, what do you think the main topic of this lesson will be?

2-3 Group

If you've been paying attention to the news at all for the last several years, you know that the unemployment rate in the United States has been a big story. But what did unemployment look like before the economic crisis of the late 2000s? These tables display the average annual unemployment rate for the years from 1992 to 2011.

Year	'92	'93	'94	'95	'96	'97	'98	'99	'00	'01
Rate (%)	7.5	6.9	6.1	5.6	5.4	4.9	4.5	4.2	4.0	4.7

Year	'02	'03	'04	'05	'06	'07	'08	'09	'10	'11
Rate (%)	5.8	6.0	5.5	5.1	4.6	4.6	5.8	9.3	9.6	8.9

Give students time to answer Question 1 on their own, then discuss as a class.

1. Use the table to write a verbal description of trends in the unemployment rate over that 20-year period.

The rate went down pretty steadily from 1992 to 2000. It then went up pretty quickly between 2000 and 2003 before dropping between 2003 and 2006. After staying steady for one year, it went up some in 2008, then increased dramatically in 2009. It went up a bit more in 2010, then came down some in 2011.

With enough effort, you were probably able to write a reasonable description. But because there's so much data in the table, spotting the trends isn't exactly a simple thing to do. Next, let's look at the same data displayed in graphical form.

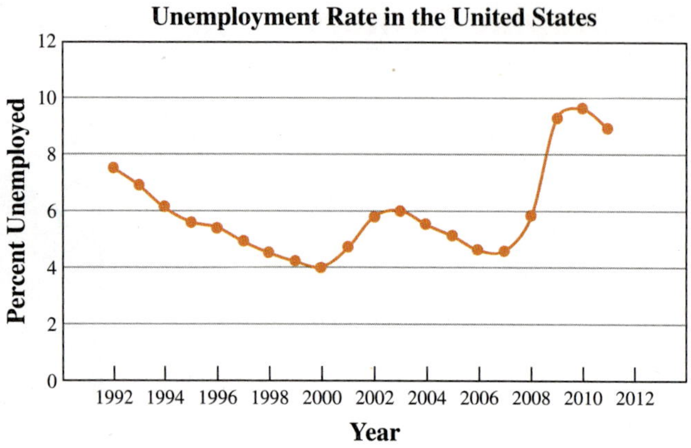

When I look at this graph, two things occur to me: It's a lot easier to see the trends than it was from looking at the table, and the graph looks a whole lot like a dinosaur, which is totally irrelevant but still pretty cool.

2. Use the graph to write a verbal description of trends in the unemployment rate, then explain why the graph makes it easier than the table did.

Unemployment fell steadily between 1992 and 2000, then went up and back down between 2000 and 2007. Then it skyrocketed before falling back a bit in 2011. By just focusing on the height of the graph, we can see when the rate went up and down and how rapidly very quickly.

Again, it's important to give students time to answer Question 2 on their own, then discuss as a class.

Without the numbers running along the bottom side of the graph and down the left side, we wouldn't be able to understand any of the information the graph provides. Those numbers provide the **scale** for the graph, and they're ALWAYS crucial in drawing a graph. Each of the number lines that we write the scale on is called an **axis** (the plural of this word is **axes**).

In preliminary numbers released in early 2013, the average unemployment rate for 2012 was listed as 8.1%. We can add that piece of information to the graph by finding 2012 on the horizontal axis and 8.1 on the vertical axis, then drawing imaginary lines up from 2012 and right from 8.1 until the lines meet: that's where we put the point corresponding to 2012 and 8.1%.

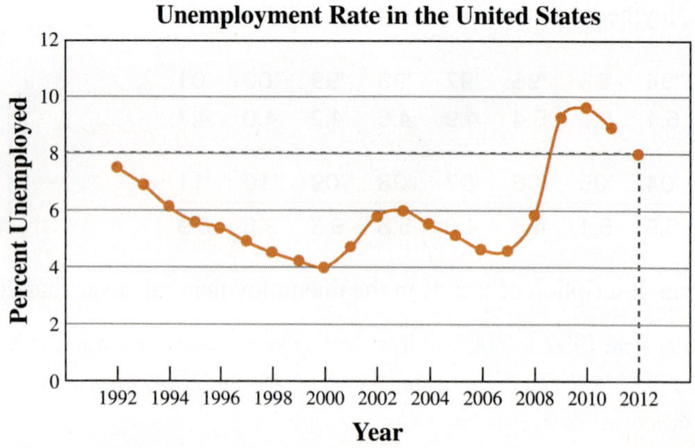

Notice that when we found the location to indicate that the unemployment rate was 8.1% in 2012, the imaginary lines we drew formed a rectangle with the two axes. That's why we call this system of graphing a **rectangular coordinate system**. Each of the numbers we used to locate that point are called **coordinates**. The horizontal axis is usually called the ***x* axis** and the vertical axis is usually called the ***y* axis**. The point where the two axes meet is called the **origin**.

Since we didn't need to worry about negative years or negative unemployment rates, the graph we drew earlier only showed positive values along each axis. But there are plenty of examples of data where negative values make perfect sense, so a rectangular coordinate system is often set up like this:

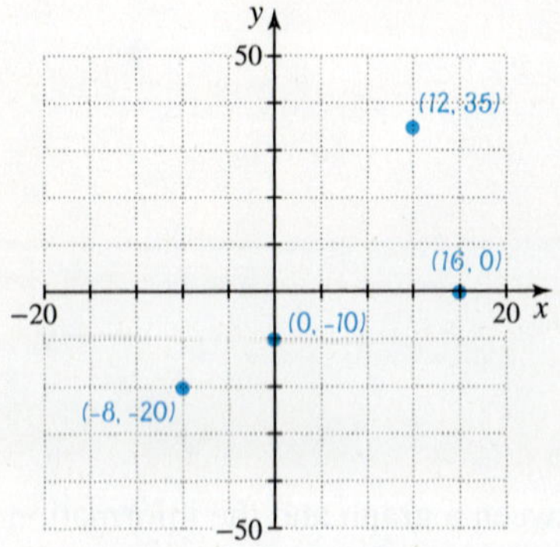

The process of locating information on a rectangular coordinate system, like the 8.1% unemployment rate in 2012, is called **plotting points**. We identify locations by writing the two coordinates together inside parentheses, like this: (2012, 8.1).

3. What are the coordinates of the origin in the rectangular coordinate system above?

(0, 0)

4. Look carefully at the numbers on each axis. What distance does each box on the grid represent along the *x* axis? What about the *y* axis? These distances are known as the **scale** for each axis, and are *always* carefully chosen so that all of the points fit nicely into the space provided. Choosing a reasonable scale is the first step in drawing ANY graph.

4 on the x axis, 10 on the y axis

5. Plot each of the following points on the rectangular coordinate system above. Label the coordinates of each point.

a. (16, 0)
b. (0, −10)
c. (−8, −20)
d. (12, 35)

Points plotted in blue.

Did You Get It

Try these problems to see if you understand the concepts we just studied. The answer can be found at the bottom of the portfolio page.

1. List the coordinates of each point graphed to the right.
2. If you wanted to plot the points (−3, 110), (6, −250), (1, −90), and (−5, 170), what would be a good scale to use on the two axes? Note that this is two questions, not one.

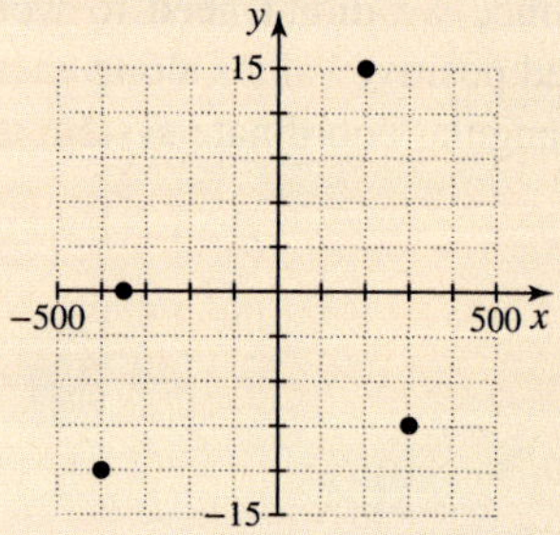

Being able to understand the connection between a graph and the information that it illustrates is not just the most important skill in graphing: It's the reason that graphing was developed. If you can't interpret the meaning of a graph, it's just really bad art!

6. A small plane takes off from a regional airport; its altitude at various times is recorded in the table below. Write ordered pairs of the form (Flight time, Altitude) for each pair of values, then plot the points on the graph.

Flight time x (minutes)	Altitude y (feet)	Ordered pair *(x, y)*
0	0	(0, 0)
10	4,000	(10, 4,000)
20	3,000	(20, 3,000)
30	2,000	(30, 2,000)
40	2,000	(40, 2,000)
50	1,000	(50, 1,000)
60	0	(60, 0)

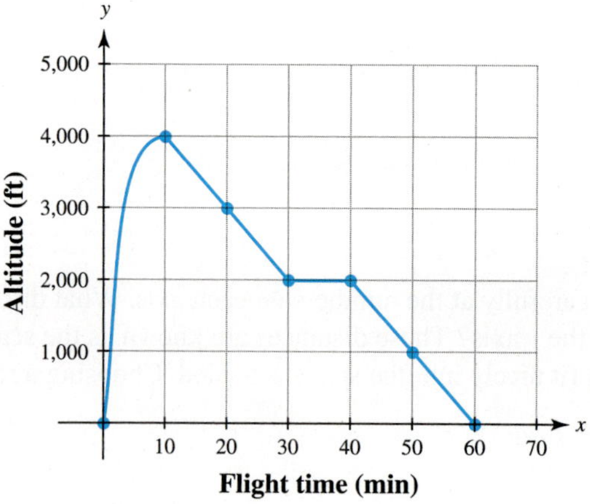

7. Connect the points you plotted to draw a graph, and describe what that graph illustrates. Then add a verbal label to each axis that describes the information it represents.

The graph illustrates the altitude of the plane in feet based on the number of minutes since the flight began. Emphasize that this is probably the simplest graph that fits the given information, but because we only know the height every ten minutes, we're just making an educated guess on how to connect the points on the graph.

For Questions 8–13, explain how you got your answer using BOTH the table and the graph. The point is to make sure you focus on the connection between the information in the table and the height of points on the graph.

8. What was the highest altitude reached by the plane? (Is the answer based on your graph different? Why?)

Both are 4,000 feet: 4,000 is the largest altitude listed in the table, and the highest point on the graph I drew is also at 4,000 feet. The graph could have been drawn to go above 4,000 feet either before or after time ten minutes.

9. How long after the flight began did the plane reach its highest altitude? (Again, you may have two slightly different answers.)

Ten minutes: altitude 4,000 in the table corresponds to flight time ten minutes: the highest point on the graph is at (10, 4,000), with the ten corresponding to ten minutes of flight time.

10. When was the plane flying level?

Between 30 and 40 minutes after takeoff, the altitude isn't changing: the height stays at 2,000 feet for two consecutive table entries, and the height of the graph stays the same.

11. How long did the flight last?

Sixty minutes: after that amount of time, the plane's altitude was back down to zero, which means it landed.

As you walk around, remind students to explain how they got each answer, and not just to write down a numerical answer.

12. When was the altitude of the plane increasing?

For the first ten minutes: after ten minutes, the next altitude in the table is lower than the one at ten minutes. Between times zero and ten, the graph is getting higher from left to right.

13. When was the altitude of the plane decreasing?

Between 10 and 30 minutes into the flight, and between 40 and 60 minutes. Explanations should be similar to Question 12.

14. Explain why we didn't bother to include negative values along either axis when drawing the rectangular coordinate system for the altitude graph. What would it mean if the graph ever has a height below the *x* axis?

The x axis corresponds to flight times, and negative times don't make sense. The y axis corresponds to altitudes, and barring a high-speed disaster, the plane will never be at a negative altitude. If there were any points below the x axis, that would indicate the plane being underground, which clearly would be a very bad result indeed.

When we plot points that correspond to pairs of data on a coordinate system, we call the result a **scatter diagram** or **scatter plot**.

15. According to taxifarefinder.com, a cab ride from the airport in Las Vegas will cost you an initial charge of $5.10, plus $2.60 per mile. Use this information to fill in the rest of the table describing the total cost for various distances. Then create a scatter diagram, using the distance as *x* coordinate and the cost as *y* coordinate. Make sure you label a scale on each axis.

© Stockbyte/Punchstock RF

Distance (mi)	Cost ($)
0	$5.10
1	$7.70
2	$10.30
3	$12.90
4	$15.50
5	$18.10

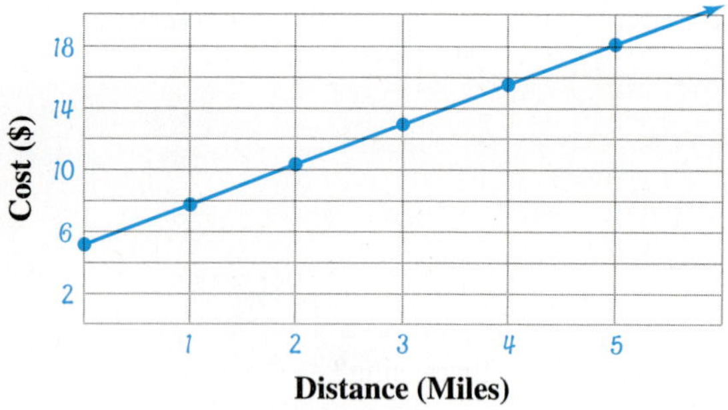

16. What do you notice about the pattern that the points on your scatter diagram are making? Use that observation about a pattern to connect the points, drawing a graph that represents the cost of a cab ride based on distance.

Make a big deal about the arrow on the end of the graph! Students need to fundamentally understand why the arrow is important in modeling a situation.

We'll be drawing quite a few scatter plots in the remainder of this unit, and will use Excel often. The box below explains how to use a spreadsheet to create a scatter plot.

Using Technology: Creating a Scatter Plot

To create a scatter plot in Excel:

1. Type the values that will go on the *x* axis in one column. (This would correspond to the flight time in Question 6 and the number of miles in Question 15.)
2. Type the values that will go on the *y* axis in one column. (This would correspond to the altitude in Question 6 and the value of the car in Question 15.)
3. Use the mouse to drag and select all the data in those two columns.
4. With the appropriate cells selected, click the Insert tab, then Charts, and click on Scatter. Then choose the type of scatter diagram you want. Options include plotting only the points, connecting the points with curves, and connecting the points with line segments.

You can add titles and change colors and other formatting elements by right-clicking on certain elements, or using the options on the Charts menu. Try some options and see what you can learn!

See the Lesson 2-3-1 video in class resources for further information.

Did You Get It

Try this problem to see if you understand the concepts we just studied. The answer can be found at the bottom of the portfolio page.

3. Draw a scatter plot for the data in the table, which represents the U.S. national surplus/deficit for even-numbered years for the period from 1998 to 2014. Positive number indicate the government taking in more money than it spent: negative numbers indicate the government spending more than it took in. All numbers are in billions of dollars. (Seriously. Billions.) Use years after 1998 on the horizontal axis, not the actual year.

Year	Surplus/Deficit
1998	70
2000	236
2002	−158
2004	−413
2006	−248
2008	−459
2010	−1,294
2012	−1,100
2014	−492
2016*	−455

* Projected

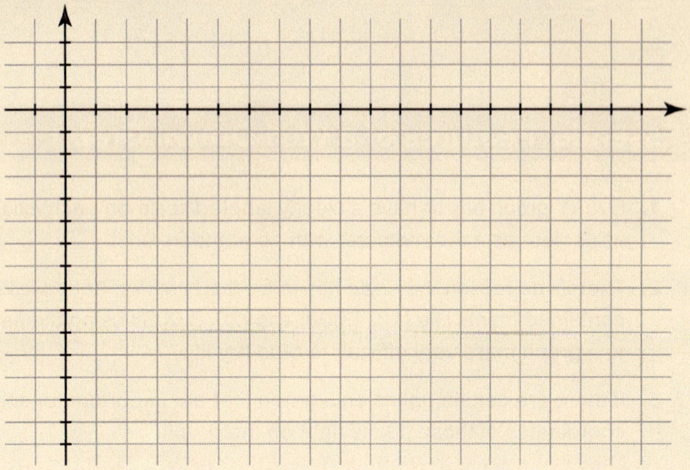

We'll return to our cab ride graph in a bit. But first, let's talk about drawing graphs in general.

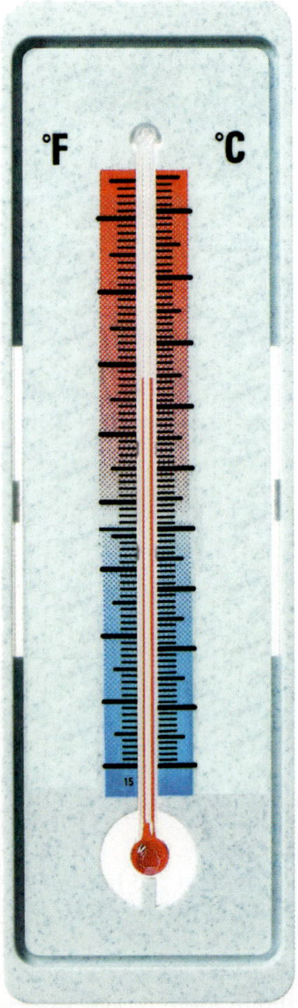

17. What temperature is this thermometer displaying?

It's impossible to tell.

18. Explain why the thermometer is pretty much useless.

There are no numbers on it! There are tick marks, but with no numbers to tell us what the tick marks mean, we can't read any temperatures.

19. What are the coordinates of the point drawn on the graph?

It's impossible to tell.

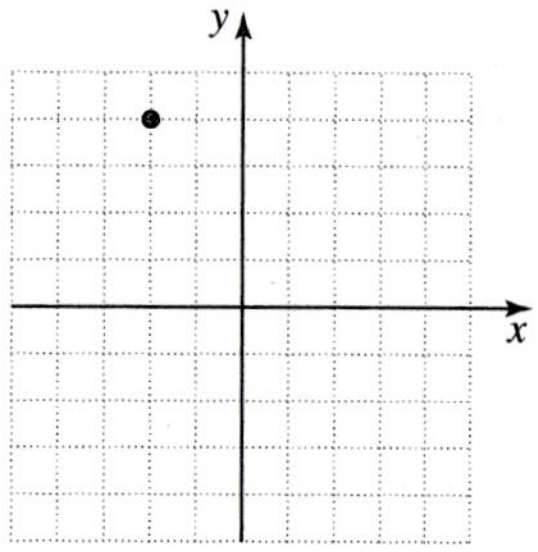

20. Explain why the graph is pretty much useless.

The graph is supposed to show the relationship between two sets of numbers, but with no scale listed on the axes, we can't tell what any of the ordered pairs are.

Essential Features of a Well-Drawn Graph

1. EVERY graph has to have a clearly labeled scale on each axis. A graph with no scale labeled is every bit as useless as a thermometer with no numbers on it.
2. There's no reason the scale has to be the same on both axes. The first graph in this group activity contained points like (10, 4,000) and (20, 3,000); using the same scale on the *x* and *y* axes would lead to a graph that's very difficult to read (try it!).
3. If there are certain points on a graph that are important for some reason, you should label the coordinates of those points right on the graph.

2-3 Class

Now it's time to refer back to the table and graph on page 138.

1. The point where any graph crosses the *y* axis is called the ***y* intercept** for the graph. Write the coordinates of the *y* intercept for the graph that you drew. More importantly, what information does the *y* intercept represent?

   The coordinates are (0, 5.10). This means that it will cost \$5.10 just to start the cab ride, even before you've gone any distance.

2. How much more will you pay for a 3-mile ride than for a 2-mile ride?

   \$12.90 − \$10.30 = \$2.60

3. How much more will you pay for a 5-mile ride than a 2-mile ride? Explain why that makes sense based on your answer to Question 2.

   \$18.10 − \$10.30 = \$7.80. The answer to Question 2 shows that you're paying \$2.60 per mile. So for a 3-mile trip, you'd pay 3 × \$2.60, which is \$7.80.

4. Pick any two points on the graph that you drew, and subtract the second coordinates. Then divide the result by the difference of the first coordinates for those same points. What does the result represent about the cab ride?

   Using the points (5, 18.10) and (2, 10.30), we get

   $$\frac{18.10 - 10.30}{5 - 2} = \frac{7.80}{3} = 2.60$$

   This shows us that the cost increases by \$2.60 for every extra mile traveled.

Hopefully, you recognize the calculation in Question 4 as a rate of change: it uses division to compare the change in cost to the change in miles driven. So specifically, it's the rate of change of cost with respect to distance. When applied to the graph of a line, the number that we found represents the change in *y* coordinate of the graph (i.e., change in height) divided by the change in *x* coordinate. This number, which measures how steep a line is, is called the *slope* of the line.

For any two points on a line, (x_1, y_1) and (x_2, y_2), the formula below, which compares the change in height to the horizontal change, is called the **slope** of the line. It's usually represented by the letter *m*.

$$m = \frac{y_2 - y_1}{x_2 - x_1}$$

When *y* represents some quantity, like the cost of a cab ride, the slope tells us the rate of change of that quantity.

5. Estimate the distance of a \$16 cab ride. You can use either the table or the graph, but make sure you explain how you got your answer.

The point on the graph at height 16 (which represents a cost of \$16) is just a little bit to the right of $x = 4$ on the horizontal axis. That means you can get a little more than 4 miles for \$16.

6. Estimate the cost of a two-and-a-half-mile cab ride. Again, explain how you got your answer.

The point on the graph halfway between $x = 2$ and $x = 3$ (which represents a distance of 2.5 miles) looks like it's close to height 12. So the ride should cost about \$12.

7. We know that the cost of the cab ride is going up by \$2.60 for each extra mile. So how much would you pay strictly for mileage if the ride is 10 miles? (Ignore the \$5.10 you'd pay just for the privilege of getting into the cab.)

$2.6 \cdot 10 = \$26$

8. How much would you pay strictly for mileage if the ride is *x* miles? Refer to Question 7 if you need help. If you're really desperate, ask your professor. (Just kidding. That should never require desperation. He or she gets paid to do stuff like help you learn.)

$2.6x$ dollars

9. Now, the key question: given that the initial fee is \$5.10, and on top of that you pay for the mileage, as described in Question 8, write an equation that describes the cost of a cab ride (y) in terms of the number of miles driven (x).

$y = 5.10 + 2.6x$

10. How could you use your equation to check your estimate of the distance of a \$16 cab ride? Do it.

Substitute 16 in for y, and solve the equation for x.

$16 = 5.10 + 2.6x \Rightarrow 2.6x = 10.90 \Rightarrow x = 10.90/2.6 \approx 4.2$ miles

11. How could you use your equation to check your estimate of the cost of a 2.5-mile cab ride? Do it.

Substitute 2.5 in for x and find the value of y.

$y = 5.10 + 2.6(2.5) = 11.6$, so the cost is \$11.60.

There's a very useful lesson to be learned from Questions 8 and 9.

- The \$2.60 that it costs for each mile is the rate of change of your cost as the number of miles increases. And we know that a constant rate of change for some quantity represents the slope of a line when you graph that quantity.
- The \$5.10 is the amount you'll pay just for starting the cab ride. So zero miles corresponds to \$5.10, and the point (0, 5.10) would be on the graph describing total cost. In fact, it would be the *y* intercept.
- The equation describing the cost can be written as $y = 2.6x + 5.10$, where x is the number of miles driven. Notice that the **coefficient** of x, which is the number x is multiplied by, is the slope of the line. Also, the **constant term**, which is the term without a variable, is the second coordinate of the y intercept.

Conclusion?

Writing the Equation of a Line

When a line has slope m and the second coordinate of its y intercept is the number b, then the equation of that line is $y = mx + b$.

12. A small business had just 12 employees when they opened their doors, but have been adding employees at the rate of 4 per month since then. Write a linear equation that describes the number of employees in terms of the number of months after the company was founded.

$E = 12 + 4m$, where E is the number of employees and m is the number of months.

Hugely important point of emphasis: When writing algebraic expressions to represent quantities, it's critical to identify what quantity each of the variables represents!

13. Use a graphing calculator to create both a table of values and a graph for the equation you wrote in Question 12. The technology boxes on page 144 will walk you through the process.

X	Y_1
0	12
3	24
6	36
9	48
12	60
15	72
18	84

X=0

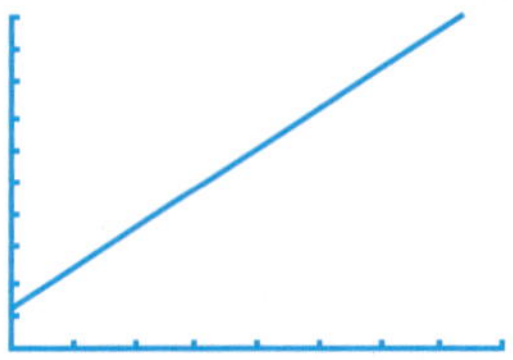

WINDOW
Xmin=0
Xmax=24
Xscl=3
Ymin=0
Ymax=100
Yscl=10
Xres=1

Did You Get It

Try this problem to see if you understand the concepts we just studied. The answer can be found at the bottom of the portfolio page.

4. Suppose that you're in desperate need of some quick cash, and a local restaurant owner offers you twenty bucks to stand on a street corner and hand out coupons to her place. You also get a \$0.50 bonus for every one of the coupons that gets turned in to the restaurant on that day. Write an equation that describes the total amount of money you'll make (y) if x coupons are turned in.

Using Technology: Making a Table on a Graphing Calculator

To make a table of values for an expression on a TI-84 Plus:

1. Press Y= to get to the equation editor screen.
2. Enter the formula you want to find values for, using X,T,θ,n for the variable.
3. Press 2nd WINDOW to get to the table setup screen. Enter the first input value you want an output for next to TblStart, and the distance between input values next to ΔTbl. (In the case of the cab ride, we chose 0 and 2 respectively to get the cost at 2-mile increments.)
4. Press 2nd GRAPH to display the table of values.

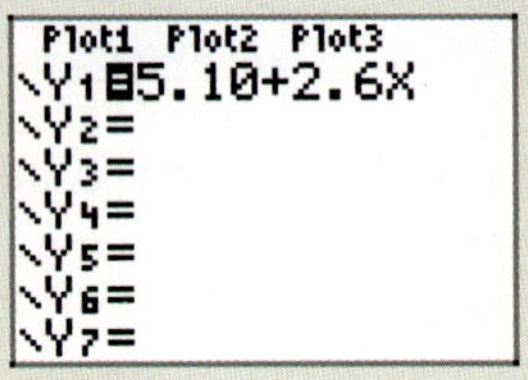

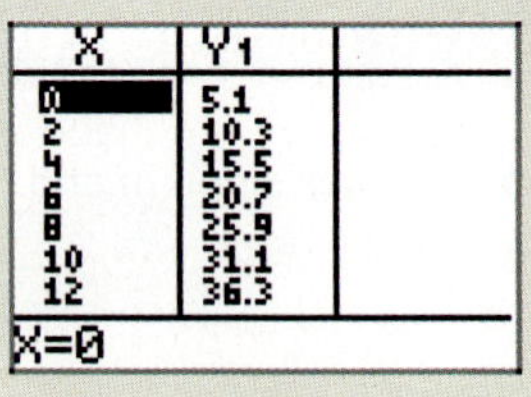

See the Lesson 2-3-2 video in class resources for further instruction.

Using Technology: Graphing an Equation on a Graphing Calculator

Starting with an equation with two variables *x* and *y*, with *y* alone on one side:

1. Press Y= to get to the equation editor screen.
2. Enter the side of the equation containing *x*, using X,T,θ,n for the variable.
3. Press WINDOW to get to the window screen.
4. Enter the lowest value you want to display along the *x* axis after Xmin, and the highest value you want to display along the *x* axis after Xmax. After Xscl, enter the distance that you want between each tick mark on the *x* axis. If you make this too small, the tick marks will be indistinct on your graph.
5. Enter the lowest value you want to display along the *y* axis after Ymin, and the highest value you want to display along the *y* axis after Ymax. After Yscl, enter the distance that you want between tick marks on the *y* axis.
6. Press GRAPH to display the graph. If you're not seeing a graph, you probably need to adjust the values you entered in the window screen.

See the Lesson 2-3-3 video in class resources for further information.

2-3 Portfolio

Name ______________________________

Check each box when you've completed the task. Remember that your instructor will want you to turn in the portfolio pages you create.

Technology

1. ☐ Use Excel to create two different scatter diagrams for the ordered pairs in the table on the next page. The first should just have the points; the second should connect the points with curves. A template to help you get started can be found in the online resources for this lesson.

Online Practice

1. ☐ Include any written work from the online assignment along with any notes or questions about this lesson's content.

Applications

1. ☐ Complete the applications problems.

Reflections

Type a short answer to each question.

1. ☐ If someone says that the point of graphing is plotting points and connecting the dots, how would you explain to them how very, very wrong they are? It'll be tough, but try to be nice.

2. ☐ Describe what the slope of a line tells us about the quantity represented by the height of that line.

3. ☐ Explain the advantages of graphed data over data in table form.

4. ☐ Take another look at your answer to Question 0 at the beginning of this lesson. Would you change your answer now that you've completed the lesson? How would you summarize the topic of the lesson now?

5. ☐ What questions do you have about this lesson?

Looking Ahead

1. ☐ Complete the Lesson 2-4 Prep Skills and read the opening paragraph in Lesson 2-4 carefully, then answer Question 0 in preparation for that lesson.

Answers to "Did You Get It?"

1. From highest to lowest: (200, 15), (−350, 0), (300, −9), (−400, −12) **2.** There's no right answer here, but I would go with either one or two per tick mark on the x axis and 50 per tick mark on the y axis. **3.** The scatter plot is available in the online resources for this lesson. **4.** $y = 0.50x + 20$

2-3 Applications

Name ______________________________

The hourly temperatures for Champaign, Illinois, on October 8, 2012 are given in the table. Use hours after midnight (NOT the actual time) as first coordinates, and the corresponding temperatures as second coordinates.

Time	Temp.	Ordered Pair
12:00 AM	34°	(0, 34)
1:00 AM	34°	(1, 34)
2:00 AM	34°	(2, 34)
3:00 AM	36°	(3, 36)
4:00 AM	35°	(4, 35)
5:00 AM	33°	(5, 33)
6:00 AM	32°	(6, 32)
7:00 AM	33°	(7, 33)
8:00 AM	37°	(8, 37)
9:00 AM	42°	(9, 42)
10:00 AM	48°	(10, 48)
11:00 AM	51°	(11, 51)
12:00 PM	54°	(12, 54)
1:00 PM	55°	(13, 55)
2:00 PM	57°	(14, 57)
3:00 PM	57°	(15, 57)
4:00 PM	57°	(16, 57)
5:00 PM	55°	(17, 55)
6:00 PM	52°	(18, 52)
7:00 PM	49°	(19, 49)
8:00 PM	47°	(20, 47)
9:00 PM	46°	(21, 46)
10:00 PM	45°	(22, 45)
11:00 PM	44°	(23, 44)
12:00 AM	43°	(24, 43)

1. Write an ordered pair for each time and temperature pairing. Remember, *the first coordinate is hours after midnight.*

2. Decide on an appropriate scale for each axis and create a scatter plot for these data on graph paper. Then connect the points with a smooth curve. If you don't have graph paper, you can easily find printable graph paper online. Make sure that the lowest height on the graph corresponds to zero degrees.

See online resources.

3. What do you need to locate in the table to find when the temperature was highest during the day? What about on the graph?

Table: The time corresponding to the largest number in the second column, or to the largest second coordinate. Graph: The highest point on the graph.

4. Use your graph to estimate ALL time spans when the temperature was increasing, and all times when the temperature was decreasing.

Increasing: 2 AM to 3 AM, 6 AM to 2 PM
Decreasing: 3 AM to 6 AM, 4 PM to 12 AM

5. Use your graph to write a verbal description of the change in temperature over that 24-hour period.

The temperature was pretty stable from midnight to 4 AM, then dropped to a low just above 30° at 6 AM. After that it rose steadily until reaching a high around 57° in mid-afternoon. After 4 PM, the temperature started to decrease fairly steadily.

2-3 Applications

Name ____________________

6. Draw a second graph for the data: this time make the lowest height on the graph correspond to 30°. (Important: Make your graph cover the same number of blocks on the graph paper as the first graph. You'll need to change the scale on the *y* axis to make this happen.) Why is this second graph deceiving in terms of how much the temperature varies?

See online resources for graph. Since the same amount of vertical space now corresponds to a smaller temperature range, it makes the changes in temperature look much more extreme than they really are.

Everyone knows that after you buy a car, in most cases its value decreases as it gets older. Suppose you bought a used car for \$15,000 a few years ago, and its value has been decreasing at the rate of \$2,000 per year since then.

Time (yrs)	Value ($)
0	\$15,000
1	\$13,000
2	\$11,000
3	\$9,000
4	\$7,000
5	\$5,000

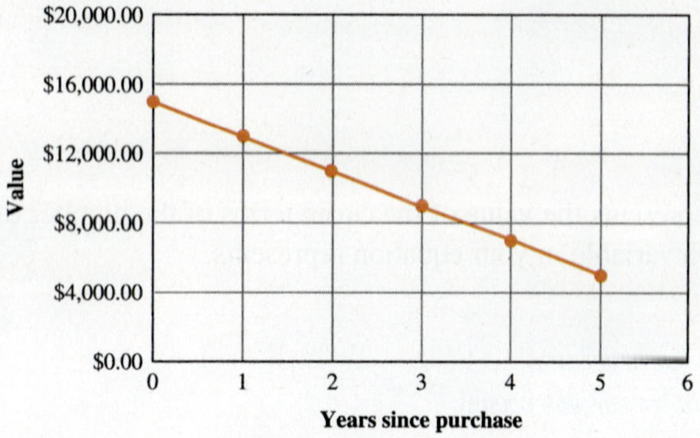

7. Find the *y* intercept of the graph and explain what it means. (Remember, the *y* intercept is a point, not a number.)

(0, 15,000); this means that the value of the car at the time you bought it (zero years after purchase) was \$15,000.

8. Find the slope of the line and explain what it means. In particular, what's the significance of the sign?

$$\frac{13{,}000 - 15{,}000}{1 - 0} = -2{,}000$$

This means that the value of the car was decreasing at the rate of \$2,000 per year.

2-3 Applications

Name ______________________________

9. Estimate the value of the car after 42 months. (Careful about time units!)

42 months is 3.5 years; there's a point on the graph with coordinates (3.5, 8,000), so the estimated value after 42 months is $8,000.

10. Estimate the number of years it took for the car to reach a value of $6,000.

The point with height 6,000 is about halfway between 4 and 5 on the horizontal axis, so it took about 4.5 years.

11. Write an equation that represents the value of the car in terms of the number of years since it was bought. Make sure you clearly identify what each variable in your equation represents.

$y = 15{,}000 - 2{,}000x$
y represents the value of the car in dollars.
x is the number of years since the car was bought.

12. Enter the expression from Question 11 next to Y1 in the equation editor of a graphing calculator, then create a table and a graph.

X	Y1
0	15000
1	13000
2	11000
3	9000
4	7000
5	5000
6	3000

X=0

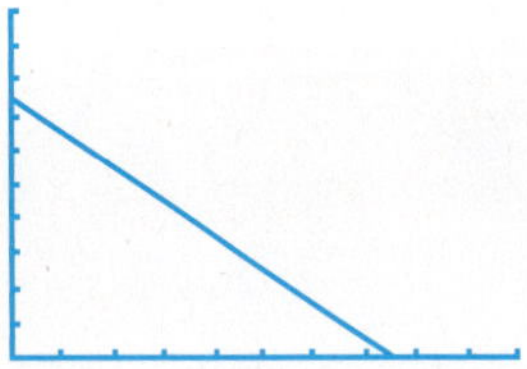

2-3 Applications

Name ____________________

13. Make up a practical problem that could correspond to each of these graphs. Include as much detail as possible, and make sure to pay close attention to the sign of the *y* coordinate. Be creative!

a.

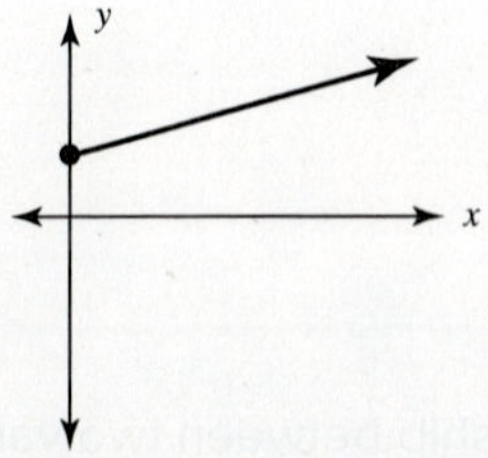

If these answers don't vary widely, I have bad news: your students are cheating. :)

b.

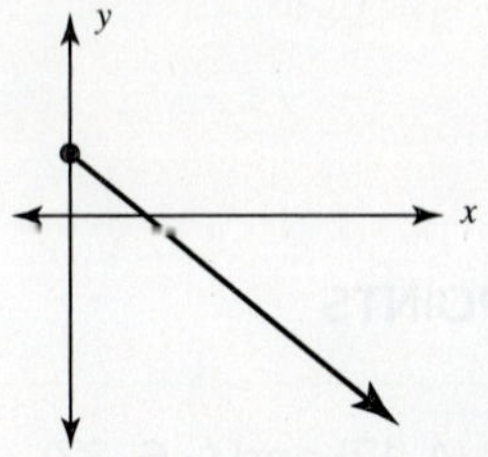

c.

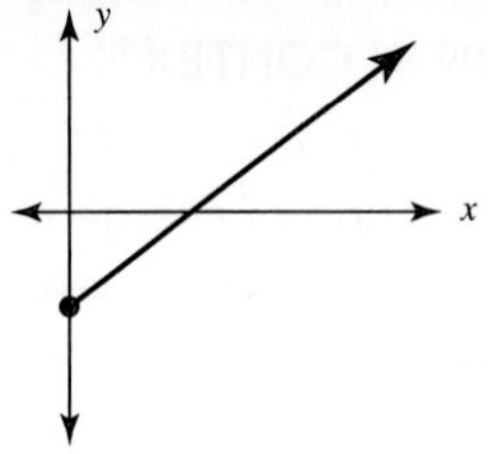

Lesson 2-4 Prep Skills

This is a short review of skills that will come in handy in the next lesson. In each case, answer the question, then rate your confidence level by checking one of the boxes. If you feel like you're struggling with these skills, consult the online resources provided by your instructor for extra practice.

I KNOW IT | THINK SO | UNSURE | NO IDEA

SKILL 1: IDENTIFY A LINEAR RELATIONSHIP

1. Which of the two tables shows a linear relationship between two variables?

x	y
2	35
4	55
6	80
8	110
10	145

x	y
10	5
20	9
30	13
40	17
50	21

SKILL 2: FIND SLOPE GIVEN COORDINATES OF POINTS

2. Find the slope of the line connecting the points (4, 12) and (−6, 37).

SKILL 3: FIND VALUES FROM AN EQUATION

3. Evaluate the equation $y = -2.3x + 11.1$ for $x = -10$, $x = 10$, and $x = 25$.

SKILL 4: FIND INFORMATION FROM AN EQUATION IN CONTEXT

4. The number of people that come to a monthly meeting of a large student organization can be modeled by the equation $y = 5.5x + 37$, where x is the number of months since the group started a large membership drive.

   a. How many people came to the meeting the month the drive started?
   b. Evaluate the equation for $x = 4$ and $x = 10$, and describe what each result tells us.

Lesson 2-4 The Great Tech Battle (Linear Relationships and Lines of Best Fit)

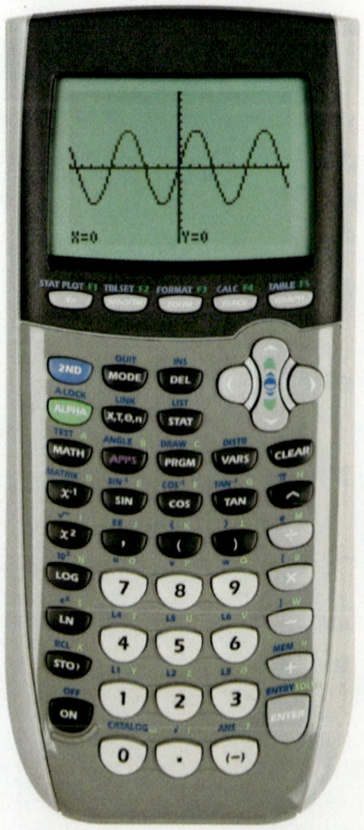

© DonNichols/Getty Images RF

LEARNING OBJECTIVES

- ☐ 1. Decide if two data sets are linearly related.
- ☐ 2. Find lines of best fit for data using spreadsheets and calculators.
- ☐ 3. Use lines of best fit to analyze data.
- ☐ 4. Interpret correlation coefficients.

One machine can do the work of fifty ordinary men. No machine can do the work of one extraordinary man.

—Elbert Hubbard

That quote says an awful lot about the dangers of too much reliance on technology, a peril with the potential to affect every modern student. The most interesting thing about it, though, is that it was written in 1903! When used correctly in math, technology allows us to focus more on understanding and interpretation, and less on computation. That's great. But when you try to use any technology as a substitute for thinking, nothing good is likely to happen. In this lesson, we'll learn how technology can help us to decide when a given data set is likely to be modeled well with a linear equation. Better still, calculators and spreadsheets will be able to find the best linear model for such situations, freeing us to focus on interpreting the model and using it to further study the data.

0. After reading the opening paragraph, what do you think the main topic of this lesson will be?

This class portion requires more guidance than most, so it works best as a teacher-led class discussion.

2-4 Class

In Lesson 2-3, we examined the relationship between time of day and outside temperature. While we needed to study data to model the exact nature of the relationship, it doesn't exactly take a genius to figure out that the two quantities are related in some way. There are many quantities that *seem* like they might be related, but it's not clear that they really are, or if so, how. So it would be nice to have a way to study relationships (if they even exist) between data sets. Fortunately, the lovely folks that program spreadsheets and graphing calculators have done most of the heavy lifting for us. Still, in our continuing quest to not rely on technology more than our brains, it's a good idea to think about when quantities are likely to be related in some way.

1. Decide if you think the two quantities are likely to be related in some way, and explain your answer.

a. Dollars spent by a political candidate and the number of votes he gets

Unfortunately, in our system spending a lot of money does have an influence on the number of votes.

b. An adult test subject's height and their IQ

There's no reason to believe that height has any connection to IQ.

c. A person's height and their shoe size

Larger people do tend to have bigger feet.

d. The age of a car and its resale value

Most cars are worth less as they get older.

2. Is there a relationship between the total fat in a fast-food sandwich and the total calories? Begin your study of this question by drawing a scatter plot based on the table.

Sandwich	Grams of fat	Calories
Hamburger	9	260
Cheeseburger	13	320
Quarter Pounder	21	420
Quarter Pounder with Cheese	30	530
Big Mac	31	560
Arch Special	31	550
Arch Special with Bacon	34	590
Crispy Chicken	25	500
Fish Filet	28	560
Grilled Chicken	20	440
Grilled Chicken Light	5	300

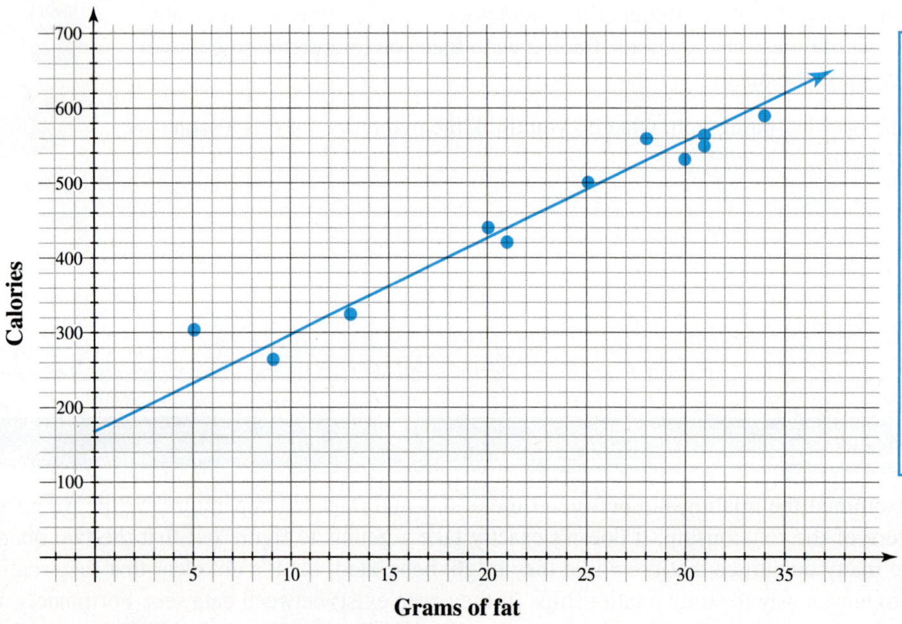

Make sure you emphasize the idea that the line of best fit may not go through any of the points. Give students a general idea of how to draw one, then walk around and see how they did. Also, show students how to make a simple straightedge by folding a sheet of paper in half.

3. The points on your scatter plot don't form a straight line, but there should be a linear trend you can observe. Use a straightedge to draw a line that you think looks like the best fit for your scatter plot. Note that it's possible that the line that fits the data best doesn't go through any of the points.

Line of best fit will vary. Sample shown.

Now that we've been able to draw a line that seems like a reasonable fit for data, it would be nice to find the equation of that line so we can do some calculations with it. We'd really like to use all of the available points to find a line to model our data, but we'll need some technology to pull that off. In the meantime, we can use a relatively simple but useful procedure to find the equation of a line when you know any two points on the line.

In the last lesson, we used the form $y = mx + b$ to describe the equation of a line. This allows us to find the equation of a line when we know the slope (m) and the second coordinate of the y intercept (b). But what if we don't know the y intercept? We can still find the equation: we just need a little ingenuity.

A Procedure for Finding the Equation of a Line Through Two Points

Step 1: Find the slope using the slope formula: difference of y coordinates divided by difference of x coordinates.

Step 2: Substitute the slope, and the x and y coordinates of one of the points in for x and y in the equation $y = mx + b$. This results in an equation you can solve to find b.

4. Pick two points on your line. (Watch out! Don't pick points from the table or the scatter plot. Actually look at the line you drew, and find points on the LINE.) Write your points here.

I chose (10, 300) and (35, 620).

5. Use your points from Question 4 to find the slope of the line you drew. What does it mean?

$$m = \frac{620 - 300}{35 - 10} = \frac{320}{25} = 12.8$$

This means that the calories go up by about 12.8 for each extra gram of fat.

6. Pick one of your points from Question 4. Using $y = mx + b$, substitute in your slope for m, the x coordinate of your chosen point for x, and the y coordinate of your chosen point for y. Solve the resulting equation to find b. What does it mean? Remember, b is the y coordinate of the y intercept.

Using the point (10, 300):

$300 = 12.8(10) + b$

$300 = 128 + b$

$b = 172$

This tells us that we would expect a sandwich with zero grams of fat to contain 172 calories.

Math Note

To see if a point is a solution to an equation with two variables, substitute the first coordinate for x and the second coordinate for y. If the resulting equation is true, the point is on the line.

7. Use your answer from Questions 5 and 6 to write the equation of the line you drew in the form $y = mx + b$.

$y = 12.8x + 172$

Students often get confused about why the values they substituted for x and y now "disappear." Remind them that any (x, y) on the line should result in a true statement, so we don't want the final answer to include specific x and y values.

8. Use your equation to predict the number of calories in a Whopper, which has 39 grams of fat.

$y = 12.8(39) + 172 = 671.2$ calories

Did You Get It

Try this problem to see if you understand the concepts we just studied. The answer can be found at the bottom of the portfolio page.

1. Find the equation of the line that goes through the points (−3, 4) and (9, 7).

So far, we've used our eyeballs and perception to draw a line that we think fits the data best, then found an equation for that line. But is your line the line of BEST fit? There's a pretty good chance that everyone in the room drew a slightly different line, and has a slightly different equation. Graphing calculators are programmed to take all data into consideration and find the line that best fits the data. That line is called the **line of best fit,** which is not the most creative name ever, but is descriptive.

Using Technology: Finding the Line of Best Fit with a Graphing Calculator

To find a line of best fit for two data sets that appear to have an approximate linear relationship:

1. Press STAT ENTER to get to the list editor.
2. Enter the data set you want to use as inputs (*x* values) under **L1.**
3. Use the right arrow key 〉 to access **L2,** then enter the data set you want to use as outputs (*y* values). When entering data from a table, make sure you enter the second list in the same order as the first.
4. Turn on **Plot1** by pressing 2nd Y= 1. Set up the screen as shown below.
5. Press Y= and if the **Y** = screen isn't blank, move the cursor over any entered equations and press CLEAR.
6. Press ZOOM 9, which is the **Zoom Stat** option; this automatically sets a graphing window that displays all of the plotted points.
7. Press STAT followed by the right arrow key to access the **STAT CALC** menu, and choose the **LinReg** (for linear regression) option, which is choice 4. Then press ENTER to calculate the line of best fit.
8. Press Y= and enter the equation of the line of best fit, then press GRAPH to display the scatter plot along with the line of best fit.

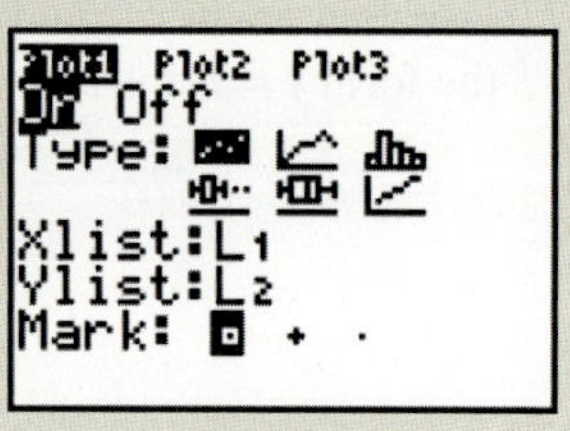

See the Lesson 2-4-1 video in class resources for further instruction.

9. Use the procedure in the Using Technology box to find the equation of the line of best fit using a graphing calculator. Round to one decimal place.

$y = 11.7x + 193.9$

10. Use your answer to Question 9 to predict the number of calories in a Whopper, which has 39 grams of fat. How does the answer compare to the number of calories predicted by your own line of best fit in Question 8? The actual number of calories in a Whopper is 660.

$y = 11.7(39) + 193.9 = 650.2$ calories
Our line gave us 671.2 calories.
The line from the calculator predicts 21 calories less.

Have a discussion about how close the predictions were, and how they could make their lines of "best fit" better. Emphasize that from now on, when they are asked for "line of best fit," they should use technology rather than finding an equation by hand.

When finding the line of best fit using a graphing calculator, the last line in the display should be a value for a quantity labeled *r*, which is known as the **correlation coefficient.** This number measures how well the line seems to fit the data. The closer *r* is to 1 (positive slope) or −1 (negative slope), the more accurately the data you entered can be modeled using a linear equation. Here are some examples, based on data from two of my classes last semester:

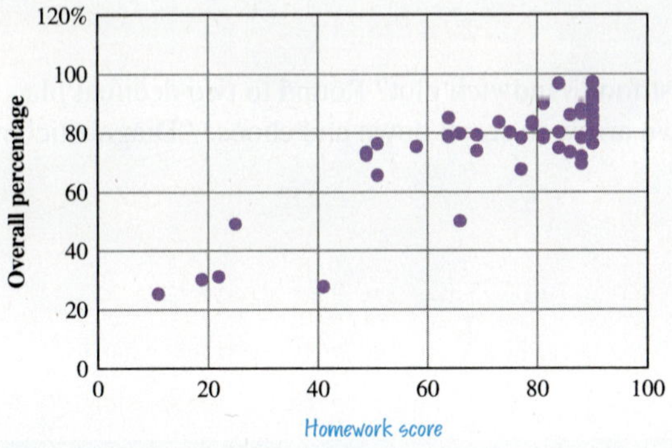

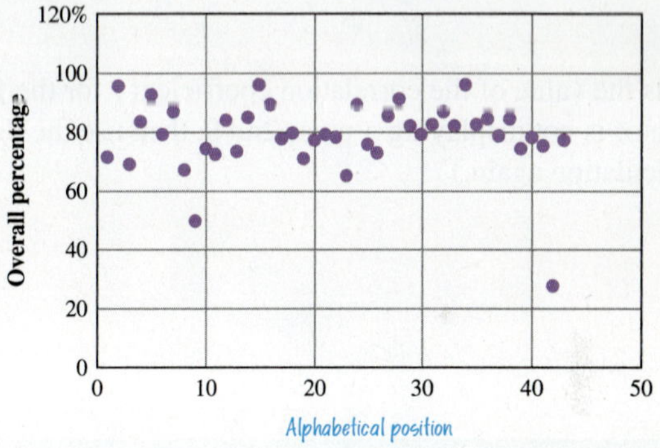

11. One of these scatter plots has a correlation coefficient of 0.12; the other has a correlation coefficient of 0.72. Discuss which you think is which, and why.

The one on the left is probably the one with coefficient 0.72. You can see a general trend from lower left to upper right: Most of the lower points are to the left, and most of the higher points are to the right. On the second graph, the points appear to be scattered in a much more random pattern.

Give students time to discuss Questions 11 and 12 in pairs, then discuss as a class.

12. One of the scatter plots has each student's homework score for the semester on the x axis and the corresponding overall percentage in the course (from 0 to 100) on the y axis. The other again has overall percentage on the y axis, but along the x axis is the student's position alphabetically in the class. (The student whose name comes first alphabetically is 1, the next alphabetically is 2, and so on.) Which plot is which? Label the x axis for each graph. Again, discuss your reasons.

The first plot has to be the relationship between homework score and overall percentage. It stands to reason that doing more homework will help you to learn, and your overall grade will improve. There's no reason to believe that the first letter of a student's last name will impact their grade in any way.

Did You Get It

Try this problem to see if you understand the concepts we just studied. The answer can be found at the bottom of the portfolio page.

2. If you drew a scatter plot that has the amount of money a small business spends on advertising on the horizontal axis and the average number of customers they have per week on the vertical axis, would you expect the correlation coefficient for this data to be closer to zero or one? Why?

13. What is the value of the correlation coefficient r for the fast food sandwich plot? Round to two decimal places. (If your calculator is not displaying r, press 2nd 0, then use the down arrow to scroll down and choose "DiagnosticOn" and try the calculation again.)

0.97

Using Technology: Finding the Line of Best Fit with a Spreadsheet

To find a line of best fit and correlation coefficient on a spreadsheet:

1. Enter the data in two columns and create a scatter plot.
2. Click on one of the points on the scatter plot, then choose "Add Trendline" from the **Chart** menu.
3. In the formatting dialog box that appears, click "Options," then click the checkboxes for "Display equation on chart" and "Display r^2 value on chart."
4. Calculate the square root of the r^2 value that appears on the chart. This is the correlation coefficient.

See the Lesson 2-4-2 video in class resources for further instruction.

14. Use the procedure in the Using Technology box to find the equations of the line of best fit and the correlation coefficient using Excel. How does each compare to the results you got from your graphing calculator?

$y = 11.731x + 193.85$; $r = 0.97$. With the same level of rounding, the results are exactly the same.

If students have laptops, they can follow along. Otherwise, use a computer projector to demonstrate.

15. Based on all of the work done in this activity, how strong do you think the relationship is between the number of grams of fat in a fast food sandwich and the number of calories in it? Make sure you justify your answer.

Very strong. The scatter plot shows a clear pattern from lower left to upper right. We were able to find a line of best fit that matches most of the data points pretty well, and the correlation coefficient is close to one, meaning there's a strong linear relationship.

An Important Note on Lines of Best Fit

In Question 10, we saw that the line of best fit for the number of calories in a sandwich doesn't give the exact correct value for a Whopper. This will almost always be the case when finding a line of best fit for a data set with more than two values. There's a good reason we call it "the line of best fit," not "the line of perfect fit"! The equations we're finding are *models* for data that allows us to make approximations and predictions; they're not *exact representations* of data.

16. How close is the actual number of calories in a Big Mac to the number predicted by the line of best fit? Is the prediction too high or too low?

$y = 11.7(31) + 193.9 = 556.6$; the prediction is really close–just 3.4 calories below the actual value.

Did You Get It

Try this problem to see if you understand the concepts we just studied. The answer can be found at the bottom of the portfolio page.

3. For the data points listed below, find the line of best fit and the correlation coefficient.
(4, 190), (8, 246), (12, 384), (16, 362), (20, 461), (24, 501), (28, 705)

2-4 Group

1. Do you think there's a strong connection between a person's shoe size and their height? Answer yes or no, explain why you feel that way, and make a guess as to the value of the correlation coefficient for these two data sets.

It seems reasonable that there's a strong connection, but of course answers will vary for this entire group of questions due to the data specific to the individual class.

Shoe Size	Height (inches)	Predicted Height

2. Fill out the first two columns in the chart for all of the students in your class, your instructor, and anyone else who happens to be walking by.

3. Draw a scatter plot with shoe size on the x axis and height in inches on the y axis. Axes are provided on the next page.

4. Use a calculator or spreadsheet to draw a scatter plot for the data, then find and write the equation of the line of best fit.

5. Add the line to your hand graph.

6. Find the correlation coefficient for the data. How closely are the two quantities related? Write a sentence to describe the connection.

7. Use your equation to predict the height of a person with size 17 shoes.

> If all goes really well, one of your students might ask about the discrepancy in sizing for men and women. Discuss whether or not you think that should be taken into account. I had the students convert women's sizes into men's equivalents, then did it myself without converting for comparison.

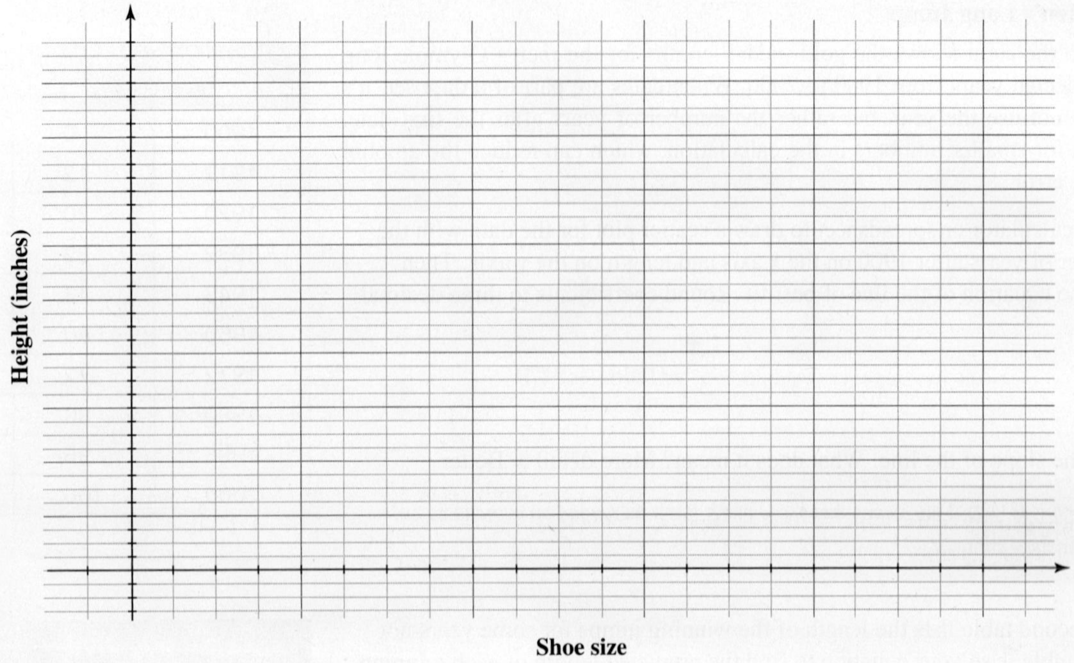

8. Use your equation and either the table feature on a graphing calculator or a spreadsheet to calculate the predicted heights for everyone in your chart based on shoe size, then put that data into the chart.

9. What percentage of the people in your chart are within 2 inches of their predicted height? What can you conclude?

Olympic Men's Long Jump

The table to the right shows the gold-medal lengths for the men's Olympic long jump for selected years from 1900 to 2000. When dates are part of a data set, it's common to not use the year, but rather the number of years after the first date. This allows for smaller numbers in the calculation, which can reduce the amount of rounding error.

Year	x	Length (meters)
1900	0	7.18
1912	12	7.60
1920	20	7.15
1932	32	7.64
1948	48	7.82
1960	60	8.12
1972	72	8.24
1980	80	8.54
1996	96	8.50
2000	100	8.55

10. Use a calculator or spreadsheet to draw a scatter plot for the data with the number of years after 1900 on the x axis and length on the y axis. Then find the equation of the line of best fit. Round coefficients to three decimal places.

See online resources for graph. The equation is $y = 0.015x + 7.179$.

11. Find the slope of the line. What does it mean? More detail = Better.

The slope is 0.015. This means that the winning jump increased on average by 0.015 meters per year.

12. The second table lists the length of the winning jumps for some years not in the table. Use your equation to find the predicted length of each winning jump.

Year	x	Length (meters)
1984	84	8.54
2004	104	8.59
2008	108	8.34
2012	112	8.31

1984: $y = 0.015(84) + 7.179 = 8.44$ m

2004: $y = 0.015(104) + 7.179 = 8.74$ m

2008: $y = 0.015(108) + 7.179 = 8.80$ m

2012: $y = 0.015(112) + 7.179 = 8.86$ m

13. How do the predictions compare to the actual jumps? What can you conclude?

The predictions are all quite a bit more than the actual jumps once you get past 2000. That tells us that the winning length increased steadily during the 20th century, but that trend changed after 2000.

14. Use your equation to predict the year that the men's long jump record will reach 9.0 meters. (Remember, the Olympics are only held every four years!)

$9 = 0.015x + 7.179 \Rightarrow 0.015x = 1.821 \Rightarrow x = 121.4$ *years. This corresponds to the year 2021, but the next Olympics after that will be in 2024.*

15. Do an Internet search for "1968 men's long jump." After reading about the event that year, explain why we purposely left that year out of the table.

The winning jump that year was staggeringly long–many consider it to be the single greatest athletic feat of all time. It's so aberrational that it doesn't fit an otherwise gradual trend, and can mess up the accuracy of the line of best fit.

2-4 Portfolio

Name ______________________________

Check each box when you've completed the task. Remember that your instructor will want you to turn in the portfolio pages you create.

Technology

1. ☐ You kind of get a break on this one . . . since this lesson is all about using technology, you'll take care of the technology part in doing the Applications portion. Cool!

Online Practice

1. ☐ Include any written work from the online assignment along with any notes or questions about this lesson's content.

Applications

1. ☐ Complete the applications problems.

Reflections

Type a short answer to each question.

1. ☐ What is a line of best fit for two sets of data?
2. ☐ Does the line of best fit provide useful information about every pair of data sets? Why or why not? Your answer should probably mention the correlation coefficient.
3. ☐ Take another look at your answer to Question 0 at the beginning of this lesson. Would you change your answer now that you've completed the lesson? How would you summarize the topic of this lesson now?
4. ☐ What questions do you have about this lesson?

Looking Ahead

1. ☐ Complete the Lesson 2-5 Prep Skills and read the opening paragraph in Lesson 2-5 carefully, then answer Question 0 in preparation for that lesson.

Answers to "Did You Get It?"

1. $y = 0.25x + 4.75$ **2.** Closer to one. It's likely that the number of customers has a reasonably strong relationship to how much is spent on advertising.
3. $y = 19.04x + 102.43$, $r = 0.96$

2-4 Applications

Name ____________________

The table to the right shows the average annual cost for tuition, room, and board at all colleges in the United States between the 2000 and 2010 school years.

School Year	x	Cost ($)	Increase
2000–2001	0	10,820	—
2001–2002	1	11,380	560
2002–2003	2	12,014	634
2003–2004	3	12,953	939
2004–2005	4	13,793	840
2005–2006	5	14,634	841
2006–2007	6	15,483	849
2007–2008	7	16,231	748
2008–2009	8	17,092	861
2009–2010	9	17,649	557
2010–2011	10	18,497	848

1. Calculate the increase over the previous year for each school year and put the value in the table. Based on the results, do you think we could model these data accurately with a linear equation? Why or why not?

   The rate of increase isn't constant, but most of the increases were in the 700-800 range, so a linear equation should do a pretty good job of modeling the costs.

2. Draw a scatter plot on the axes provided below. The x values are given in the table; use values from the Cost column as y values. Does the result seem to match your answer to Question 1?

   The trend does look pretty linear.

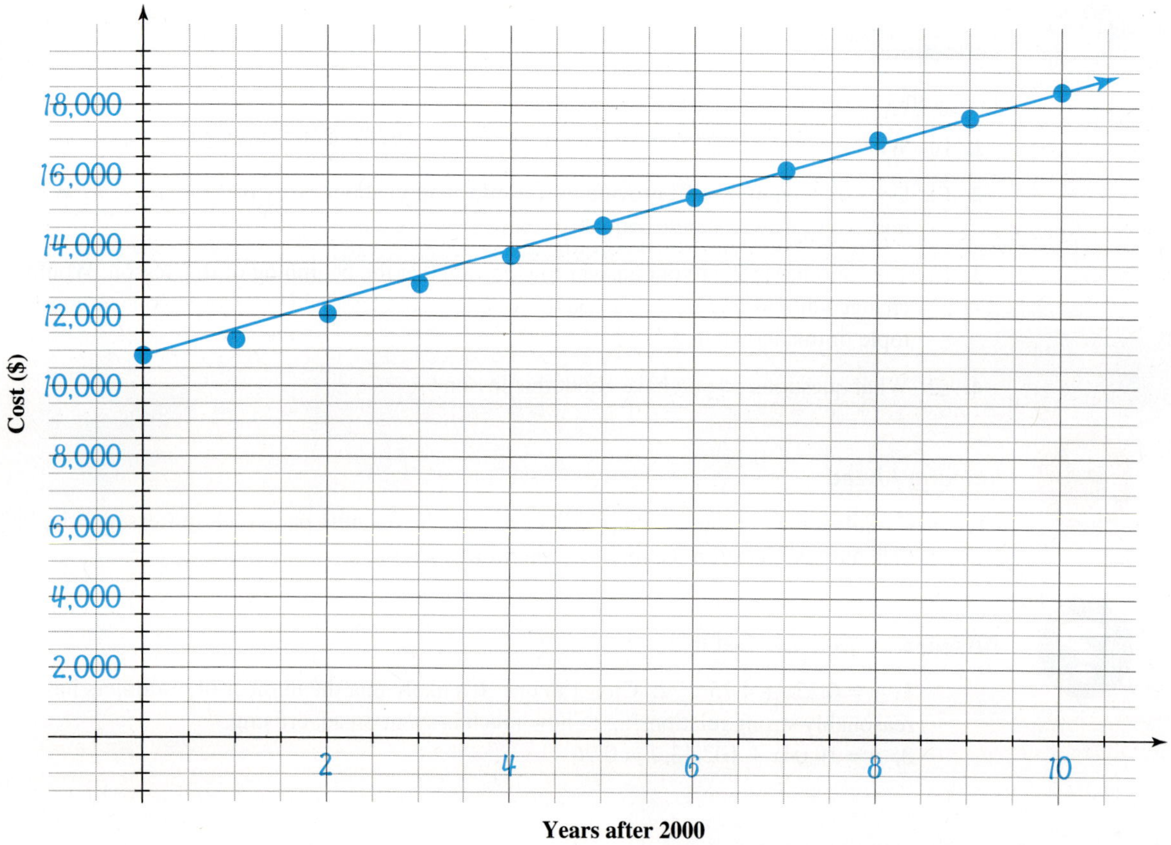

Lesson 2-5 Sit Back and Watch Your Money Grow (Exponential Growth Equations)

LEARNING OBJECTIVES

- ☐ 1. Illustrate exponential growth with a graph.
- ☐ 2. Solve problems using graphs representing exponential growth and decay.
- ☐ 3. Identify the significance of the parameters in an exponential equation.
- ☐ 4. Find and interpret exponential models.

If I take 30 steps linearly, I get to 30. If I take 30 steps exponentially, I get to a billion.

—Ray Kurzweil

Raise your hand if you hope to be rich someday. There's a pretty good chance that your hand is up right now: financial independence is a goal for a lot of people. We've already studied the differences between quantities that grow linearly and those that grow exponentially. Now that our skills at interpreting graphs have grown, well, exponentially, it's a good time to revisit exponential growth to learn a little bit more about its magic. We already know that the long-term growth of money is an ideal way to study exponential growth, so we'll start there, and on the way start you on the path to riches!

0. After reading the opening paragraph, what do you think the main topic of this lesson will be?

2-5 Class

1. Describe the differences between linear and exponential growth based on what we know from earlier in the course.

With linear growth, the increase is steady; that is, the rate of change is constant. The new amount comes from ADDING the same thing repeatedly to the previous amount. With exponential growth, the rate of increase keeps getting bigger. The new amount comes from MULTIPLYING the previous amount by the same thing.

2. What do you think is meant by the "relative change" in a quantity? How do you compute relative change? (Referring to the topic of relative difference in Lesson 1-6 should help.)

It refers to the size of the change compared to the original size of the quantity. So you divide the change by the original quantity.

If you're like a lot of people, you'd go on a spending spree if you won $10,000 on a game show. But what if you decided to invest that money rather than spend it, and what if you were patient enough to let it grow for 40 years in a savings account? Let's take a look. Make sure that you answer every question with a complete sentence.

Time (years)	Value ($)
0	$10,000.00
1	$10,700.00
2	$11,449.00
3	$12,250.43
4	$13,107.96
5	$14,025.52
6	$15,007.30
7	$16,057.81

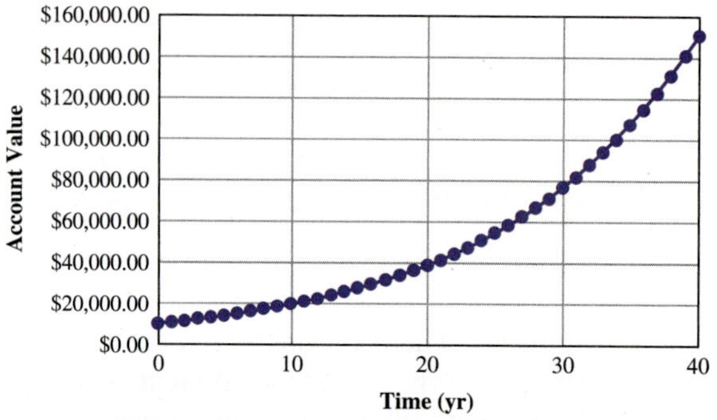

3. For each value in the table after the initial investment, divide the amount by the value of the account the previous year. What do you notice?

In each case, the result is either 1.07 or very close.

4. What does your result in Question 3 say about the type of growth this account exhibits?

This shows that the value of the account is multiplied by 1.07 each year, which makes it exponential growth.

5. Find the relative change in value for each year in the table compared to the previous year. What can you conclude?

The relative change is 7% for each year. This shows that the value of the account grows by 7% per year.

Here's a quick summary of what we learned on page 166: the account starts at \$10,000 and grows by 7% each year. Mathematically, this is accomplished by multiplying the previous year's amount by 1.07. So after one year, the amount is \$10,000(1.07), after two years \$10,000(1.07)(1.07), and so on. The result is that an equation that describes the value after x years is

$$y = 10{,}000(1.07)^x$$

And now we know why this type of growth is called "exponential growth": we can use an exponent to represent the repeated multiplications that we get when a quantity is multiplied by the same amount over and over.

6. Verify that the above equation generates the account values in the table. Ideally, you would use either the table feature on a graphing calculator or spreadsheet, but in any case, make sure you describe how you got your values.

Answers vary depending on the method used. But in every case, of course, the students should find that the equation does in fact match the values in the table.

Math Note

We'll study this method of building exponential models in greater depth when we study consumer math in Unit 3.

Did You Get It

Try this problem to see if you understand the concepts we just studied. The answer can be found at the bottom of the portfolio page.

1. Elaine keeps an investment whose value x years after opening the account can be described by the equation $y = 12{,}500(1.05)^x$. What was her initial investment? By what percentage does it grow each year?

7. Look back at the graph that illustrates the value of the account. Trace along the graph slowly from left to right, and describe what the changing steepness of the graph tells you about how the value is growing.

The graph starts out very gradually, which indicates that the value of the account is not growing very quickly. But it starts to get noticeably steeper around year 15 or so, and by the time you get out to 30 years it's gotten very steep, which indicates that the value of the account is growing rapidly.

8. How much interest will you earn from 0 to 5 years into the investment? This will be the difference in the value after 5 years and the initial investment. First, estimate the answer by looking at the graph. Then perform a calculation using the table provided.

From the graph, it looks like about \$5,000 or so. The actual amount is \$4,025.52.

9. How much interest will you earn between years 35 and 40? Again, estimate from the graph first. Then use the equation you wrote to get an exact value.

From the graph, it looks like about \$45,000 or so. The actual amount is \$42,978.76.

10. Describe what you can learn from comparing the answers to Questions 8 and 9. Also describe how that fits in with your answer to Question 7.

The amount of interest earned over a 5-year period gets a LOT bigger when the money has been invested for a longer period of time. This perfectly matches what we observed from the graph: The big change in steepness indicates the value growing much faster after 35 years than after 5.

My all-time favorite TV show was *Highlander.* It featured a race of immortals whose destiny is to battle each other over hundreds and in some cases thousands of years. The fact that most of them appear to be fabulously wealthy strikes a lot of people as hokey, but based on what we've learned in this lesson, it actually makes a great deal of sense. Suppose that you could travel back in time to the year 1700 and invest the very modest sum of \$10. Suppose further that the investment draws a very modest return of just 5% each year.

© AF archive/Alamy Stock Photo

11. Comparing to the equation at the top of page 167, write an equation that would describe the value of this account after x years.

$y = 10(1.05)^x$

12. Use your equation to find the value of that \$10 investment now. Remember, your variable x represents years after 1700.

$y = 10(1.05)^{316} = \$49{,}638{,}483.63$ as of 2016.

Did You Get It

Try this problem to see if you understand the concepts we just studied. The answer can be found at the bottom of the portfolio page.

2. Elaine from Did You Get It 1 is quite mortal, but she does have a daughter. Suppose she never touches the money in that account, and it gets passed down through three generations. After 94 years in the bank, how much would the account be worth?

2-5 Group

We've now studied exponential growth a couple of different times in this unit. But there's a flip side to exponential growth, known as **exponential decay**. The thing that distinguishes exponential growth is that the growth starts off relatively slowly, then speeds up dramatically as time passes. Let's see how exponential decay compares.

Coffee is a popular morning drink because the caffeine it contains acts as a temporary stimulant. Once the caffeine gets into your system, the amount dissipates exponentially (assuming that you don't add more, of course). Answer every question (except for 2 and 6) with a full sentence.

© Ingram Publishing/ SuperStock RF

Time (hours)	Caffeine (mg)
0	180.0
1	144.0
2	115.2
3	92.2
4	73.7
5	59.0
6	47.2
7	37.7

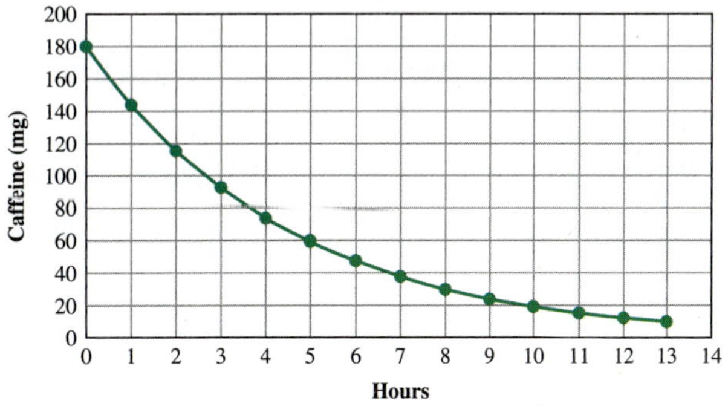

1. If you're the coffee drinker represented by the table and graph, how much caffeine was initially in your system?

I initially had 180 mg in my system.

2. Find the relative change in caffeine for each one-hour period listed in the table below. Round to the nearest full percent.

Hours	Relative change
0–1	−20%
1–2	−20%
2–3	−20%
3–4	−20%
4–5	−20%
5–6	−20%
6–7	−20%

3. Complete this important statement about exponential change: A quantity either grows or decays exponentially when the __relative__ __change__ in that quantity is __constant__.

4. Looking at the graph of the caffeine remaining in your system, describe what exponential decay looks like compared to exponential growth.

The amount decreases rapidly at first, then starts to level off. That's pretty much the opposite of exponential growth, which starts off pretty level, then starts to increase more rapidly.

5. What's the multiplication factor that you'd need to multiply by a previous hour's amount to get the next hour's amount? (Hint: If 10% of something goes away each hour, 90% remains.)

You would need to multiply by 0.8 because each hour the value is 20% less than the previous hour, which means 80% remains.

Pay attention to make sure that students understand that you subtract from 100% to find the multiplication factor for exponential decay. As you circulate, discuss the differences between growth and decay with the groups.

6. Use your answer to Question 5 to write an equation of the form "$y =$ expression," where the expression describes the amount of caffeine left in your system after x hours. Your answer to Question 11 in the Class portion is a good place to look if you need help.

$y = 180(0.8)^x$

Did You Get It

Try this problem to see if you understand the concepts we just studied. The answer can be found an the bottom of the portfolio page.

3. The radioactive elements used in medical imaging are good examples of substances that decay exponentially. Iodine-123 is a radioactive substance used in diagnosing thyroid conditions. This particular isotope decays in such a way that 95% of the initial sample remains after each one-hour period. If an initial sample is 200 mg, write an equation that describes the amount remaining after x hours.

7. Use a graphing calculator to verify that the equation you wrote in Question 6 gives you the values in the table, and the given graph. Describe your results.

The results in the table match the calculator table exactly to the number of decimal places given, and the graph looks identical.

8. Use the TRACE feature on your graphing calculator to find the amount of caffeine left in your system after 90 minutes. (Pay attention to units, and don't forget that we're still answering in complete sentences.)

The amount of caffeine left after 1.5 hours is about 129 mg.

9. Use the graph to estimate the amount of time needed for the amount of caffeine to drop to 40 mg. Describe how you found your answer.

It should take a little bit less than 7 hours for the amount to reach 40 mg. This comes from identifying the point at height 40, then visually estimating the x coordinate of that point.

10. Use either TABLE or TRACE commands on your calculator to try and come up with a more accurate estimate of the time needed for the caffeine level to reach 40 mg.

It will take about 6 hours and 45 minutes.

Did You Get It

Try this problem to see if you understand the concepts we just studied. The answer can be found at the bottom of the portfolio page.

4. Use the equation you wrote in Did You Get It 3, along with a graphing calculator, to answer each question.
   a. How much of the initial 200 mg sample remains after 8 hours? One week?
   b. About how long does it take until half of the original sample remains?

11. Make up a situation that could be modeled by each of the following exponential graphs. As always, extra points for creativity!

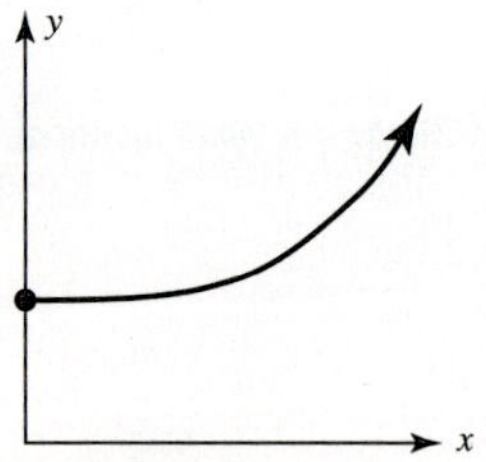

Answers vary. (Legal disclaimer: If you needed us to point that out, you should not be teaching a college math course.)

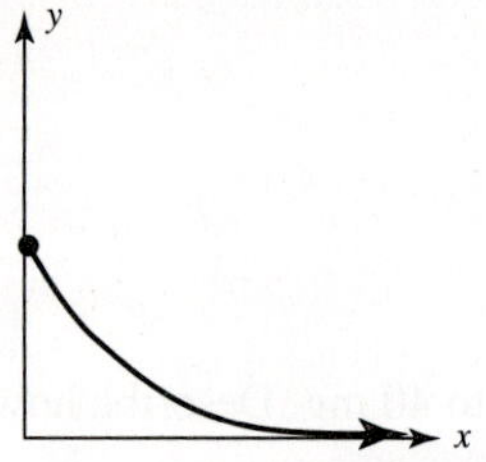

2-5 Portfolio

Name ____

Check each box when you've completed the task. Remember that your instructor will want you to turn in the portfolio pages you create.

Technology

1. ☐ Use an Excel spreadsheet to extend both the table and the graph for the caffeine problem (page 169) out to the point where there's less than 1 mg left in your system. How long does this take? A template to help you get started can be found in the online resources for this lesson.

Online Practice

1. ☐ Include any written work from the online assignment along with any notes or questions about this lesson's content.

Applications

1. ☐ Complete the applications problems.

Reflections

Type a short answer to each question.

1. ☐ Compare exponential growth and exponential decay. What do they have in common? How are they different? A really good answer will say something about relative change.

2. ☐ What did you learn in this lesson about the value of leaving money invested for long periods of time? How can you apply this to retirement?

3. ☐ Take another look at your answer to Question 0 at the beginning of this lesson. Would you change your answer now that you've completed the lesson? How would you summarize the topic of this lesson now?

4. ☐ What questions do you have about this lesson?

Looking Ahead

1. ☐ Complete the Lesson 2-6 Prep Skills and read the opening paragraph in Lesson 2-6 carefully, then answer Question 0 in preparation for that lesson.

Answers to "Did You Get It?"

1. \$12,500, 5% **2.** \$1,226,603.29 **3.** $y = 200(0.95)^x$
4. a. 8 hours: 132.7 mg, 1 week: 0.036 mg **b.** 13.5 hours

2-5 Applications

Name ______________________________

We'll begin this assignment by looking at the savings plan for Joe, a 30-year-old with a steady job making $30,000 a year, with an annual raise of 4%. Looking ahead to the future, Joe commits to saving 12% of his earnings every year. He chooses a fixed fund investment that returns 9% annually. The following graph shows Joe's earnings, the amount of money he spends, and the amount of money that is generated in interest on his investment each year. The purple curve isn't the amount he's saved: it's the amount of interest earned each year. The values on the *x* axis represent Joe's age in years.

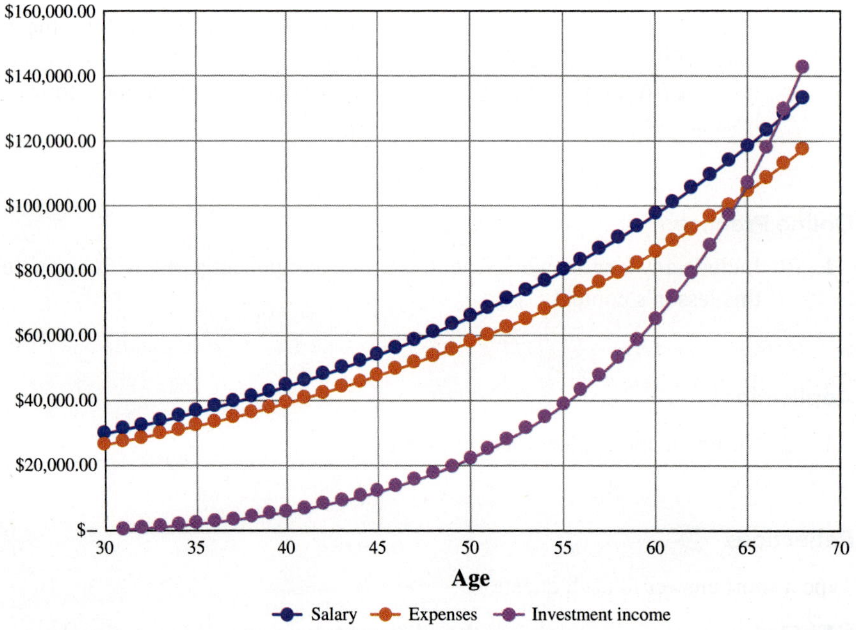

The investment income graph displays the true magic of exponential growth: the income is limited when Joe's a young guy, but as he gets closer to thinking about retirement, the income really takes off. Notice that at age 65, something awesome happens: the income that Joe's making from his investments is more than his living expenses. At that point, he can retire and live as well as ever without, you know, working for a living. Sweet! This age is known in financial circles as the **crossover point**.

Use the information in the paragraphs above to answer Questions 1–5.

1. How much money is Joe earning when he's 30?

   $30,000 *per year*

2-5 Applications

Name ____________________

2. How much money is he allowing himself to spend when he's 30? Show a calculation, then describe whether or not your answer seems correct based on the graph.

88% of \$30,000 = \$26,400: that seems pretty reasonable based on the yellow curve in the graph.

3. How much is Joe earning when he's 35? Show a calculation based on the annual relative growth rate of 4%. Then describe whether or not your answer seems correct based on the graph.

$\$30{,}000(1.04)^5 = \$36{,}499.59$: the blue curve looks to be about 3/4 of the way between 20,000 and 40,000 at $x = 35$, so that seems about right.

4. How much money is Joe allowing himself to spend when he's 35? Show a calculation, then describe whether or not your answer seems correct based on the graph.

88% of \$36,499.59 = \$32,119.64: the yellow curve is a little more than halfway between 20,000 and 40,000 at $x = 35$, which is about right.

5. Carefully explain the meaning of the crossover point.

The money you have invested for retirement earns more money in interest, and the amount it earns grows due to exponential growth. The crossover point is the time when the amount of money you're earning from your retirement investment passes the amount of money you're spending.

Now turn your attention to the graph to answer Questions 6–8.

6. How much is Joe earning in salary when he's 60?

A little less than \$100,000 per year

7. How much is Joe allowing himself to spend when he's 60?

About \$85,000 per year

2-5 Applications

Name ______________________________

8. How much is Joe earning from his investments at 60?

Around \$65,000 per year

9. Write an equation of the form "y = expression" where the expression describes the amount Joe earns x years after age 30. You'll find Question 11 in the Class portion of this lesson helpful.

$y = 30{,}000(1.04)^x$

10. Use your answer to Question 9 to find how much Joe earns at age 60. How does it compare to the estimate you got from the graph?

$y = 30{,}000(1.04)^{30} = \$97{,}301.93$; that's maybe a tad lower than the height of the graph, but pretty close.

11. Write an equation of the form "y = expression" where the expression describes the amount Joe spends x years after age 30.

$y = 0.88 \cdot 30{,}000(1.04)^x = 26{,}400(1.04)^x$

12. Use your answer to Question 11 to find how much Joe spends at age 60. How does it compare to the estimate you got from the graph?

$y = 26{,}400(1.04)^{30} = \$85{,}625.69$; that's pretty much right on what the graph shows.

2-5 Applications

Name ____

Now let's look at a coworker of Joe's named Fran. She has the same job as Joe, and in an amazing coincidence happens to be the same age. Sounds like her and Joe should get together and go bowling or something. Anyhow, Fran is much more frugal than Joe: she commits to spending only 70% of her salary and saving the rest in the same retirement plan that Joe is invested in. Let's see what Fran's graph looks like. Again, the horizontal axis is her age.

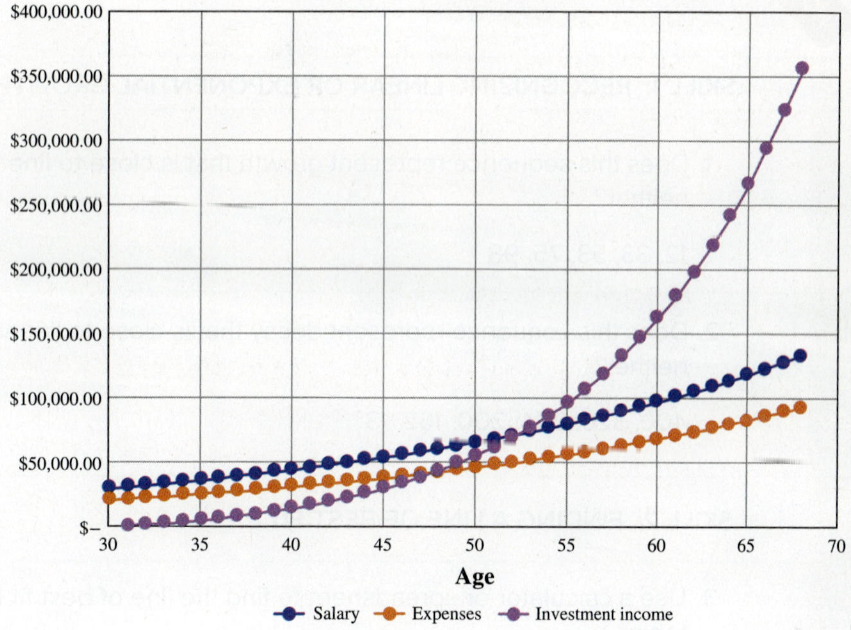

13. How much less money does Fran spend at age 30 than Joe?

70% of $30,000 = $21,000. This is $5,400 less than Joe.

14. How much earlier does Fran reach her crossover point?

Fran's crossover point is at age 49, which is 16 years earlier than Joe. Wow.

15. Describe what impact this article might have on your saving vs. spending plan in the future.

Answers vary.

Lesson 2-6 Prep Skills

This is a short review of skills that will come in handy in the next lesson. In each case, answer the question, then rate your confidence level by checking one of the boxes. If you feel like you're struggling with these skills, consult the online resources provided by your instructor for extra practice.

I KNOW IT | THINK SO | UNSURE | NO IDEA

SKILL 1: RECOGNIZING LINEAR OR EXPONENTIAL GROWTH/DECAY

1. Does this sequence represent growth that is close to linear, exponential, or neither?

   12, 33, 53, 75, 98

2. Does this sequence represent decay that is close to linear, exponential, or neither?

   400, 320, 254, 200, 162, 13

SKILL 2: FINDING A LINE OF BEST FIT

3. Use a calculator or spreadsheet to find the line of best fit for the data in the table.

x	12	24	36	48	60	72
y	\$312	\$600	\$1,040	\$1,279	\$1,703	\$2,041

SKILL 3: INTERPRET PARAMETERS IN AN EXPONENTIAL EQUATION

4. A quantity is described by the equation $y = 26(1.2)^x$. What was the initial size of the quantity? By what percentage is it increasing each time period?

SKILL 4: FIND RELATIVE CHANGE

5. The first time a new course is offered at a college, 73 people signed up for it. The following semester, only 60 signed up. Find the relative change.

Lesson 2-6 Follow the Bouncing Golf Ball (Exponential Curve Fitting)

LEARNING OBJECTIVES

- ☐ 1. Gather and organize data from an experiment.
- ☐ 2. Find an exponential curve of best fit for a real data set.
- ☐ 3. Study the rate of decay for an exponential equation.

It's just a glorious day. The only way to ruin a day like this would be to play golf on it.

—David Feherty

Courtesy of David Sobecki

Avid golfers have a classic love-hate relationship with the game. They enjoy it enough to spend a lot of time and money pursuing it, but they also know how downright maddening it can be. Golf is a whole lot harder than it looks, and sometimes it seems like the ball just moves and bounces with a mind of its own—an evil mind intent on ruining the golfer's day.

Even so, like everything else in our world, golf balls are bound by the laws of physics and their motion actually is predictable—at least in a lab setting, if not on the golf course. In this lesson, we'll follow a bouncing golf ball to help us further our study of exponential growth and decay, relying on technology to find exponential equations that best fit a data set.

0. After reading the opening paragraph, what do you think the main topic of this lesson will be?

2-6 Group

Here's the list of supplies needed for our golf ball experiment:

1. A golf ball; without this, it's not a golf ball experiment, is it?

2. A meter stick or tape measure with metric units

3. A chair or short stepladder

4. Masking tape and a marking pen

5. A hard, smooth floor area next to a wall

6. A graphing calculator with a regression feature

You can have one set of supplies for the whole class, or one set for each group (recommended). Have one student drop the ball, and the rest watch for its bounce height. They may need extra bounces before they agree on what the bounce height was. Encourage them to work together to get the most accurate possible readings. If you can, use different types of golf balls, then compare results.

Part 1: The Bounce Lab

Location is everything here: You'll need a relatively high ceiling, and a smooth concrete or other hard floor. Tile with grout lines won't work: The grout lines will affect the bounces of the ball. The key is to get consistent bounces straight up with no deflection. Have the tallest member of your group stand on a chair or short stepladder, and run a strip of masking tape vertically along the wall from the floor to the highest point they can reach. (Ideally, the starting point will be at least 250 cm above the floor.) Using your meter stick/tape measure and the marking pen, label heights on the masking tape every 10 cm starting with 0 cm at floor level.

Now you're ready to bounce. Hold the golf ball a few inches from the wall with the bottom of the ball at the highest height marked on the tape. Drop the ball and have someone in the group note the approximate height the ball reaches at the top of its first bounce. This will give you a rough idea of where to look.

Now repeat the drop, and have the person who watched the height of the first bounce put their finger on the tape at the height the ball reaches on the first bounce (the bottom of the ball, that is). Then mark that height with the pen. Repeat twice more FROM THE SAME HEIGHT, then calculate and record the *average* of the three bounce heights. (Use the table below to record your results.)

Next, we'd like to calculate the height of the second bounce. But rather than having the ball bounce twice, use the height of the first bounce that you calculated as the new release point. Clever, right? Repeat three times, recording the bounce heights in the table and calculating the average. Then use that average height as release point to calculate the height of the third bounce, and so on. Continue until you have heights for the first 8–10 bounces.

Sample data from one student group shown.

1. Record the data from your experiment in the table.

x Bounce #	Trial 1 bounce height (cm)	Trial 2 bounce height (cm)	Trial 3 bounce height (cm)	Total bounce height (cm)	*y* Average bounce height (cm)
0					200
1					160
2					125
3					100
4					80
5					65
6					55
7					45
8					35
9					30
10					25

2. Use these data to create a scatter plot. Use the average bounce height as your *y* coordinates.

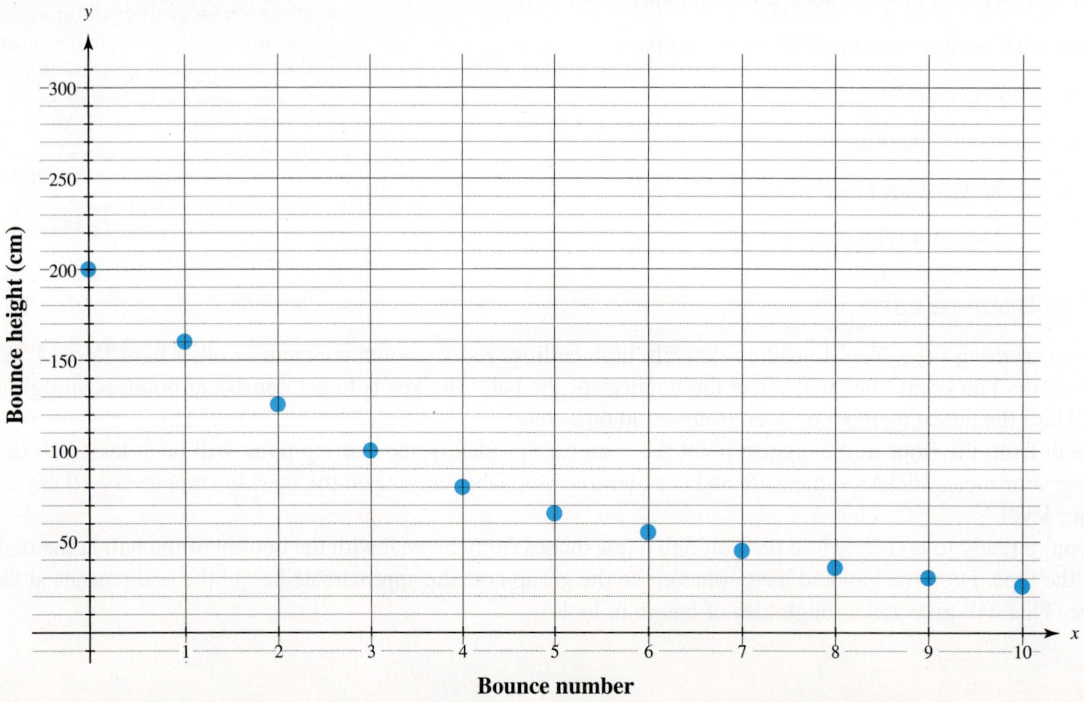

Part 2: Modeling Your Data

3. Using your graphing calculator, find an exponential equation of best fit for your data. The command for this is usually **ExpReg** on the **STAT-CALC** menu; your equation should be of the form $y = ab^x$. Round to two decimal places.

$y =$ $191.34(0.81)^x$

4. Use your equation to approximate the height of the 7th bounce. How does the result compare to the value in your table? Does this surprise you?

$y = 191.34(0.81)^7 = 43.8$ cm. The value in the table for the 7th bounce is 45 cm, which is pretty close.

Emphasize that models arent perfect, so the results you get from a model are estimates.

5. Evaluate your equation for $x = 0$. What does your answer mean?

$y = 191.34(0.81)^0 = 191.34$ cm. It means that the ball was originally dropped from a height of 191.34 cm; this is also reasonably close to the actual height of 200 cm.

Did You Get It

Try this problem to see if you understand the concepts we just studied. The answer can be found at the bottom of the portfolio page.

1. When trying this experiment at my campus, I got the model $y = 263(0.84)^x$.
   a. What was the approximate height I dropped the ball from originally?
   b. About how high was the seventh bounce?

Now let's analyze the significance of every symbol in our equation.

6. What does the variable x represent?

The bounce number

7. In the equation $y = ab^x$ that models your data, what is the value of a? What does it tell us about the experiment? Your answers to Questions 5 and 6 should help. Does it match the results in your table? Why do you think that is?

$a = 191.34$ is the model's version of the height the ball was bounced from.

8. In the equation $y = ab^x$ that models your data, what is the value of b? What does it tell us about the experiment?

$b = 0.81$ is the average percentage that each bounce height is of the previous bounce height.

As you talk to the groups, emphasize the fact that b is the multiplication factor they have used before with exponential growth and decay in Lesson 2-5.

Did You Get It ?

Try this problem to see if you understand the concepts we just studied. The answer can be found at the bottom of the portfolio page.

2. Look back at the bounce height equation I got in Did You Get It 1. What is the significance of the 0.84 in that equation? What does it tell you about the difference between my golf ball experiment and yours?

Part 3: The Connection Between Exponential Decay and the Rate of Decay

We've put a considerable amount of energy into studying exponential growth or decay. The independent variables we've worked with have typically been whole numbers: years after an account was opened, hours since caffeine was introduced to the bloodstream, number of bounces for a golf ball, etc. In each case, the distinguishing feature is that each new output comes from multiplying the previous by the same fixed number. In the bounce lab, we see that same pattern: each bounce is a certain percentage of the height the ball was dropped from. The result is an equation where the input is an exponent, and that's why we call this type of equation an **exponential equation**.

We decided to classify this entire activity as Group, but students will almost surely need more guidance with Part 3 than the first two parts.

Now let's study the rate of decay.

9. Using data from your original table, fill in this new table. The drop height corresponds to the height of the previous bounce; start with the original height the ball was dropped from. Use the average bounce heights from the original table. Note: The difference between this table and the previous one is that the input is no longer the bounce number: It's the *height of the previous bounce.*

x Drop height (cm)	y Bounce height (cm)
200	160
160	125
125	100
100	80
80	65
65	55

x Drop height (cm)	y Bounce height (cm)
55	45
45	35
35	30
30	25
25	—

10. Use these data to create a second scatter plot.

Point out that since we have no data on the bounce height for a 25-cm drop, there's no point to plot.

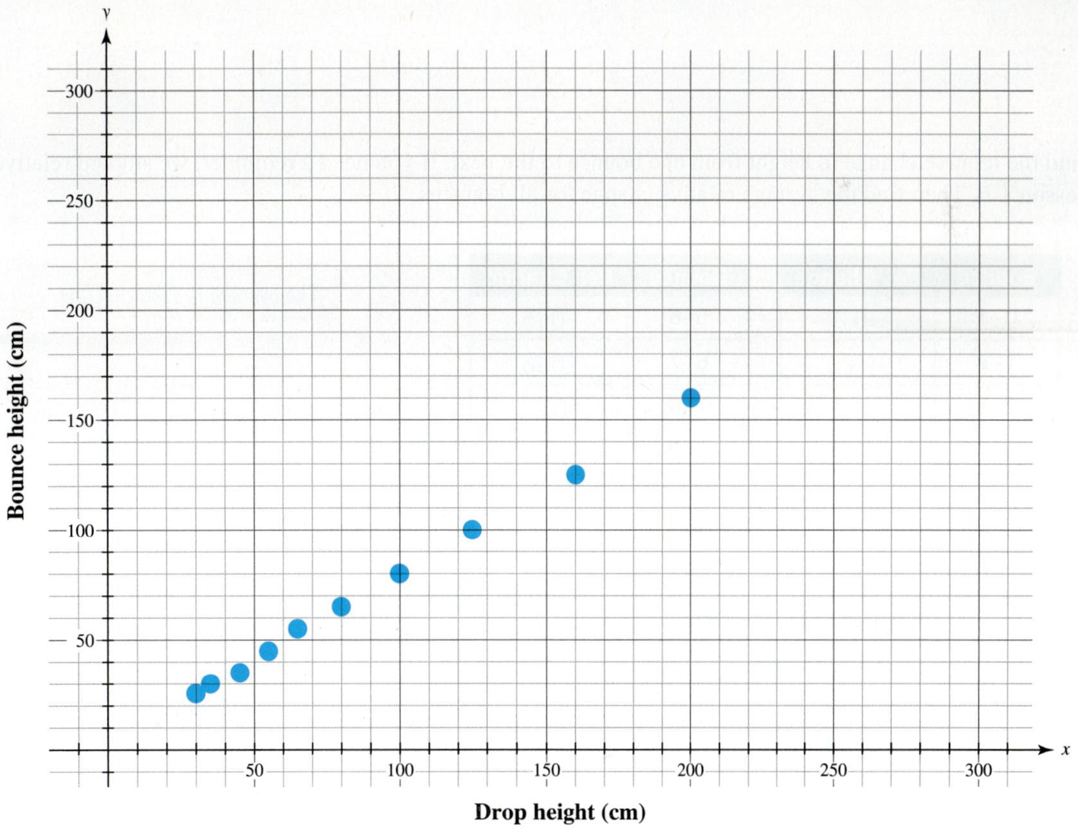

11. Based on the scatter plot, what kind of equation seems to be a good choice for modeling this data?

It looks pretty linear.

I can almost guarantee that students will want to say exponential here, since that's the stated topic of the lesson. Ask them to put a straightedge to the graph before deciding.

12. Use your graphing calculator to find the equation of best fit. Round to two decimal places.

$y =$ 0.78x + 1.80

13. Your equation has a variable in it, but also has **parameters:** These are constants in an equation of a certain form that are determined by the data for a particular model. For example, when you found an equation of the form $y = ab^x$ earlier, x was the variable, while the values you found for a and b were parameters. Identify the variables and parameters in your answer to Question 12, then describe what each means in this setting.

x is the independent variable, and it represents the height the ball is falling from. y is the dependent variable, and it represents the height of the bounce when dropped from x cm. 0.78 is a parameter that represents the slope of the line; in this case, it's the average rate of change of the bounce height. 1.80 is a parameter that ideally would represent the height of a bounce when the ball is dropped from height zero.

14. If you were to drop the golf ball from your experiment out of a second story window 20 feet above a concrete patio, how high would you expect it to bounce?

16 feet; If a drop from 200 cm bounces to 160 cm, a drop from 20 feet would be expected to bounce to 16 feet.

15. Find the relative change in height from one bounce to the next. If you need a reminder, we studied relative change in Lesson 1-6. Then find the average relative change for all bounces.

Bounce	Rel. Change
0–1	-0.2
1–2	-0.22
2–3	-0.2
3–4	-0.2
4–5	-0.19

Bounce	Rel. Change
5–6	-0.15
6–7	-0.18
7–8	-0.22
8–9	-0.14
9–10	-0.17

The average is about −0.19.

16. (This one requires some thought, but is the key question.) How does your answer to Question 15 relate to EACH of the best-fit equations found in this lesson?

The relative changes average to −0.19; this is exactly the percent of the height that is lost on each bounce in our exponential model! The factor of 0.81 means that about 19% is lost on each bounce. Notice that the slope of the linear model is 0.78; if the data were perfect, that slope would have been 0.81, as it's a comparison between the new bounce height and the previous bounce height.

Did You Get It ?

Try this problem to see if you understand the concepts we just studied. The answer can be found at the bottom of the portfolio page.

3. Recall that the equation giving the height of the bounces in terms of bounce number (from Did You Get It 1) was $y = 263(0.84)^x$.
   a. What should the relative change in bounce heights work out to be? In other words, if I'd made a table comparing the drop height to the bounce height, like in Question 9, what should I have gotten for the average relative change (Question 15)?
   b. If my experiment had worked out ideally, with perfectly consistent bounces and completely accurate measurements, what would I have gotten for the equation of the line of best fit for the data comparing drop heights to bounce heights? Your answers from Questions 13 and 16 will save the day here.

17. List as many factors as you can think of that contribute to the data from this experiment not working out to be a perfect exponential equation.

Answers, of course, vary. Some of the most likely candidates are:

- Error in measuring the bounce heights
- Inconsistent drop heights
- Inconsistency in the bounce surface-loose tile, uneven concrete
- Golf balls have dimples on them, so there's by nature inconsistency in the way they bounce.
- It gets harder to measure as the bounce heights get smaller.

According to the International Federation of Tennis, a regulation tennis ball should bounce about 55 inches when dropped from a height of 100 inches.

18. What's the bounce height as a percentage of the drop height for a regulation tennis ball?

55%

19. Write an expression that calculates the bounce height for a regulation tennis ball dropped from 200 inches.

$(0.55)(200)$

20. Write an equation that describes the bounce height (y) of a regulation tennis ball when it's dropped from x inches.

$y = 0.55x$

21. Using your answer to Question 18, it's possible to write an equation that describes the height of the xth bounce when a regulation tennis ball is dropped from a height of 100 inches. Do it! The things we learned in the golf ball experiment will definitely come in handy.

$y = 100(0.55)^x$

22. If you were to toss a tennis ball a bit forward from 10 feet above the ground, draw a graph of what you think the path of the ball would look like. The horizontal axis should measure lateral distance from where it was tossed, and the vertical axis should measure height. In essence, your graph should be a side view of watching the ball bounce forward.

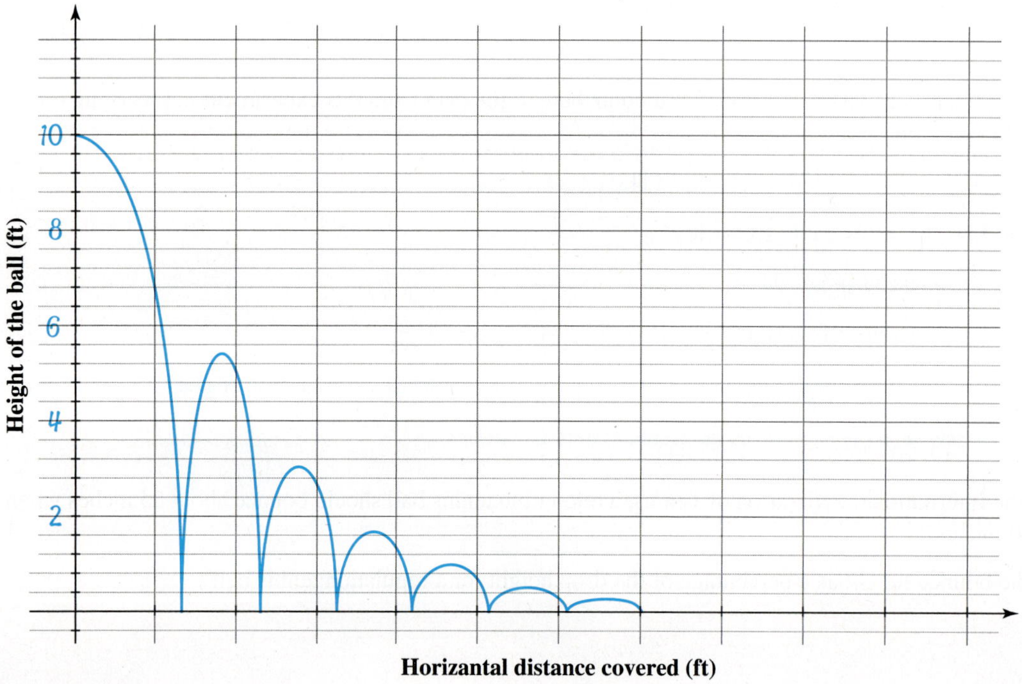

Sample answer shown. This is a good opportunity to point out that since we have no information on how far forward the ball is going, there's no real reason to put a scale on the horizontal axis. But if they choose to make a guess, that's fine too. The key here is the heights of the bounces being roughly 55% of the previous.

2-6 Portfolio

Name ______________________________

Check each box when you've completed the task. Remember that your instructor will want you to turn in the portfolio pages you create.

Technology

1. ☐ Questions 1, 2, and 3 in the Group section are great examples of the three ways we've modeled data: numerically, graphically, and algebraically. Build an Excel spreadsheet that illustrates each of those models, using a table, connected scatter plot, and a formula. A template to help you get started is available in the online resources for this lesson. (Note: The exponential regression equation supplied by Excel will be in a different form than the one you got from your graphing calculator. The Excel version uses the number e as its base: This number is about 2.7. To write the equation in the form we're accustomed to, find the base b by raising 2.7 to a power. The power is the number part of the exponent in the equation that Excel supplied.) Make a table of values and a graph for each of the two equations to verify that they're equivalent.

You should probably go over in class how to change the Excel exponential model to our form.

Online Practice

1. ☐ Include any written work from the online assignment along with any notes or questions about this lesson's content.

Applications

1. ☐ Complete the applications problems.

Reflections

Type a short answer to each question.

1. ☐ We've studied linear and exponential lines of best fit in this course. If you're looking at a set of data, how can you decide which of those would be most appropriate for the data?
2. ☐ How could you tell if neither of those types of equations is a good fit for a data set?
3. ☐ Take another look at your answer to Question 0 at the beginning of this lesson. Would you change your answer now that you've completed the lesson? How would you summarize the topic of this lesson now?
4. ☐ What questions do you have about this lesson?

Looking Ahead

1. ☐ Complete the Lesson 2-7 Prep Skills and read the opening paragraph in Lesson 2-7 carefully, then answer Question 0 in preparation for that lesson.

Answers to "Did You Get It?"

1. a. 263 cm **b.** 77.6 cm **2.** It's an average of what percent of the previous height each bounce reaches. By comparing to your result, you can tell if my golf ball bounces more or less than yours. **3. a.** −0.16 **b.** $y = 0.84x$

2-6 Applications

Name ____

1. The equation $y = 81.15(1.013)^x$ can be used to model the population of the United States in millions of people based on the number of years since 1900. Identify the variables and the parameters in this equation, and describe what the significance of each is in terms of population.

Independent variable: x, which represents the number of years since 1900. Dependent variable: y, which represents the population of the United States in millions.

Parameters: 81.15, which tells us that the population in 1900 was about 81.15 million; 1.013, which tells us that the population grew by an average of 1.3% per year.

2. Use the equation in Question 1 to complete the table, then use the table to draw a graph of the equation. Don't forget to label an appropriate scale on each axis.

Years after 1980	Population (Millions)
0	81.15
20	105.07
40	136.04
60	176.14
80	228.06
100	295.28

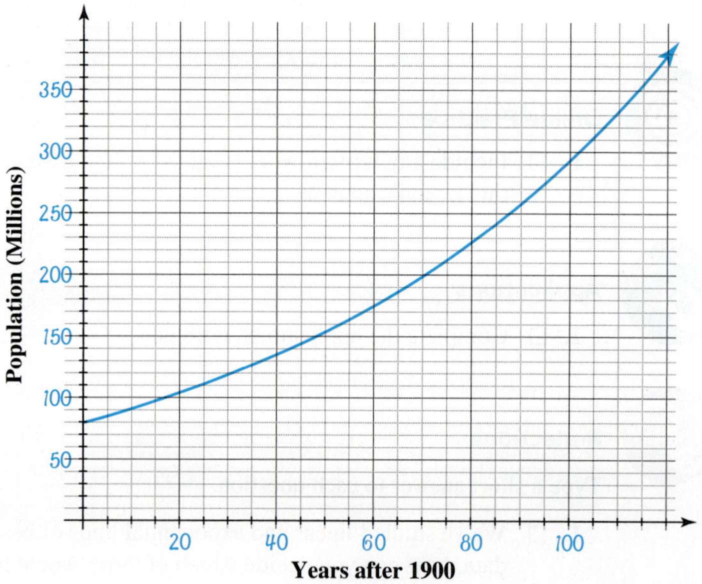

3. The equation $y = 81.15(1.013)^x$ illustrates ___exponential___ growth. The population was growing by ___1.3___% each year.

4. The model used in Questions 1–3 was obtained using official data for every census from 1900 to 2010. (The census is conducted every 10 years.) Use your model to predict the population in 2015. Then find the population of the United States in 2015 online and compare to your result. How accurate is it?

The estimate that I found online for 2015 is 325.13 million. The model:

$y = 81.15(1.013)^{115} = 358.41$ million. This is off by quite a lot-over 30 million people!

2-6 Applications

Name ____________________

5. What do you think the result of your answers to Question 4 say about population growth in the 21st century?

It must have slowed down. The model was built from data going back over 100 years. Since the actual population is so much less than the prediction, the population must not be growing as quickly as the model anticipated.

6. What if we use only more recent data? Since the census is conducted only every 10 years, we'll have to use a different source. I used estimates from a site called Worldometers, which are based on multiple data sources that are likely to make them more accurate than the census. In fact, the census is historically not the best measure of the true population for a wide variety of reasons. The table below has data for years from 1990 to 2010. Use the data to find the exponential equation of best fit. Include FIVE digits in all parameters. *Note that x now represents years after 1990.*

Year	Yrs. after 1990	Population (Millions)
1990	0	254.51
1995	5	268.04
2000	10	284.59
2005	15	298.16
2010	20	312.25

The equation of best fit is
$y = 255.06(1.0104)^x$

7. Based just on the base of the exponential in your equation, do you think this equation is likely to give a better estimate for the 2015 population than our first model? Why or why not?

The growth rate for this equation is 1.04%, compared to 1.3% for the first equation. This lower growth rate probably should help to keep the estimate for 2015 from being way too large, like it was with the first equation.

8. Use your equation to estimate the population in 2015 and 2050. Compare the 2015 estimate to the population you already found on the Internet. Then search for a site that does population projections and see how your prediction compares. What does the comparison say about that site's population growth forecast?

$y = 255.06(1.0104)^{25} = 330.35$ million. This is much closer.

$y = 255.06(1.0104)^{60} = 474.51$ million. Worldometer predicts a population of 400.85 million in 2050. That's quite a bit less than our projection, which indicates that the good folks at Worldometer anticipate that population growth in the U.S. will slow down even more.

Lesson 2-7 Prep Skills

This is a short review of skills that will come in handy in the next lesson. In each case, answer the question, then rate your confidence level by checking one of the boxes. If you feel like you're struggling with these skills, consult the online resources provided by your instructor for extra practice.

SKILL 1: CHOOSE AN APPROPRIATE TYPE OF MODEL

1. For each pair of scatter plots, decide if a linear or exponential model would be a better choice.

a.

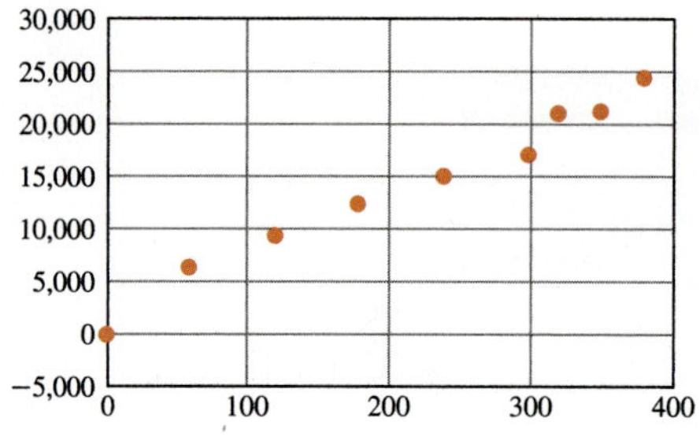

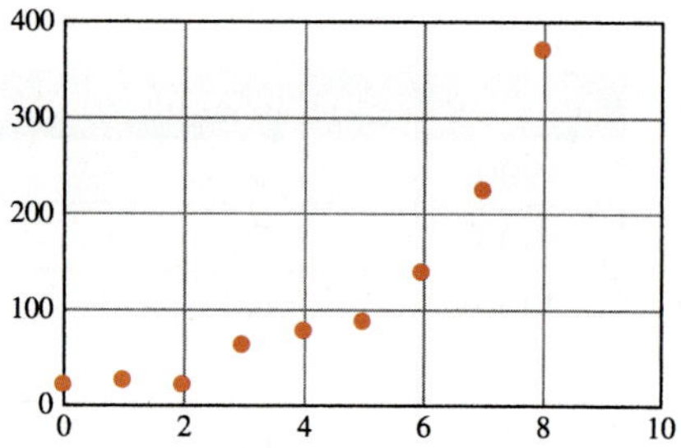

b.

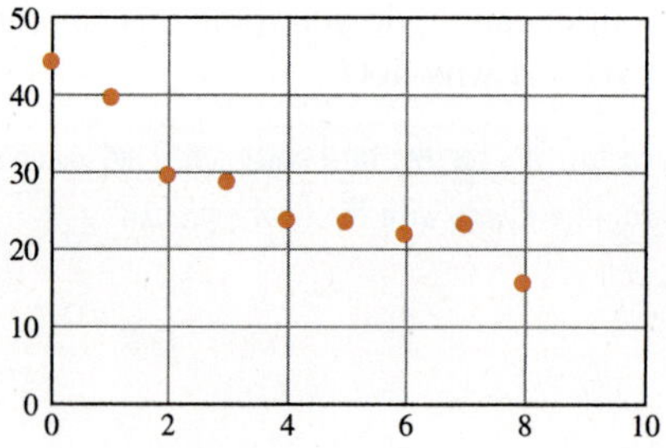

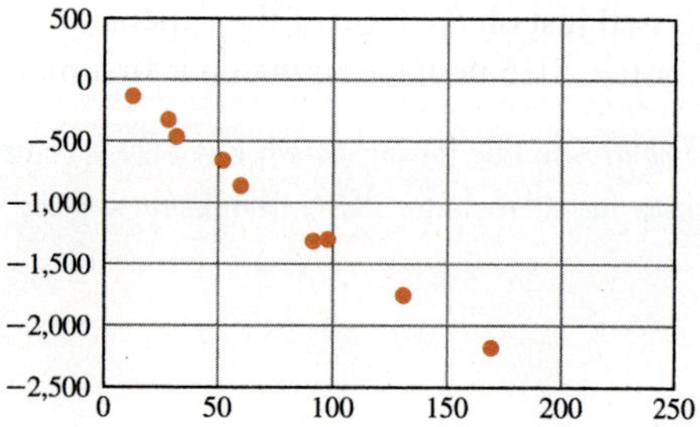

SKILL 2: COMPUTE RELATIVE ERROR

2. Carpet is ordered for a room with an actual area of 149 square feet, and the estimate is 165 square feet. What is the relative error in the estimate?

SKILL 3: INTERPRET CORRELATION COEFFICIENT

3. For one particular data set, both a line and exponential equation of best fit are found. For the line, the value of r^2 is 0.3229, and for the exponential the value of r^2 is 0.984. Which type of equation is most likely the better choice for this data?

Lesson 2-7 Keeping Current (Modeling Current Data with Linear or Exponential Equations)

LEARNING OBJECTIVES

- ☐ 1. Decide on an appropriate type of model for a given data set.
- ☐ 2. Model data with a linear or exponential equation.
- ☐ 3. Research current data to compare to values predicted by a model.

Source: CNN

To me, the real "state of the union" is found in how Americans react to current events.

—Henry Rollins

Studying history is important, as eloquently described by Sir Winston Churchill: "Those who fail to learn from history are doomed to repeat it." (Giving credit where it's due, Churchill was paraphrasing the philosopher and poet George Santayana.) But a true understanding of our world requires that we keep abreast of current events as well, and one of the key ways to do that well is to practice finding facts online. There are literally millions of people out there intent on passing off their opinions as facts, so the saying "Don't believe everything you hear" is more relevant than ever. In this lesson, we'll be modeling some data with linear and exponential equations, then doing some Internet research to see how effective these models are at predicting the most current data available. As you do so, keep in mind that it's the responsibility of good citizens to find facts and use them to form opinions—not the other way around.

0. After reading the opening paragraph, what do you think the main topic of this lesson will be?

2-7 Class

If you're like many students, you're concerned with the cost of attending college. It's certainly not a secret that college tuition has been rising rapidly over the last 20 to 30 years. But what are the *facts*? For that, we need some data. The table below lists the average tuition costs for all public American colleges and universities for a variety of school years starting in 1981.

School Year	Cost
1981	$2,663
1991	$5,138
2001	$8,022
2002	$8,502
2003	$9,247
2004	$9,864
2005	$10,454

School Year	Cost
2006	$11,049
2007	$11,573
2008	$12,256
2009	$12,804
2010	$13,564
2011	$14,292

Source: U.S. Department of Education, National Center for Education Statistics. (2013). Digest of Education Statistics, 2012 (NCES 2014-015), Chapter 3.

To analyze the data and look for patterns, we've seen that scatter plots can be useful tools. We've also seen that when modeling data based on time, it's a good idea to choose input values based on time passed since the first data value we have. In this case, we should use inputs that correspond to the number of years since 1981.

1. Fill in the table of values using the information in the last table. Your x values should be years since 1981, and your y values should be the average tuition cost for that year.

x	y
0	$2,663
10	$5,138
20	$8,022
21	$8,502
22	$9,247
23	$9,864
24	$10,454

x	y
25	$11,049
26	$11,573
27	$12,256
28	$12,804
29	$13,564
30	$14,292

Scale is very important in drawing a good scatter plot for these data, so before students get too far into plotting the points, make sure they chose a good scale.

2. Draw a scatter plot of the data in your table. Make sure you choose an appropriate scale for each axis, and clearly mark that scale on your graph.

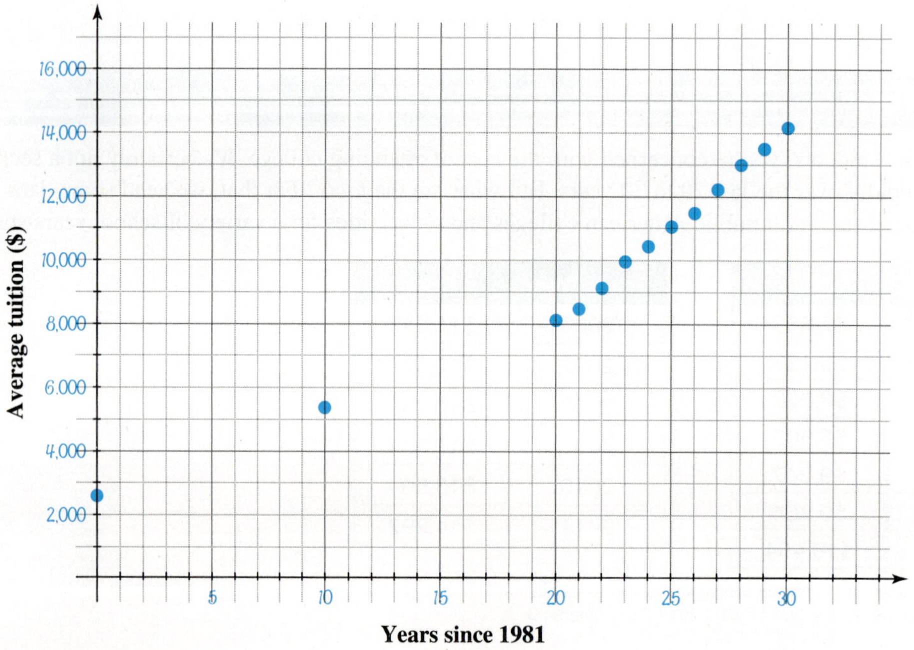

3. Based on your scatter plot, which type of equation do you think would most appropriately model the tuition data, linear or exponential? Explain your answer, please.

This is an opinion question, so of course answers will vary.

4. Using a graphing calculator or a spreadsheet, find the linear equation of best fit for your data. Round all coefficients so that they have 5 digits total.

$y = 385.66x + 1{,}501.2$

5. Graph the line of best fit and your scatter plot on screen. Based on the graph, how well do you think your model fits the data? Explain.

Answers vary, but a lot of people think that linear is perfect for this equation because the points starting with $x = 20$ line up nicely, but the trendlline shows that the later data don't align very well with the first 20 years.

6. What is the correlation coefficient for your equation of best fit? Does this change your perspective on how accurately the model fits the data? Explain. (Recall that if your calculator is not displaying a value for *r*, you should discard it and buy a new one. Just kidding: You should press 2nd 0, then use the down arrow to scroll down and choose "DiagnosticOn" and try the calculation again.)

$r = 0.97093$; *Explanations will vary.*

7. Write a description of what the slope of your linear model tells us about the growth of tuition since 1981. Be very specific.

It tells us that the average rate of change in tuition over that 30-year span was $385.66. More specifically, it says that if the linear model is accurate, then tuition grew on average $385.66 per year.

8. Now use your calculator or a spreadsheet to find the exponential equation of best fit for the data. Again, round all coefficients so that they have five total digits.

$y = 2{,}757.0(1.0566)^x$

9. Graph the exponential equation of best fit and your scatter plot on screen. Based on the graph, how well do you think your model fits the data? Explain.

Answers vary, but this graph surely appears to fit the data points better than the line.

10. Write a description of what the base of the exponent in your exponential model tells us about the growth of tuition since 1981. Again, more specific and detailed is better.

It tells us that the growth factor, or relative growth rate, was about 5.66% over that time span. More specifically, it says that tuition went up, on average, by about 5.66% per year.

11. What is the correlation coefficient for your exponential equation of best fit? Does this change your perspective on how accurately the model fits the data? Explain.

$r = 0.99826$; Explanations will vary.

12. Based on all of the work we've done in this portion of the lesson, which model do you think is the best choice for modeling the tuition data in the table? Why do you feel that way?

Answers vary, but both the visual evidence and the correlation coefficient support the exponential being the better fit. A really great answer would consider the idea of ignoring the first two data points and finding a line of best fit based on the data from 2001 on.

Did You Get It

Try this problem to see if you understand the concepts we just studied. The answer can be found at the bottom of the portfolio page.

1. The table below shows the number of women serving in the U.S. House of Representatives for selected years from 1980 to 2008. Find both a linear and an exponential model for the data, write the correlation coefficients, then decide which model you think is a better fit for the data. Round coefficients so that they have five total digits.

Year	x	Women in House
1980	0	19
1984	4	22
1988	8	25
1992	12	47
1996	16	51
2000	20	59
2004	24	68
2008	28	78

2-7 Group

Now that we've analyzed the tuition data and tried out some equations to model it, let's do some further analysis and see how well each model does in predicting current tuition costs.

1. Do an Internet search to find the most recent data you can on tuition costs. Keep in mind that in order to keep consistent with the model, you'll need to find tuition cost data compiled from public institutions. (Many sources will break the data down by public and private, since private schools tend to be more expensive than public.) Also, many sources break costs down into 2- and 4-year schools, so you'll need a source that combines them. Write the cost you found, as well as the year it's for and make sure you list your source as well.

   Answers vary for much of this activity due to the nature of gathering current data.

   You may want to search for some data in advance and suggest some sources to make the student's search more efficient.

2. Keeping in mind that the input of your modeling equations from the Class portion is years since 1981, use each equation to predict the tuition for the year you listed in Question 1.

   The linear model predicts tuition of $ __________ for the school year beginning in __________.

   The exponential model predicts tuition of $ __________ for the school year beginning in __________.

3. Find the error in each prediction, and the relative error. (If you need a refresher, relative error was covered in Lesson 1-6.)

4. After answering Questions 1–3, which of the models do you think is the best fit for the actual data? Explain. Is this the same answer you gave in Question 12 of the Class portion of this lesson?

Math Note

The variation in the predictions made by our regression equations shows exactly why you have to be really careful when using a regression equation based on past data to make future predictions. To say the least, it's not an exact science.

5. Discuss some potential reasons for the error in the tuition predicted by your model. Don't just think about the mathematics of it: consider sources and other factors as well.

Did You Get It

Try this problem to see if you understand the concepts we just studied. The answer can be found at the bottom of the portfolio page.

2. Use each of the models you found in Did You Get It 1 to predict the number of women in the House of Representatives in 2012 and 2016. Then look up the actual numbers and describe if that changes your opinion on which model is better.

6. Sometimes tweaking a model can help to make it work better to predict current and future trends. In this case, it's possible that the data from 1981 and 1991 is too old to really say anything relevant about the recent trend. Look back at your scatter plot from the Class portion of this lesson. If you ignore the first two data points, what type of model do you think will be a good fit? Explain.

As mentioned earlier, if you throw out the first two points, the data looks pretty close to linear.

7. Recalculate the linear and exponential models using only the data points from 2001 on. Don't change the input values though: x should still represent years after 1981.

Linear: $y = 619.75x - 4{,}436.6$

Exponential: $y = 2{,}617.8(1.0586)^x$

8. What are the correlation coefficients for each model now?

Linear: $r = 0.99930$

Exponential: $r = 0.99734$

9. Use these new models to make predictions for the more recent tuition data you found in Question 1.

The new linear model predicts tuition of $ __________ for the school year beginning in __________.

The new exponential model predicts tuition of $ __________ for the school year beginning in __________.

10. What does the slope of the new linear model tell us about the growth of tuition?

It tells us that the average rate of change in tuition over that 10-year span was \$619.75. More specifically, it says that if the linear model is accurate, then tuition grew on average \$619.75 per year. Comparing to the earlier model, this shows that the growth in tuition increased quite a bit sometime after 1991.

11. What does the base of the new exponential model tell us about the growth of tuition?

It tells us that the growth factor, or relative growth rate, was about 5.86% over that time span. More specifically, it says that tuition went up, on average, by about 5.86% per year. This is remarkably consistent with the growth rate we got by considering all the data, which tends to support the exponential model as being most appropriate.

12. Did either of these new models improve the prediction? Which of the four models do you now think provides the best overall representation of the data? Explain.

Answers vary.

2-7 Portfolio

Name ______________________________

Check each box when you've completed the task. Remember that your instructor will want you to turn in the portfolio pages you create.

Technology

1. ☐ Do an Internet search on some topic that you find interesting, and find data that you may be able to model with a linear or exponential equation. Write a brief description of the topic and the source. Then describe the data and why you think a linear or exponential model might be appropriate.

Online Practice

1. ☐ Include any written work from the online assignment along with any notes or questions about this lesson's content.

Applications

1. ☐ Complete the applications problems.

Reflections

Type a short answer to each question.

1. ☐ Write a few sentences on the difference between a mathematical model fitting past data well, and being useful to predict current or future data.
2. ☐ What did you learn about tuition costs in this lesson? Were you surprised by anything in particular?
3. ☐ Take another look at your answer to Question 0 at the beginning of this lesson. Would you change your answer now that you've completed the lesson? How would you summarize the topic of this lesson now?
4. ☐ What questions do you have about this lesson?

Looking Ahead

1. ☐ Complete the Lesson 3-1 Prep Skills and read the opening paragraph in Lesson 3-1 carefully, then answer Question 0 in preparation for that lesson.

Answers to "Did You Get It?"

1. Linear: $y = 2.2292x + 14.917$, $r = 0.98398$; Exponential: $y = 19.173(1.0556)^x$, $r = 0.97003$. The linear appears to be the better choice here.
2. The linear model predicts 86 in 2012 and 95 in 2016. The exponential model predicts 108 in 2012 and 134 in 2016. The actual numbers are 79 in 2012 and 84 in 2016.

2-7 Applications

Name ____________________

1. Do an Internet search to find a prediction for the cost of tuition in the future. We're not going to specify a particular year, but it should be 2020 or later. Remember, we're looking for the cost of all public institutions, both two- and four-year combined. Note the year and the predicted cost, as well as the source.

Answers will vary for Questions 1-3.

2. In the Group portion of this lesson, you chose one of four models used to represent tuition data. Use that model to predict the average cost of tuition for a public college or university in the year from Question 1. How close is the prediction?

3. Use the other three models to predict tuition in that year. Do any of them come close to matching the prediction you found online?

4. The next table provides the average tuition costs for all two-year American colleges for a variety of school years starting in 1981. Fill in the x column with years after 1981, which will act as input for some modeling equations on the next page.

School Year	x	Cost
1981	0	$2,224
1991	10	$3,623
2001	20	$5,137
2002	21	$5,601
2003	22	$6,012
2004	23	$6,375
2005	24	$6,492

School Year	x	Cost
2006	25	$6,815
2007	26	$6,975
2008	27	$7,568
2009	28	$7,703
2010	29	$8,085
2011	30	$8,561

2-7 Applications

Name ______________________________

5. Draw a scatter plot of the data in your table. Make sure you choose an appropriate scale for each axis, and clearly mark that scale on your graph.

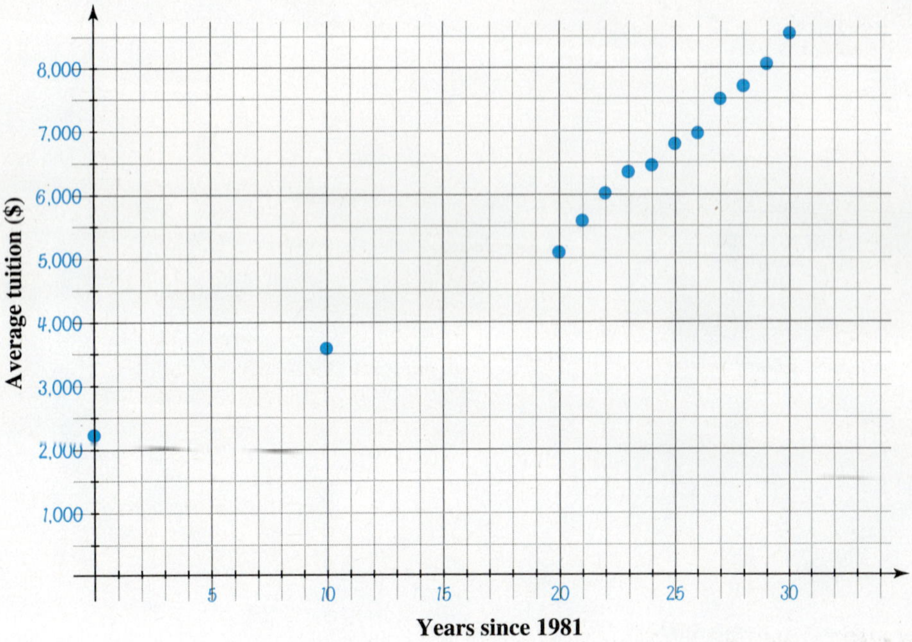

6. We found four types of models for the tuition data in the Class and Group portions of this lesson: linear and exponential equations using all of the data, then linear and exponential equations based only on the data beginning in 2001. Use your scatter plot to make an educated guess (what a math person would call a **conjecture**) as to which of those four model types would be the best choice for the two-year tuition data. Explain why you made your choice, then use a graphing calculator or spreadsheet to find the model.

Answers will vary.

Unit 3
Managing Your Money

Outline

Lesson 3-1 Prep Skills

This is a short review of skills that will come in handy in the next lesson. In each case, answer the question, then rate your confidence level by checking one of the boxes. If you feel like you're struggling with these skills, consult the online resources provided by your instructor for extra practice.

I KNOW IT	THINK SO	UNSURE	NO IDEA	
				SKILL 1: WRITE PERCENTS IN DECIMAL FORM
				1. Write the decimal form of each percent: 65%
				4.25%
				SKILL 2: COMPUTE PERCENTAGE OF AN AMOUNT
				2. What is 6.5% of $4,500?
				3. What is 104% of $1,250?
				4. Mel's salary was $42,300 last year, and he's getting a 5.5% raise this year. How much will he make this year?
				SKILL 3: EVALUATE ALGEBRAIC FORMULAS
				5. Evaluate the formula Prt for $P = 500$, $r = 0.9$, and $t = 4$.
				6. For the formula $A = P(1 + rt)$, find A if $P = 9{,}500$, $r = 0.025$, and $t = 10$.
				SKILL 4: SOLVE BASIC EQUATIONS
				7. Solve for t: $1{,}460 = 1{,}200(0.06)t$
				8. Solve for r: $21{,}126 = 17{,}750r(5)$

Lesson 3-1 A Topic of Interest (Simple Interest)

LEARNING OBJECTIVES

- ☐ 1. Define interest and understand related terminology.
- ☐ 2. Develop simple interest formulas.
- ☐ 3. Use simple interest formulas to analyze financial issues.

© Photodisc/Getty Images RF

An investment in knowledge pays the best interest.
—Benjamin Franklin

The topic of the next two lessons is of interest to anyone who plans to buy a house or a car, have a credit card, invest money, have a savings account—in short, pretty much everyone. This interesting topic is *interest*—a description of how fees are calculated when money is borrowed, and how your money grows when you save. If you prefer not to be separated from your hard-earned money, this is a topic you should be eager to understand well. The key to understanding isn't to just focus on a bunch of formulas: it's about understanding what interest is about and how it's calculated. This will give you the best chance to make sound financial decisions throughout your lifetime. So unless you can grow money, this should indeed be a topic of interest.

0. After reading the opening paragraph, what do you think the main topic of this lesson will be?

3-1 Group

New group time again! If it doesn't make you uncomfortable, exchange the following information with the classmates in your Unit 3 group. This will be your small group for the third unit. It would be a good idea to schedule a time for the group to meet to go over homework, ask/answer questions, or prepare for exams. You can use this table to help schedule a mutually agreeable time.

Name	Phone Number	Email	Available times

Interest is a fee paid for the use of money. For example, if you borrow money from a bank to buy a car, you must not only pay back the amount of money that you borrowed, but also an additional amount, called the interest, for the use of the bank's money. On the other hand, if you deposit money in a savings account, the bank will pay you interest for saving money since it will be using your money to provide loans, mortgages, etc. to people who are borrowing money. The stated rate of interest is generally given as a yearly percentage of the amount borrowed or deposited.

There are two kinds of interest. *Simple interest* is paid strictly as a percentage of the original amount of money. *Compound interest* is a percentage of an original amount, as well as a percentage of the new amount including previously calculated interest. We will study simple interest in this lesson, and compound interest in the next.

1. Let's say that you get a $600 tax refund, and you wisely decide to put it into a long-term investment that pays 5% in simple interest each year. How much money would that add to the account each year? (Remember, with simple interest a certain percentage of an original amount is added to the account.)

   5% of $600 is $30, so $30 would be added to the account each year.

2. Write out the multiplication you did to get your answer to Question 1. The next goal is to write a general formula for computing the amount of interest earned. Substitute the letter *P* for the original amount invested ($600) and the letter *r* for the percentage you multiplied by (written in decimal form, of course).

   $\$600 \times 0.05 = \$30$

   $P \times r = \text{Interest}$

A lot of people think that simple interest isn't used in "real" financial transactions, but that is absolutely not the case, as we'll see working through this chapter. You might do a search for "simple interest" for the class to find examples.

Did You Get It

Try this problem to see if you understand the concepts we just studied. The answer can be found at the bottom of the portfolio page.

1. Suppose that you take out a loan to buy new furniture, borrowing $1,700. If the interest charge will be 7.9% of the borrowed amount per year, how much interest will you pay each year?

3. Fill in the following table, which shows the growth of the account over several years, as well as the total amount of interest earned. You'll need a correct answer to Question 1, and remember that this is simple interest, so the interest earned each year is always calculated on the original value of the investment.

Time	Total Interest Earned	Value of Account
Start	–	$600
After 1 year	$30	$630
After 2 years	$60	$660
After 3 years	$90	$690
After 4 years	$120	$720
After 5 years	$150	$750
After 10 years	$300	$900

4. Look carefully at the first and second columns of your table. What is the relationship between the number of years that have passed and the total interest earned? Describe it verbally.

The total amount of interest earned is the product of the amount earned in one year and the number of years that have passed.

3-1 Class

In the Group portion of this lesson, you actually developed formulas for computing simple interest—whether you realized it or not! Now let's formalize what we discovered.

In order to compute simple interest, we need exactly three pieces of information: the *principal,* the *rate,* and the *time.* The box below describes the terminology we'll use, as well as the letters we'll use to represent the key data.

Interest (I) is a fee charged for the use of money.

Principal (P) is the amount of money borrowed or placed into a savings account.

Rate (r) is the percent of the principal paid for having money loaned, or earned for investing money. Unless indicated otherwise, rates are given as a percent for a term of 1 year.

Time (t) or term is the length of time that the money is being borrowed or invested. When the rate is given as a percent per year, time has to be written in years.

Future value (A) is the amount of the loan or investment plus the interest paid or earned.

In the work we did in the Group portion, the principal was $P = \$600$; the rate was 5% written as a decimal, which is $r = 0.05$. The amount of time varied when filling out the table of interest and future value.

1. In Question 2, you should have discovered that the formula for computing the interest in one year is $P \times r$. Use this, and your answer to Question 4, to write a formula for finding the amount of interest I that accumulates in t years:

   $I = P \times r \times t$

2. The total value of the account after t years, which we're calling future value and representing with the letter A, is calculated by adding the total amount of interest to the principal amount. Use your answer to Question 1 above to write a formula for future value in terms of P, r, and t.

   $A = P + P \times r \times t$

3. Use your formula from Question 2 to find the future value of your original $600 investment at 5% interest in 30 years.

   $A = 600 + 600(0.05)(30) = \$1{,}500$

Did You Get It

Try this problem to see if you understand the concepts we just studied. The answer can be found at the bottom of the portfolio page.

2. Find the future value of a loan if $3,600 is borrowed for 3 years at a simple interest rate of 8% per year.

4. The right side of your answer to Question 2 should be a sum, meaning that it has two terms. Use some algebra to rewrite the formula so that it has just one term. The result should look like the principal amount times some quantity.

   $A = P + Prt = P(1 + rt)$

Most groups will need a little nudge on the algebra here.

Here's a summary of the formulas we've built to describe simple interest:

Formulas for Computing Simple Interest and Future Value

1. Interest = principal × rate × time:

$$I = Prt$$

2. Future value = principal + interest:

$$A = P + I \quad \text{or} \quad A = P(1 + rt)$$

5. To meet payroll during a down period, United Ceramics Inc. needed to borrow $2,000 at 4% simple interest for 3 months. Without doing any calculations, make an educated guess as to how much interest you think would be due.

Answers vary.

6. Now use formulas to find the interest they paid, and the future value at the end of three months. (Hint: How many years does three months correspond to?)

$I = \$2{,}000(0.04)(1/4) = \$20$

Future value: $2,020

Did You Get It

Try this problem to see if you understand the concepts we just studied. The answer can be found at the bottom of the portfolio page.

3. Marta needs some quick cash for books at the beginning of spring semester, so she borrows $600 at 11% simple interest for 2 months. How much interest will she pay, and what is the future value of the loan?

It's very common for simple interest loans to be paid off in monthly installments. This is how many loans for things like furniture and electronics work. Let's look at finding the monthly payment for a simple-interest loan.

A biotech startup borrowed $68,000 to buy lab equipment at 9% simple interest for 1½ years.

7. Find the future value of the loan.

$A = \$68{,}000(1 + 0.09(1.5)) = \$77{,}180$

8. Divide the future value by the number of months to find the monthly payment.

$$\frac{\$77{,}180}{18} = \$4{,}287.78$$

Did You Get It

Try this problem to see if you understand the concepts we just studied. The answer can be found at the bottom of the portfolio page.

4. The Lookout Restaurant took out a loan for $5,000. The simple interest rate was 6.5%, and the term of the loan was 3 years. Find the interest, future value, and monthly payment.

3-1 Group (Again)

Notice that the simple interest formula contains four variable quantities: interest, principal, rate, and time. Because we know how to solve equations from algebra, if we know three of those four quantities for a given account, we can solve to find the one we don't know.

The Problem: Phillips Health and Beauty Spa is replacing one of its workstations. The interest on a loan secured by the spa was $93.50. The money was borrowed at 5.5% simple interest for 2 years. What was the amount that they borrowed?

1. What information is provided in the problem? Write out each piece of information as an equation. For example, if the loan had been taken out for 8 years, you would write $t = 8$.

$I = \$93.50$
$t = 2$
$r = 0.055$

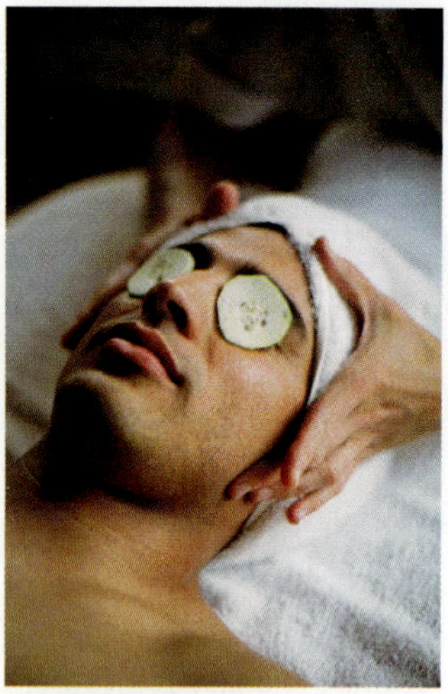

© Royalty-Free/Corbis

2. What is it you're being asked to find? Answer in a complete sentence, please.

We are asked to find the principal of the loan.

3. Use the simple interest formula to set up and solve an equation to solve the problem. Show all work!

$I = Prt \Rightarrow \$93.50 = P(0.055)(2)$

$0.11P = 93.50$

$P = \frac{93.50}{0.11} = \$850$

Questions 1–3 illustrate a really useful framework for solving this type of problem. When you're using an algebraic formula, it's ALWAYS a good idea to clearly identify the information that you're given, as well as what it is you're being asked to find. Then you can plug that information into the known formula and solve the resulting equation.

4. After receiving an inheritance, a newly married couple invested $15,250 for 10 years and received $9,150 in simple interest. What was the rate that the investment paid? Use the framework from Questions 1–3, and show all of your work. Write your answer in the form of a sentence.

$P = \$15{,}250,\ t = 10,\ I = \$9{,}150$

$I = Prt \Rightarrow \$9{,}150 = \$15{,}250r(10)$

$\$152{,}500r = \$9{,}150 \Rightarrow r = 0.06$

The investment paid 6% interest.

Point out to students that they'll be using a lot of formulas in this unit, and in most cases the difference between failure and success is clearly identifying all of the information needed to substitute into a formula.

5. If that couple had been able to find an investment that paid 9% interest, how much more quickly would they have made that $9,150 in simple interest? Use the framework from Questions 1–3, and show all of your work. Write your answer in the form of a sentence.

$P = \$15{,}250,\ r = 0.09,\ I = \$9{,}150$

$I = Prt \Rightarrow \$9{,}150 = \$15{,}250(0.09)t$

$\$1{,}372.50t = \$9{,}150 \Rightarrow t = 6\frac{2}{3}$

They would have made the amount in 3 years and 4 months less time.

Did You Get It

Try this problem to see if you understand the concepts we just studied. The answer can be found at the bottom of the portfolio page.

5. If you invest $8,000 for 30 months and receive $1,000 in simple interest, what was the rate?
6. A pawn shop offers to finance a guitar costing $750 at 4% simple interest. The total interest charged will be $150. What is the term of the loan and the monthly payment?

If you look at it a certain way, simple interest in a savings account doesn't seem terribly fair. Remember, interest is a fee charged for using someone's money. So when you put money into a bank account, the bank is essentially paying you to borrow your money, because they can use it to increase their business in other areas. So far, so good.

But when you keep that money in the account for a long time, things start to go bad. Let's say that you invest $1,000 into a simple-interest account at 3% interest per year. Then according to the simple interest formula, after one year you would have earned $I = (\$1{,}000)(0.03)(1)$, or $30. Don't spend it all in once place! In any case, at the end of that first year, you now have $1,030 in your account, not $1,000. Yet when interest is simple, the amount of interest paid per year remains constant over the life of the investment. Why is this bad for you? Let's see.

6. Forget about the original amount, and just think of the $1,030 in your account at the end of year one as a new investment. If you get $30 in interest over the next year, what percentage did you actually get? Show work, please.

$I = Prt \Rightarrow \$30 = \$1{,}030r(1)$

$$r = \frac{30}{1{,}030} = 0.02913$$

You actually got about 2.9%, not 3%.

This closing material presents an important introduction to the idea of compound interest.

7. If you'd actually been paid the full 3% on your $1,030 in the second year, how much more interest would you have earned? Again, show work!

3% of $1,030 = 0.03 × $1,030 = $30.90; you'd earn 90 cents more.

We'll take up this discussion in the next lesson. For now, refer back to these questions when answering the first question in the Looking Ahead section of the Portfolio.

3-1 Portfolio

Name ______________________________

Check each box when you've completed the task. Remember that your instructor will want you to turn in the portfolio pages you create.

Technology

1. ☐ A quick glance at the table you filled out in Question 3 at the beginning of this lesson should be enough to convince you that spreadsheets are particularly well-suited to interest calculations. First, recreate that table using a spreadsheet. Of course, the point is to use formulas to do the calculations, not just key in the results. Then use the "copy down" feature to calculate the future value of the account and the total interest earned for years 11 through 30. A template to help you get started is included in the online resources for this lesson.

   Next, suppose that your goal was for the future value of the investment to be $1,000 after ten years. Use trial and error, changing the amounts in cells A2 and B2, to find (1) how much more you'd need to invest up front, and (2) what interest rate you'd need to earn to achieve that goal if you keep the original investment at $600.

Online Practice

1. ☐ Include any written work from the online assignment along with any notes or questions about this lesson's content.

Applications

1. ☐ Complete the applications problems.

Reflections

Type a short answer to each question.

1. ☐ Why is it reasonable that a bank would pay interest on savings, or charge interest on a loan?
2. ☐ What is the key feature of simple interest?
3. ☐ Why is it important to understand the concept of interest?
4. ☐ Take another look at your answer to Question 0 at the beginning of this lesson. Would you change your answer now that you've completed the lesson? How would you summarize the topic of this lesson now?
5. ☐ What questions do you have about this lesson?

Looking Ahead

1. ☐ Look back at page 212. Write a brief description of why simple interest is somewhat unfair when applied to a savings account.
2. ☐ Complete the Lesson 3-2 Prep Skills and read the opening paragraph in Lesson 3-2 carefully, then answer Question 0 in preparation for that lesson.

Answers to "Did You Get It?"

1. $134.30 **2.** $4,464 **3.** $11, $611 **4.** $975, $5,975, $165.97 **5.** 5%
6. 5 years; $15

3-1 Applications

Name ______________________________

Some of the most common (and most dangerous) examples of simple interest are short-term payday loans, which some lenders also call "cash advances." (This is not to be confused with a cash advance on a credit card.) Let's say that you're going to get paid in ten days, and you need some cash for a car repair now. A payday lender might lend you \$350 now, and you'll be asked to pay them back when your paycheck comes. Of course, you'll have to pay interest. The median fee charged by these types of lenders is \$15 per every \$100 borrowed.

1. How much would you have to pay back in ten days?

   Fee: \$15 + \$15 + \$15 + \$7.50 = \$52.50

   Amount paid back: \$402.50

2. What is the percent interest you're charged?

   $\frac{\$52.50}{350} = 0.15$; 15% interest

   It will be interesting here to see how many students recognize that they can simply do 15/100 to get the percentage interest. Not many would be my guess. But the really clever ones will.

3. What portion of a year is the ten-day period for this loan? Write your answer as both a fraction, and a decimal.

   $\frac{10}{365} \approx 0.027397$ year

4. As you know, one month is 1/12 of a year. If you paid 2% simple interest in one month on a loan, what percentage would you pay in a full year?

   24%, which of course is 2% times 12.

5. In order to find the annual interest rate, you'll need to find rate as a percent per year. Use your answers from Questions 2 and 3 to find the annual interest rate for this loan. Discuss how good or bad of a deal you're getting as the borrower. You can use your calculation from Question 4 as a guide.

   $\frac{15\%}{0.027397 \text{ year}} = 547.5\%$ per year

   This is a TERRIBLE deal for the borrower-about as bad as it gets without risking having your thumbs broken by a loan shark.

3-1 Applications

Name ______________________________

Another example of simple interest is a certificate of deposit, or CD as they're more commonly known. This is an investment where you agree to leave your deposit untouched for a fixed amount of time, and in exchange you are guaranteed a certain percentage interest. Below are some CD rates listed in December of 2015 on Bankrate.com for CDs with a 1-year term.

6. For a \$2,000 investment, how much more would you earn if you chose Synchrony over CIT?

Synchrony: $I = \$2{,}000(0.0125)(1) = \$25$

CIT: $I = \$2{,}000(0.0122)(1) = \$24.40$

Synchrony pays a big 60 cents more.

1 yr CD

Institution	APY	Rate	Min Deposit
SYNCHRONY Bank MEMBER FDIC Great Rates + Safety = Peace of Mind	1.25% Wed Dec 9	1.24%	\$2,000
CIT Bank Member FDIC Great rates, low minimum deposits, Member FDIC.	1.22% Wed Dec 9	1.21%	\$1,000
My e-BAnC BAC Florida Bank Member FDIC Make sure to check our Jumbo Rates	1.25% Wed Dec 9	1.24%	\$1,500

7. If your goal was to earn \$40 in interest on that \$2,000 investment, what interest rate would you need to find?

$I = Prt \Rightarrow \$40 = \$2{,}000r(1)$

$r = \dfrac{40}{2{,}000} = 0.02$; 2% interest would be required.

8. If you got the same rate that Synchrony is offering for a longer length of time, how long would you need to invest for to get that \$40 interest you were shooting for?

$I = Prt \Rightarrow \$40 = \$2{,}000(0.0125)t$

$t = \dfrac{40}{2{,}000(0.0125)} = 1.6$; 1.6 years would be required, or a little more than a year and seven months.

9. As you can see, CDs don't return a ton of interest. What they DO offer is a guaranteed return. Suppose that you also had the opportunity to invest that \$2,000 in a business venture that promised a 12.5% return in one year, unless the company failed to turn a profit, in which case you would get nothing at all back. Calculate the amount of interest you'd get, and discuss whether you'd prefer the risky investment over the CD.

$I = \$2{,}000(0.125)(1) = \$250$

Answer to risk vs. reward question, of course, will vary.

Lesson 3-2 Prep Skills

This is a short review of skills that will come in handy in the next lesson. In each case, answer the question, then rate your confidence level by checking one of the boxes. If you feel like you're struggling with these skills, consult the online resources provided by your instructor for extra practice.

I KNOW IT | THINK SO | UNSURE | NO IDEA

SKILL 1: WRITE PERCENTAGES IN DECIMAL FORM

1. Write each percent in decimal form.

   47%

   9.4%

SKILL 2: COMPUTE SIMPLE INTEREST

2. If you borrow $1,200 for furniture at 11% simple interest for two years, how much will you pay in interest?

SKILL 3: INTERPRETING EXPONENTS AS REPEATED MULTIPLICATION

3. Write the expression below in a shorter form using an exponent:

   $500(1 + 0.03)(1 + 0.03)(1 + 0.03)(1 + 0.03)$

SKILL 4: PERFORM CALCULATIONS INVOLVING ORDER OF OPERATIONS

4. Perform each calculation:

   $12(8 + 4(13))$

   $400\left(1+\frac{0.05}{12}\right)^{60}$

SKILL 5: SUBSTITUTE INTO AN EXPRESSION

5. For the expression $a(1 + b)^c$, find the value when $a = 20$, $b = 0.09$, and $c = 6$.

Lesson 3-2 Like a Snowball Rolling Downhill (Compound Interest)

LEARNING OBJECTIVES

- ☐ 1. Describe how compound interest differs from simple interest.
- ☐ 2. Develop compound interest formulas.
- ☐ 3. Use compound interest formulas to analyze financial issues.

© McGraw Hill Education

My wealth has come from a combination of living in America, some lucky genes, and compound interest.

—Warren Buffett

If you've ever made a snowman, you're familiar with how it starts. You make a little snowball in your hands, then plop it down and start rolling it through the snow. As you do, it picks up more snow on all sides (if you roll it well) and gets bigger. At first, not that much snow is picked up because the original ball is little. But as it gets bigger, it has more surface area to pick up more snow, and it starts to grow very quickly—to the point that before you know it, you need help pushing it to gather more snow. Push that baby down a steep hill, and you could end up with one massive snowball (and maybe a spanking or a lawsuit if something valuable gets in the way). At the very end of the last lesson, we looked into interest being calculated not just on the principal amount, but on interest that had previously been paid as well. As we'll see in this lesson, given enough time this can become the financial equivalent of a snowball rolling downhill, leading to a wonderful place we call "financial independence."

0. After reading the opening paragraph, what do you think the main topic of this lesson will be?

3-2 Group

We've already studied how an investment can grow when simple interest is calculated; the table below is borrowed from Lesson 3-1. (Borrowed . . . stolen . . . whatever.) In this case, a principal amount of $600 was invested at 5% interest.

Our first goal in this lesson is to recalculate the interest earned and value of the account, this time calculating the next year's interest on the entire value of the account (including previously earned interest), rather than just on the principal amount.

Time	Total Simple Interest Earned	Value of Account
Start	–	$600
After 1 year	$30	$630
After 2 years	$60	$660
After 3 years	$90	$690
After 4 years	$120	$720

1. Compute the amount of interest earned during the second year if that amount is 5% of both the principal amount of $600 AND the $30 in interest that was already earned in the first year. Find the new ending value as well. Show work, please.

   5% of $630 is $31.50. The new amount is $630 + $31.50 = $661.50.

2. We know from Lesson 3-1 that we can compute the amount of interest earned in one year using the formula $I = Pr$. This is using $t = 1$ in the interest formula $I = Prt$. Use this formula with $P = \$630$ to show that you get the same answer you did in Question 1.

   $I = \$630(0.05) = \$31.50$. This matches the amount of interest found in Question 1.

3. Use your answer from Question 1 to fill in the row of the table corresponding to "After 1 year." Then repeat your calculations to fill in the remainder of the table, keeping in mind that in each case interest should be calculated on the *new amount at the beginning of the year.*

Time	Total Interest Earned	Value of Account
Start	–	$600
After 1 year	$30	$630
After 2 years	$61.50	$661.50
After 3 years	$94.58	$694.58
After 4 years	$129.30	$729.30

If all went well, you should have found that at the end of 4 years, the account would have $9.30 more when interest is calculated on the new yearly amount, rather than the original principal. This type of interest is called **compound interest,** and each time the new amount is calculated (in this case, at the end of each year), we say that interest has been **compounded** at that time. Since interest is being compounded once per year in this example, we say that it is interest **compounded annually.**

You're probably not overly impressed by that extra nine bucks, but when we develop some formulas for efficiently computing compound interest, we'll be able to see that something VERY interesting happens in the long run. So let's get to work on developing a formula by looking for a pattern.

Value after 1 year = 600(1 + 0.05)

Value after 2 years = Value after 1 year(1 + 0.05) = $\underbrace{[600(1 + 0.05)]}_{\text{Value after 1 year}}(1 + 0.05) = 600(1 + 0.05)^2$

4. Now fill in the boxes in the computation to find the value after 3 years:

Value after 3 years = Value after 2 years(1 + 0.05) = $\underbrace{\boxed{600(1 + 0.05)^2}}_{\text{Value after 2 years}}(1 + 0.05) = 600(1 + 0.05)^3$

Now here comes the key observation of this lesson, so fasten your seat belts. So far, we know the following:

Value after 1 year = 600(1 + 0.05)

Value after 2 years = $600(1 + 0.05)^2$

Value after 3 years = $600(1 + 0.05)^3$

5. Make a conjecture as to the value after 4 years and 5 years.

Value after 4 years = $600(1 + 0.05)^4$

Value after 5 years = $600(1 + 0.05)^5$

6. Make a conjecture as to the value after *t* years, where *t* represents any number of years:

Value after t years = $600(1 + 0.05)^t$

Congratulations! Your conjecture (you did get it, right?) can be formalized into our first important formula for compound interest.

Formula for Compound Interest Compounded Annually

$$A = P(1 + r)^t$$

where

A is the future value.

P is the principal amount.

r is the interest rate in decimal form.

t is the term of the investment in years.

7. Use the formula for interest compounded annually to find the future value of the original $600 investment at 5% interest after 10, 20, 30, and 40 years. Then use the simple interest formula from Lesson 3-1 to find the future value if the interest is simple.

Time	Future Value, Compounded Annually	Future Value, Simple Interest
After 10 years	$977.34	$900
After 20 years	$1,591.98	$1,200
After 30 years	$2,593.17	$1,500
After 40 years	$4,223.99	$1,800

8. Use the table to draw a graph of the future value for both accounts, then write a verbal description of what you learned from the table and the graph. Make sure you choose an appropriate scale for each axis, and of course label that scale on your graph. Do something to distinguish the two graphs: Use different colors, or make one dashed or thicker. What kind of growth does each account illustrate?

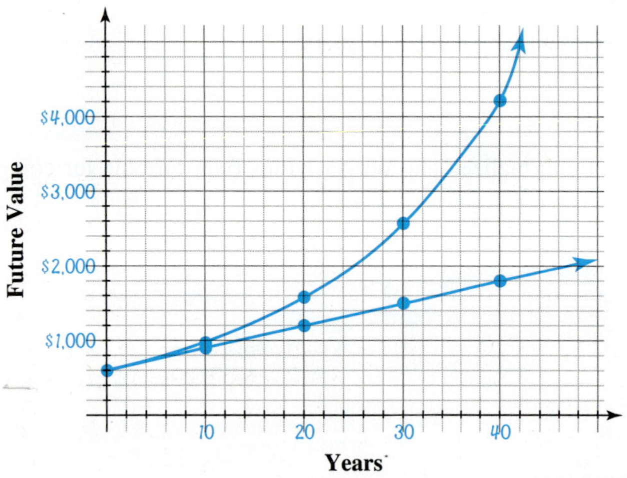

With simple interest, the graph is a line, meaning the growth is linear. With compound interest, the graph starts off growing gradually, but then starts to curve dramatically upward. This is typical of exponential growth.

Did You Get It

Try this problem to see if you understand the concepts we just studied. The answer can be found at the bottom of the portfolio page.

1. For an investment of $100,000 at 6% interest for 4 years, find (a) the simple interest, and (b) the compound interest if interest is calculated once per year.

3-2 Class

Based on what we've learned so far, if you're the one getting paid interest, compound interest is a MUCH better deal for you than simple interest. But what if you plan to invest your money for less than a year? You'd end up getting a whole lot of nothing, since the interest is compounded for the first time at the end of the first year. This would provide a competitive disadvantage for a banker, as wise investors would prefer more flexibility. At some point, bankers decided that compounding interest more often would make them more attractive to potential customers.

So let's say that interest is being compounded 4 times per year (or **quarterly**). Going back to our investment of $600 at 5% interest, the banker would not add 5% to the account 4 times per year—this would correspond to a far higher rate than the 5% they promised. Instead, they'd divide that rate by 4, and pay 1.25% every 3 months. The first calculation, after 3 months, would look like this:

$$A = 600\left(1 + \frac{0.05}{4}\right)$$

1. After 6 months, write a calculation that describes the future value, using the same idea from the Group portion of the lesson. Remember that interest is being calculated four times per year.

$$A = 600\left(1 + \frac{0.05}{4}\right)^2$$

2. By the time the first year is up, the interest would have been calculated __4__ times, so the total exponent on the parentheses in our future value calculation will be __4__, which is the number of times per year that interest has been compounded times the number of years that have passed.

3. If interest is compounded monthly, after 5 years it would be compounded __60__ times.

4. If interest is compounded n times per year, after t years it would be compounded __nt__ times.

5. Now adapt your answer from Question 1 into a formula for calculating the future value of that \$600 investment at 5% interest if interest is compounded *n* times per year for *t* years. Remember that the interest rate needs to be divided into *n* equal portions.

$$A = 600\left(1 + \frac{0.05}{n}\right)^{nt}$$

And that, friends, is how you develop a formula for interest compounded a certain number of times per year. Nicely done.

General Formula for Compound Interest

$$A = P\left(1 + \frac{r}{n}\right)^{nt}$$

where

A is the future value.

P is the principal amount.

r is the interest rate in decimal form.

t is the term of the investment in years.

n is the number of times that interest is compounded per year.

6. Use the compound interest formula to calculate the future value of our \$600 invested at 5% interest compounded quarterly for 10 years, then compare to the future value you calculated in Question 6 of the Group portion using interest compounded annually. Which results in more interest?

$P = \$600,\ r = 0.05,\ n = 4,\ t = 10$

$$A = 600\left(1 + \frac{0.05}{4}\right)^{40} \approx \$986.17$$

This is almost \$9 more in interest than when interest was compounded annually.

Math Note

Important note: This general formula makes the earlier formula for interest compounded annually obsolete. (You can just use this formula with $n = 1$.) *Don't think of it as two formulas.*

7. Repeat Question 6, this time using interest **compounded monthly** (12 times per year) and interest **compounded daily** (365 times per year). In each case, find the amount of interest that was earned (not just the future value).

Monthly: $P = \$600, r = 0.05, n = 12, t = 10$

$$A = 600\left(1 + \frac{0.05}{12}\right)^{120} \approx \$988.21 \quad \text{Interest: } \$388.21$$

Daily: $P = \$600, r = 0.05, n = 365, t = 10$

$$A = 600\left(1 + \frac{0.05}{365}\right)^{3,650} \approx \$989.20 \quad \text{Interest: } \$389.20$$

Help students with Question 8. Thinking about that one will really help foster an understanding of the magic of compound interest.

8. Draw a general conclusion about what happens to the amount of interest as the number of times compounding yearly goes up. Then (and this question really requires some thought) see if you can come up with a reason why this makes perfect sense. A true understanding of the difference between simple and compound interest will really help.

In general, as the number of times compounded goes up, so does the future value. This actually makes sense for the same reason that compound interest in general leads to more interest than simple interest: Every time that interest is compounded, its getting a percentage of the interest that has been previously added to the account. So the more often that happens, the greater the amount of interest should be.

Did You Get It ?

Try this problem to see if you understand the concepts we just studied. The answer can be found at the bottom of the portfolio page.

2. Find the interest on an investment of $75,000 that pays 4.5% compounded weekly for 8 years.

Effective Interest Rate

We've now seen that the number of times interest is compounded per year has an effect on how much interest is actually earned. It's not a *huge* effect, mind you, but is at least worth studying. The bottom line is that when interest is compounded more than one time per year, the actual percentage earned is a bit higher than the stated interest rate. That's good if you're the lender, not so good if you're the borrower. In any case, there's a formula that can be used to measure this effect. First, the terminology.

The **effective rate** (also known as the **annual yield)** is the simple interest rate which would yield the same future value over 1 year as the given compound interest rate. The compound interest rate is sometimes known as the **nominal rate.**

Now, the formula.

Formula for Effective Interest Rate

$$E = \left(1 + \frac{r}{n}\right)^n - 1$$

where

E is the effective rate.

n is the number of times that interest is compounded per year.

r is the nominal interest rate.

9. A loan has a nominal interest rate of 4%, and the interest is compounded weekly. Without doing any calculations, would you expect the effective interest rate to be higher or lower than 4%? Explain.

Higher. We know that compound interest leads to more interest than simple, so to get the same return with simple interest, which is what the effective interest rate measures, you'd need a higher rate.

10. Refer to Question 9. Calculate the effective interest rate.

$$E = \left(1 + \frac{0.04}{52}\right)^{52} - 1 = 0.04079477$$

The effective rate us 4.079477%

11. Find the future value after one year for an investment of \$10,000 at 4% compounded weekly.

$$A = \$10{,}000\left(1 + \frac{0.04}{52}\right)^{52(1)} = \$10{,}407.95$$

12. Divide your answer to Question 11 by 10,000. What do you notice?

$$\frac{\$10{,}407.95}{\$10{,}000} = 1.040795$$

The decimal part of the result is the effective interest rate.

13. Comparing investments is a very important part of any financial strategy. Which savings account is a better investment: 6.2% compounded daily or 6.25% compounded semiannually? There are different ways you could approach this, so make sure you show all work and explain how you made your decision.

Students may want to pick a principal amount and calculate the actual return, which is fine. But the simplest thing to do is simply compute the effective interest rate for each. As we can see below, the 6.2% compounded daily has a higher effective interest rate.

$$E = \left(1 + \frac{0.062}{365}\right)^{365} - 1 \approx 0.06396$$

$$E = \left(1 + \frac{0.0625}{2}\right)^{2} - 1 \approx 0.06348$$

Did You Get It

Try this problem to see if you understand the concepts we just studied. The answer can be found at the bottom of the portfolio page.

3. Which is a better investment, 3% compounded monthly or 3.25% compounded semiannually?

The remainder of this lesson can be considered optional if you don't want to cover interest compounded continuously. If you do chose to cover it, students will need some help with the algebra on the next two pages. You can choose to sort of gloss over that part and just focus on what the compound interest formula becomes when *n* approaches infinity. In that case, you'll need to define e for the class.

Interest Compounded Continuously

The goal of the remainder of this lesson is to see if we can figure out what happens to the compound interest formula when the number of times that interest is compounded per year gets really, really large. Large to the point that in essence, interest is being compounded every instant of every day. Wow.

As you know, the compound interest formula is $A = P(1 + \frac{r}{n})^{nt}$, where n is the number of times per year that interest is compounded.

14. To rewrite this formula in a form that makes it easier to decide what happens when n gets really big, we're going to introduce a new variable u, which will be defined by $u = \frac{n}{r}$. First, solve this equation to find a new name for n.

$n = ur$

15. Now replace n in the compound interest formula with the new name for n that you found in Question 14 and simplify to show that the formula becomes $A = P\left(1 + \frac{1}{u}\right)^{urt}$.

$A = P\left(1 + \frac{r}{ur}\right)^{urt} = P\left(1 + \frac{1}{u}\right)^{urt}$

16. Explain why this can be rewritten as $A = P\left[\left(1 + \frac{1}{u}\right)^{u}\right]^{rt}$.

This is just a standard property of exponents. It's easiest to see if you look at it backward: In the ending form, we are raising an expression with an exponent to a further power. To do this, you multiply the exponents, so you get a singe exponent of u r t.

17. Remember that $u = \frac{n}{r}$. What happens to this fraction when n gets really big? Remember that r is the interest rate, and it can't change.

Since n is in the numerator of the fraction, when n gets really big the entire fraction does as well. In fact, since the number we're dividing by (the interest rate in decimal form) is always less than 1, then u is always BIGGER than n. So as n gets bigger and bigger, u gets bigger still.

Here's a summary of what we learned in Questions 14–17: the compound interest formula can be rewritten in the form below, and as we make the number of times that interest is compounded per year get larger and larger, the variable *u* in that expression gets larger and larger as well. So next we'll study what happens to the expression inside the brackets when *u* gets larger and larger, and we'll find that we can find a MUCH simpler form for this formula.

$$A = P\left[\underbrace{\left(1 + \frac{1}{u}\right)^{u}}\right]^{rt}$$

What happens to this expression when *u* gets larger and larger?

18. Using a calculator, fill in the table below. Round to three decimal places. What can you conclude about what happens to the value of $\left(1 + \frac{1}{u}\right)^{u}$ as the number of compounding periods tends to infinity? ("Tends to infinity" is the phrase we use to describe a variable getting larger and larger without bound.)

u	50	100	500	1,000	1,500	2,000	2,500
$\left(1 + \frac{1}{u}\right)^{u}$	2.692	2.705	2.716	2.717	2.717	2.718	2.718

19. The number that you should have found in Question 18 is about 2.718, and is denoted *e*. (Actually, *e* is irrational, so its decimal equivalent neither terminates nor repeats: $e \approx 2.718\ldots$). What is the result of the compound interest formula above when the number of times interest is compounded gets larger and larger? In this case, we say that interest is **compounded continuously.** You can think of this as the interest being compounded every instant of every day.

$A = P[e]^{rt}$ or, more simply $A = Pe^{rt}$

Using Technology: Evaluating Expressions Containing e

To evaluate e^3:

Standard Graphing Calculator: 2nd LN 3 ENTER

Standard Scientific Calculator: 3 2nd LN

Spreadsheet: = EXP(3)

20. Use the formula you wrote in Question 19 to find the future value after 20 years of a $5,000 account that earns 4% interest compounded continuously, then compare that to the future value of the same account if interest is compounded annually.

$$A = Pe^{rt} = \$5{,}000e^{0.04(20)} = \$11{,}127.70$$

$$A = P\left(1 + \frac{r}{n}\right)^{nt} = \$5{,}000\left(1 + \frac{0.04}{1}\right)^{20} = \$10{,}955.62$$

21. Finally, let's compare interest compounded continuously to simple interest. Suppose that $50,000 is invested in two accounts: One earns 6% simple interest, the other earns 6% compounded continuously. Fill in the table with the future value of each account after each term, rounded to the nearest dollar. Wouldn't this be a GREAT spot to use a spreadsheet, or at least the table feature on a graphing calculator?

Years	5	10	15	20	25	30
6% simple interest	$65,000	$80,000	$95,000	$110,000	$125,000	$140,000
6% compounded continuously	$67,492	$91,106	$122,980	$166,006	$224,084	$302,482

Did You Get It

Try this problem to see if you understand the concepts we just studied. The answer can be found at the bottom of the portfolio page.

4. Find the future value of a $40,000 investment with 6.9% interest compounded continuously 25 years after it's started.

3-2 Portfolio

Name ____________________

Check each box when you've completed the task. Remember that your instructor will want you to turn in the portfolio pages you create.

Technology

1. ☐ After completing this lesson, you should be aware of the fact that the formulas involved are a little complicated, so you have to be really careful when doing the computations. Wouldn't it be nice if there were some sort of machine for doing computations? Wait, what? That's what the word "computer" means? Cool. In any situation where repeated calculations may be done, programming a spreadsheet to do them is likely to be helpful. So your task is to build a compound interest calculator, setting it up like the sample below. Obviously, you should use a formula to calculate future value. Then choose an amount you'd like to invest, an interest rate, and a time period, then use your spreadsheet to find the future value for interest compounded annually, quarterly, monthly, weekly, daily, and hourly.

	A	B	C	D	E
1	Principal	Interest rate	Number of times compounded per year	Term (Yrs)	Future Value
2	$600.00	0.04	1	10	$888.15
3	$600.00	0.04	12	10	$894.50
4	$600.00	0.04	52	10	$894.96

Online Practice

1. ☐ Include any written work from the online assignment along with any notes or questions about this lesson's content.

Applications

1. ☐ Complete the applications problems.

Reflections

Type a short answer to each question.

1. ☐ Explain the difference between simple and compound interest.
2. ☐ Why is compound interest a much better deal for you if you're the investor?
3. ☐ What does the term "compounding" refer to?
4. ☐ Take another look at your answer to Question 0 at the beginning of this lesson. Would you change your answer now that you've completed the lesson? How would you summarize the topic of this lesson now?
5. ☐ What questions do you have about this lesson?

Looking Ahead

1. ☐ Complete the Lesson 3-3 Prep Skills and read the opening paragraph in Lesson 3-3 carefully, then answer Question 0 in preparation for that lesson.

Answers to "Did You Get It?"

1. a. $24,000; **b.** $26,247.70 **2.** $32,482.97 **3.** 3.25% semiannually
4. $224,500.84

3-2 Applications

Name ______________________________

Over the last two lessons, we've studied many aspects of interest. The goal of this assignment will be to compare various investments and decide which would be most beneficial. Of course, the amount of money that anyone chooses to invest at any given time is very personal and depends on too many factors to consider in a few lines of text. So we'll pick an amount and go with it, but if you'd prefer to work this assignment with a different amount, check with your instructor to see if that's okay.

Example: An investor has $12,000 to invest and is willing to keep it invested for up to five years. Each of the options below is available. Find the future value of the investment and the effective rate in each case. When you've completed all of them, write a short essay explaining which you think would be the best choice for you. Consider as many factors as you can.

1. A 1-year certificate of deposit that pays 2.1% simple interest

$A = P(1 + rt) = \$12{,}000(1 + 0.021) = \$12{,}252$

With simple interest, the effective rate is the same as the nominal rate.

2. A 5-year certificate of deposit that pays 2.9% interest compounded annually

$$A = P\left(1 + \frac{r}{n}\right)^{nt} = \$12{,}000\left(1 + \frac{0.029}{1}\right)^{5} = \$13{,}843.89$$

$$E = \left(1 + \frac{r}{n}\right)^{n} - 1 = \left(1 + \frac{0.029}{1}\right)^{1} - 1 = 0.029$$

3. You lend the money to a friend who's starting a business, and she writes a contract guaranteeing you a 12% return on your investment, to be paid in full at the end of 4 years, with an option to reinvest if the company is successful.

12% of $\$12{,}000 = 0.12 \times \$12{,}000 = \$1{,}440$

Future value $= \$13{,}440$

This is simple interest, so the effective rate is 3% per year.

4. A savings account offers 1.98% interest compounded continuously. You can remove the money any time you like, but plan to keep it in the investment for 5 years. (For interest compounded continuously, the effective rate formula is $E = e^r - 1$.)

$A = Pe^{rt} = \$12{,}000e^{0.0198(5)} = \$13{,}248.80$

$E = e^{0.0198} - 1 \approx 0.019997$

3-2 Applications

Name ____________________

5. A savings account that offers 2.18% compounded monthly: You can remove the money any time you like, but plan to keep it in the investment for 5 years.

$$A = P\left(1 + \frac{r}{n}\right)^{nt} = \$12{,}000\left(1 + \frac{0.0218}{12}\right)^{12(5)} = \$13{,}380.62$$

$$E = \left(1 + \frac{0.0218}{12}\right)^{12} - 1 \approx 0.022$$

6. A rival bank to the one in Question 5 offers the same interest rate, but they compound hourly. The plan is also to keep the money invested for 5 years.

$$A = P\left(1 + \frac{r}{n}\right)^{nt} = \$12{,}000\left(1 + \frac{0.0218}{365 \cdot 24}\right)^{365 \cdot 24 \cdot 5} = \$13{,}381.95$$

$$E = \left(1 + \frac{0.0218}{365 \cdot 24}\right)^{365 \cdot 24} - 1 \approx 0.022$$

7. So how would you invest the money? Obviously, you have to discuss you reasoning. Think about the fact that when investing, there are other potential concerns than simply the total amount at the end of the investment.

Hopefully, answers will vary!

Lesson 3-3 Prep Skills

This is a short review of skills that will come in handy in the next lesson. In each case, answer the question, then rate your confidence level by checking one of the boxes. If you feel like you're struggling with these skills, consult the online resources provided by your instructor for extra practice.

I KNOW IT | THINK SO | UNSURE | NO IDEA

SKILL 1: WRITE PERCENTAGES IN DECIMAL FORM

1. Write each percent in decimal form.

   91%

   6.48%

SKILL 2: CALCULATE A PERCENTAGE OF A WHOLE

2. What is 1.8% of $11,375?

SKILL 3: COMPUTE SIMPLE INTEREST

3. Suppose you borrowed $450 to buy an iPhone. If the interest is 14.7% simple interest for 2 years, how much would you have to pay back?

SKILL 4: USE INFORMATION FROM A TABLE

4. The table below shows the number of violent crimes committed per 100,000 citizens in various cities during 2014. Given the fact that there were about 438,500 people in Omaha in 2014, how many violent crimes were committed there?

City	Violent crimes per 100,000 citizens
Detroit	1,989
Columbus	549
Miami	1,060
Omaha	561
San Francisco	795

Lesson 3-3 Buying Stuff Without Money (Installment Buying)

LEARNING OBJECTIVES

- ☐ 1. Compute payments and charges associated with installment loans.
- ☐ 2. Identify the true cost of a loan by computing APR.
- ☐ 3. Evaluate the costs of buying items on credit.

© SW Productions/Getty Images RF

Never assume that all is well when it comes to your savings or loans. Get a printout to ensure you know exactly what is happening.

—Tagene Brown-McBean

A lot of people celebrate their college graduation by getting a new car. Unless you're very wealthy, you won't be plunking down a stack of crisp hundred dollar bills to make that happen—you'll need to get a loan. In this case, you'll be doing what is called **installment buying.** This is when an item is purchased and the buyer pays for it by making periodic partial payments, or installments.

There are natural advantages and disadvantages to installment buying. The most obvious advantage is that it allows you to buy an item that you don't have enough money to pay for, and use it while you're raising that money. The most obvious disadvantage is that you pay interest on the amount borrowed, so you end up paying more for the item—in some cases a lot more. When you buy a car, for example, you know the price of the car, and you're usually told how much the monthly payment will be. If you're buying a car for $12,260 with a $3,000 down payment and are told the payment will be $231.50 per month for 48 months, how much will you actually end up paying for that car? Work through this lesson, and soon you will know.

0. After reading the opening paragraphs, what do you think the main topic of this lesson will be?

3-3 Group

Fixed Installment Loans

A **fixed installment loan** is a loan that is repaid in equal payments. Sometimes the buyer will pay part of the cost at the time of purchase. This is known as a **down payment.** Here are some other key terms used to describe installment loans.

The **amount financed** is the amount a borrower will pay interest on.

Amount financed = Price of item − Down payment

The **total installment price** is the total amount of money the buyer will ultimately pay.

Total installment price = Sum of all payments + Down payment

The **finance charge** is the interest charged for borrowing the amount financed.

Finance charge = Total installment price − Price of item

Now let's take a look at the problem posed in this lesson's opening paragraph. The pertinent information is this:

Cost of car: \$12,260
Down payment: \$3,000
Monthly payment: \$231.50
Term of loan: 48 months

Math Note

You're often given quite a bit of information in a problem about finances. Identifying and writing out that information in an organized way, like we did here, is ALWAYS a good idea.

1. What was the amount financed? Refer to the definitions on the previous page if you need to.

Cost of car − Down payment = \$12,260 − \$3,000 = \$9,260

2. What's the total amount of all monthly payments?

48 × \$231.50 = \$11,112

3. What's the total installment price? (This is the answer to the question "How much money will you actually end up paying for the car?")

\$11,112 + \$3,000 = \$14,112

4. Find the finance charge, then describe what that amount really represents.

\$14,112 − \$12,260 = \$1,852

This is the amount of interest paid over the life of the loan.

Did You Get It

Try this problem to see if you understand the concepts we just studied. The answer can be found at the bottom of the portfolio page.

1. If you buy a used car for $8,200 with a down payment of $1,000 and 36 monthly payments of $270, find the amount financed, the total installment price, and the finance charge.

In some cases, it's a good idea to be able to calculate a monthly payment before making a purchase. Not only can this help you to decide whether or not you can afford an item, it gives you a way to make sure you're not getting cheated when you're quoted a monthly payment. So how can you find the monthly payment for a loan? Using a formula, of course. (Although to be fair, there are many loan calculators online.)

Calculating Monthly Payments on a Fixed Installment Loan

A monthly payment M can be calculated using the formula

$$M = \frac{P \cdot \frac{r}{12}}{1 - \left(1 + \frac{r}{12}\right)^{-n}}$$

where

P is the principal value of the loan.
r is the annual percentage rate (APR) in decimal form.
n is the total number of payments.

Memorize this formula, and then eat it. Hold on, we're just kidding—obviously the formula is pretty complicated. You'll need to be especially careful when performing calculations involving this formula. An important note: if you do the calculation in stages, which we don't necessarily recommend, make sure that you don't round too much. That can change the calculation considerably. A reasonable rule of thumb is at least four decimal places.

In order to calculate the monthly payment on an auto loan, we need to know the annual percentage rate, or APR. As we'll see shortly, this isn't exactly the same as the quoted interest rate for a loan. If the APR isn't quoted to you immediately when you start to talk about buying a car, ask for it! Salesmen are fond of steering the conversation toward monthly payments, but what determines how good (or bad) of a deal you're getting is the APR. You should always check local rates in advance of buying a car so you know what rate you should be getting.

5. Have you noticed that car makers never have sales? They're always "events." Weird. Anyhow, for the Ford 2015 holiday "event," they offered a 2015 Taurus with a special APR of 2.9% for 60 months. If one particular vehicle has a selling price of $27,295 and a buyer trades in an older car for a down payment of $4,500, calculate the monthly payment for this loan.

Principal value $(P) = \$27{,}295 - \$4{,}500 = \$22{,}795$; $r = 0.029$, $n = 60$

$$M = \frac{\$22{,}795 \cdot \frac{0.029}{12}}{1 - \left(1 + \frac{0.029}{12}\right)^{-60}} = \$408.58$$

If students are using a graphing calculator that can handle the entire calculation, encourage them to do so to avoid rounding error. A demonstration of how much rounding can affect a stage-by-stage calculation can be helpful.

6. If the loan was based on simple interest, the future value would be $22,795(1 + 0.029 × 5). Calculate this amount, then divide by 60 to find the monthly payment. What do you think accounts for the fact that simple interest makes it HIGHER?

$\$22,795(1 + 0.029 \cdot 5) = \$26,100.28$

$\$26,100.28/60 = \$435.00$

With the fixed installment loan, the interest being charged each month is only calculated on the remaining balance of the loan. The simple interest loan is calculating interest on the entire principal for all five years.

Did You Get It

Try this problem to see if you understand the concepts we just studied. The answer can be found at the bottom of the portfolio page.

2. A graphic design pro buys a new iMac for $1,499 with a $200 down payment, and gets manufacturer financing for 5 years at 12% APR. Find (a) the amount financed; (b) the monthly payment; (c) the total installment price, and (d) the finance charge.

Annual Percentage Rate

Many lenders add upfront fees to a loan and then spread them over the life of the loan. This has the effect of making the actual interest rate that a borrower pays higher than the quoted rate. Because this can get confusing, lenders are required by law to disclose an annual percentage rate, or APR, that reflects the true interest charged. This allows consumers to compare loans with different terms. The mathematical procedures for computing APR are extremely complicated, so tables have been compiled that help you to estimate APR for a loan. There is also a wide variety of APR calculators available online. A partial APR table is shown in the following table.

Number of Payments	Annual Percentage Rate												
	6.0%	6.5%	7.0%	7.5%	8.0%	8.5%	9.0%	9.5%	10.0%	10.5%	11.0%	11.5%	12.0%
	(Finance charge per $100 of amount financed)												
6	$1.76	$1.90	$2.05	$2.20	$2.35	$2.49	$2.64	$2.79	$2.94	$3.08	$3.23	$3.38	$3.53
12	3.28	3.56	3.83	4.11	4.39	4.66	4.94	5.22	5.50	5.78	6.06	6.34	6.62
18	4.82	5.22	5.63	6.04	6.45	6.86	7.28	7.69	8.10	8.52	8.93	9.35	9.77
24	6.37	6.91	7.45	8.00	8.54	9.09	9.64	10.19	10.75	11.30	11.86	12.42	12.98
30	7.94	8.61	9.30	9.98	10.66	11.35	12.04	12.74	13.43	14.13	14.83	15.54	16.24
36	9.52	10.34	11.16	11.98	12.81	13.64	14.48	15.32	16.16	17.01	17.86	18.71	19.57
48	12.73	13.83	14.94	16.06	17.18	18.31	19.45	20.59	21.74	22.90	24.06	25.23	26.40
60	16.00	17.40	18.81	20.23	21.66	23.10	24.55	26.01	27.48	28.96	30.45	31.96	33.47

Procedure for Computing APR

Step 1. Find the finance charge per $100 borrowed using the formula

$$\frac{\text{Finance charge}}{\text{Amount financed}} \times \$100$$

Step 2. Find the row in the table marked with the number of payments and move to the right until you find the amount closest to the number from step 1.

Step 3. The APR (to the nearest half percent) is at the top of the corresponding column.

7. Find the finance charge per $100 for the loan described in Questions 1 through 4 (step 1 in computing APR).

Finance charge = $1,852 (from Question 4), amount financed = $9,260

$$\frac{\$1{,}852}{\$9{,}260} \times \$100 = \$20$$

8. Look at the row in the table corresponding to 48 months, and find the two entries that are closest to your answer from Question 7. Then find the APR corresponding to each, and use the result to estimate the APR for that loan (steps 2 and 3 in computing APR). Describe how you got your estimate.

In the 48-month row, $20 is almost exactly halfway between the amount in the 9.0% column ($19.45) and the one in the 9.5% column ($20.59), so we can estimate that the APR is pretty close to halfway between 9.0% and 9.5%, which would be 9.25%.

Discuss interpolating between values for situations where the exact amount isn't in the table.

One way to save money on a fixed installment loan is to pay it off early. This will allow a buyer to avoid paying the entire finance charge. The amount of the finance charge that is saved when a loan is paid off early is called **unearned interest.** One way to calculate unearned interest for a loan uses the APR table we've already been introduced to: this is called the **actuarial method.**

The Actuarial Method for Computing Unearned Interest

Unearned interest can be calculated using the formula

$$u = \frac{kRh}{100 + h}$$

where

u is the unearned interest.

k is the number of payments remaining, excluding the current one.

R is the monthly payment.

h is the finance charge per $100 borrowed for a loan with the same APR and k monthly payments.

As a reminder, here are the specifics of the loan we've been working with:

Cost of car: $12,260
Down payment: $3,000
Monthly payment: $231.50
Term of loan: 48 months

9. When making your 36th payment, how many payments would remain?

12 payments remain. Don't overthink it!

10. If you decide to pay this loan off at the time of the 36th payment, how much money would you save? (You'll need to do some work to find the value of *h*. Use the 12 Payments row of the APR table.)

$k = 12$, $R = \$231.50$, $h = \$5.08$, which comes from the 12 payments row of the APR table; more specifically from splitting the difference between the amounts for 9.0% and 9.5%.

$$u = \frac{12 \cdot 231.50 \cdot 5.08}{100 + 5.08} = \$134.30$$

11. In this situation, you would surely want to know how much it would cost to pay off the loan. The first step would be to find the total cost of all the remaining payments after the 36th. So do that.

12 × $231.50 = $2,778

12. Your answer to Question 11 is what the payoff would be if you weren't getting credit for unearned interest. But you are, so subtracting the unearned interest you found will give you the payoff amount.

$2,778 − $134.30 = $2,643.70

13. Discuss the pros and cons of paying off that particular loan early. If you were in that situation, do you think it would have been worthwhile for you to pay it off early?

Pros: No more monthly payment, save about $135 overall.
Cons: you have to cough up over $2,600 all at once, which for many people is worse than paying a little more over the course of a year.

Did You Get It ?

Try this problem to see if you understand the concepts we just studied. The answer can be found at the bottom of the portfolio page.

3. A $16,500 truck loan is to be paid off in 48 monthly installments of $387.50. The borrower decides to pay off the loan on his 42nd payment. Find the amount of interest saved.

3-3 Class

So far, we've covered examples of closed-ended credit; this is credit with a fixed number of payments and a specific payoff date. We turn our attention now to **open-ended credit,** where there is no fixed number of payments or payoff date. By far the most common example of this is credit cards. Since more and more people use credit cards for convenience, an understanding of open-ended credit is more important than ever. One of the easiest ways to end up in a really bad place financially is to not understand how open-ended credit works.

The Unpaid Balance Method

One way that banks calculate finance charges on a credit card is the unpaid balance method. In this case, interest is charged only on the balance left over from the previous month.

Suppose that you had an unpaid balance of $356.75 at the beginning of August and made purchases of $436.50. A payment of $200.00 was made during the month. The interest on the unpaid balance is 1.8% per month. Our goal will be to find the finance charge and the balance on September 1. (The billing cycle for credit cards and loans can begin and end on any day of the month; however, for the examples given in this lesson, we'll assume the cycle ends on the last day of the month.)

Math Note

As of July 2015, total credit card debt in the U.S. was over nine hundred BILLION dollars, or almost $7,500 for every household in the country.

1. Find the amount of interest that would be charged on the unpaid balance. Show your work!

1.8% of $356.75 = $6.42

When studying credit cards, it's impossible to overstate how dangerous it can be to start carrying credit card balances.

2. Use your answer from Question 1 and the information about payments and purchases to find the new balance on the card. Show all work again!

Purchases + Interest = $436.50 + $6.42 = $442.92

Payment = $200

New balance = $356.75 + $442.92 − $200 = $599.67

Did You Get It ?

Try this problem to see if you understand the concepts we just studied. The answer can be found at the bottom of the portfolio page.

4. For the month of January, Christina has an unpaid balance of $846.50 from December. Purchases for the month were $532.86 and a payment of $350.00 was made during the month. If the interest on the unpaid balance is 2% per month, find the finance charge and the balance on February 1.

The Average Daily Balance Method

Because a company that issues credit cards doesn't make as much money when users pay off the balance every month, they sometimes charge an annual fee for use of the card. Another way for them to make money from all users is to use the average daily balance method for computing the finance charge. In this method, the balance for each day of the month is used to compute an average daily balance, and interest is computed on that average. So in effect, consumers start paying interest on the day they make a purchase. Ouch!

Calculations using the average daily balance method can be done in a series of steps, as illustrated by the next set of questions. The credit card statement shown here displays more detail on the account we studied in Questions 1 and 2.

August 1	Previous Balance	$356.75
August 7	Purchases	$59.95
August 12	Purchases	$223.75
August 18	Payment	$200.00
August 24	Purchases	$152.80

The goal is to find the average daily balance, the finance charge for the month, and the new balance on September 1. The interest rate is 1.8% per month on the average daily balance.

Step 1: Find the balance on each day that a transaction occurred.

3. Write the new balance next to each day listed below in the first blank. Ignore the second blank for now.

August 1 $356.75 6

August 7 $416.70 5

August 12 $640.45 6

August 18 $440.45 6

August 24 $593.25 8

Step 2: Find the number of days for each balance.

4. How many days was the balance $356.75? Using the same idea, fill in the second blank next to each balance above with the number of days that the balance was at that amount. (Make sure the total number of days adds up to 31 days, which is the number of days in August.)

Step 3: Multiply each balance by the number of days, and add these products.

5. Copy the first two columns in the table below from your answers on the previous page, then multiply to fill in the last column. Then add your results.

Date	Balance	# of days	Balance × # of days
August 1	$356.75	6	$2,140.50
August 7	$416.70	5	$2,083.50
August 12	$640.45	6	$3,842.70
August 18	$440.45	6	$2,642.70
August 24	$593.25	8	$4,746.00
		Sum	$15,455.40

Step 4: Divide by the total number of days to get the average daily balance for the month.

6. Divide your sum from Question 5 by the number of days in August. This gives you the average daily balance for the account.

$15,455.40/31 = $498.56

Step 5: Find the finance charge by calculating the amount of interest on the average daily balance.

7. Recall that interest on this card is 1.8% per month. How much interest would be due on the average daily balance?

$498.56 × 0.018 = $8.97

Step 6: Find the new ending balance by adding the finance charge to the last balance.

8. Find the final balance that would appear on the next statement.

$8.97 + $593.25 = $602.22

Did You Get It

Try this problem to see if you understand the concepts we just studied. The answer can be found at the bottom of the portfolio page.

5. A credit card statement for the month of November showed the following transactions:

November 1	Previous Balance	$937.25
November 4	Purchases	$531.62
November 13	Payment	$400.00
November 20	Purchases	$89.95
November 28	Payment	$100.00

a. Find the average daily balance.
b. Find the finance charge. The interest rate is 1.9% per month on the average daily balance.
c. Find the new balance on December 1.

Each month when you get a credit card statement, there's a minimum payment that must be paid. Before 2003, these minimum payments had shrunk to the point that they didn't even cover the interest for the month, so if a cardholder paid just the minimum payment, the balance would continue to rise. In response to a public outcry, the federal government set new regulations in 2003 requiring credit card companies to set the minimum payment high enough that it covers the interest for the month, as well as part of the principal balance. (The exact minimum payment formula varies some from company to company.)

This is not to say that making minimum payments only is a good idea. If at all possible, you should always pay off the full amount each month. This is the only foolproof way to make sure that you don't get in the habit of spending more money than you can afford. We'll close this lesson by examining the effects of making only minimum payments.

Suppose you have a $2,300 balance on a credit card with an interest rate of 1.25% per month, and the minimum payment for any month is the amount of interest plus 2% of the principal balance, or $5, whichever is higher. If you don't make any more purchases on that card and make the minimum payment for 6 months, how much will you pay down the balance? Let's find out. (By the way, the interest and minimum payment percentages in this problem reflect industry averages in 2015.)

9. This certainly sounds like we'll need to compute some interest. After all, we were told that the interest is 1.25% per month. But sometimes problems contain extraneous information. Read the problem carefully and explain why we don't need to worry about the amount of interest in calculating how much the balance decreases.

It says that the minimum payment is 2% plus the amount of interest. So each month you'd pay the interest, then 2% of the balance that remains. So the balance is decreasing by 2% each month.

10. In fact, what really matters is that the balance is going to decrease by 2% each month. Use this fact to find the new balance after 1 month.

If the balance decreases by 2%, 98% of it remains. 98% of $2,300 is $2,254.

11. Repeat your calculations from Question 10 five more times to find how much the balance would have decreased after six months.

Two months: 98% of $2,254 is $2,208.92
Three months: 98% of $2,208.92 is $2,164.74
Four months: 98% of $2,164.74 is $2,121.45
Five months: 98% of $2,121.45 is $2,079.02
Six months: 98% of $2,079.02 is $2,037.44

12. Just make a guess: How long do you think it would take to pay off the full balance at this rate? Discuss, then ask your instructor to give their thoughts on this question.

Answers vary, of course. This is a GREAT opportunity to tie in some stuff from earlier in the course: students can write an exponential decay equation, then graph it and see how many months it takes to get down to $5, at which point the minimum payment will be $5 and the account is paid off. For the record, it turns out to be 303 months, although obviously anyone with a lick of sense would go ahead and just pay it off when it gets down to twenty bucks or so, which is just 234 short months.

Math Note

In the Technology portion of the homework, we'll use Excel to calculate how long it would take to pay off this card if you make minimum payments.

3-3 Portfolio

Name ____

Check each box when you've completed the task. Remember that your instructor will want you to turn in the portfolio pages you create.

Technology

1. ☐ At the end of the Class portion of this lesson, we began to study the effects of making minimum payments on credit card debt. We can set up a spreadsheet to help calculate how long it would take to pay down the $2,300 debt. *The key is to recognize that if 2% of the debt is paid every month, then the new balance is 98% of the previous balance.* This allows you to set up a formula that calculates 98% of the previous balance. Also, the bank requires that the minimum payment each month is not less than $20, so once the balance gets below $1,000, the monthly payment will be $20 per month. At that point, change your formula so that it subtracts $20 from the previous balance. Use your spreadsheet to find how long it takes to pay off the balance, in months and in years. A template to help you get started can be found in the online resources for this lesson.

Online Practice

1. ☐ Include any written work from the online assignment along with any notes or questions about this lesson's content.

Applications

1. ☐ Complete the applications problems.

Reflections

Type a short answer to each question.

1. ☐ Explain the difference between closed-ended and open-ended credit.
2. ☐ What did you learn in this lesson about keeping balances on credit cards?
3. ☐ If making a large purchase, what factors do you need to consider?
4. ☐ Take another look at your answer to Question 0 at the beginning of this lesson. Would you change your answer now that you've completed the lesson? How would you summarize the topic of this lesson now?
5. ☐ What questions do you have about this lesson?

Looking Ahead

1. ☐ Complete the Lesson 3-4 Prep Skills and read the opening paragraph in Lesson 3-4 carefully, then answer Question 0 in preparation for that lesson.

Answers to "Did You Get It?"

1. Amount financed, $7,200; total installment price, $10,720; finance charge, $2,520
2. a. $1,299 **b.** $28.90 **c.** $1,934 **d.** $435 **3.** $40.21
4. Finance charge = $16.93; balance on Feb. 1 = $1,046.29 **5. a.** $1,198.69
b. $22.78 **c.** $1,081.60

3-3 Applications

Name ______________________________

In this assignment, we'll study various options for making a large purchase, including saving money in advance and paying cash, obtaining a fixed installment loan, and using a credit card. The goal is not just to perform calculations, but to think about all aspects and decide which you think is the best choice. We'll use furniture as an example, as it's pretty common to finance furniture purchases when you need to fill a new place. Let's say that the goal is to buy $3,500 worth of furniture.

We'll start with a fixed-installment loan from the furniture store. They're offering three-year financing with no money down at 11.99% annual interest.

1. Find the monthly payment.

$$M = \frac{\$3{,}500 \cdot \frac{0.1199}{12}}{1 - \left(1 + \frac{0.1199}{12}\right)^{-36}} = \$116.23$$

2. Find the total installment price and finance charge.

Total installment price $= 36 \cdot \$116.23 = \$4{,}184.28$

Finance charge $= \$4{,}184.28 - \$3{,}500 = \$684.28$

The store also offers 18-month financing with no payments and no interest, but if the balance isn't paid in full at the end of 18 months, 17.99% interest will apply to the entire purchase amount, and will continue on any unpaid balance after 18 months.

3. Let's say you take this option and forget to pay off the final amount before the 18 months is up. And believe me, it's easy to do—they don't exactly go out of their way to remind you of the end date. If you'd paid $100 per month up until that point, how much would you still owe after the interest charge kicks in?

At $100 per month, in 18 months you'd have paid $1,800. At that point, you incur a deferred interest charge of 17.99% of $3,500, which is $629.65. So at that point you'd still owe $\$3{,}500 - \$1{,}800 + \$629.65 = \$2{,}329.65$.

3-3 Applications

Name ______________________________

The next option is to put the entire purchase on a bank-issued Visa card with an annual interest rate of 12.49%. Let's say that you use this particular card strictly for this purchase.

4. In this case, would the amount of interest calculated vary depending on whether the bank used the unpaid balance or the average daily balance method? Explain.

It depends on when during the billing cycle your monthly payment posts. If it's the same as the beginning date of the billing cycle, and you don't make any purchases, then the average daily balance would be the same as the previous month's balance, so the two methods would lead to the same amount of interest. If your payment posts sometime during the month, you'd pay a little less with the average daily balance method since the remaining balance would go down at some point during the month.

5. This particular card has a minimum monthly payment of 3% of the balance plus any interest accrued during the month. What would your first minimum payment be?

3% of \$3,500 is \$105.00. Add to that the interest on \$3,500. Since the interest rate is 12.49% annually, one-twelfth of that applies in the first month, so interest is 0.1249/12 times \$3,500, or \$36.43. Total payment: \$105 + \$36.43 = \$141.43.

6. In this case, the equation $y = 3{,}500(0.97)^t$ would describe the remaining balance, where t is months after purchase. Graph this equation and use the graph to estimate how long it would take to pay off the purchase making only minimum payments. You can assume that once the balance reaches \$100, you'd go ahead and pay off the rest.

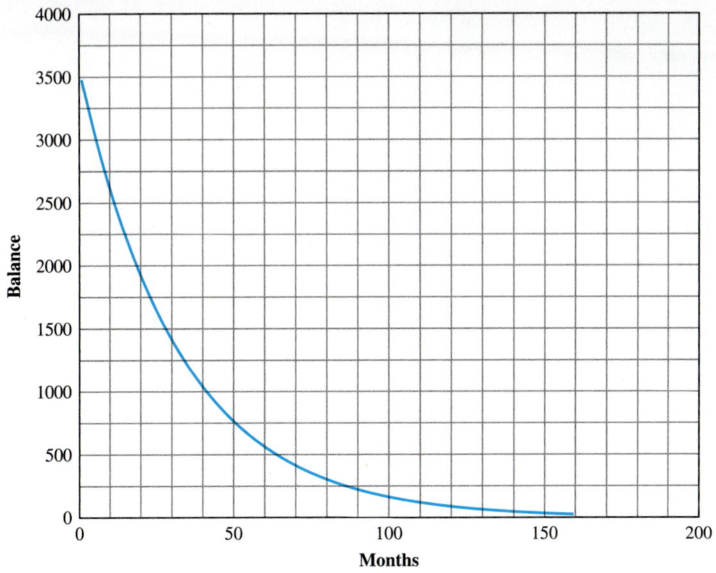

Based on the visual of the graph, it would take about 120 months, or a bit less than 10 years. The actual time is about 117 months.

3-3 Applications

Name ____________________

Suppose that you want to avoid paying interest and decide you'll only buy the furniture when you have the money to pay for it. An **annuity** is basically the opposite of a fixed-installment loan: You deposit a fixed amount each month and receive interest based on the total amount that's been saved. The future value formula is

$$A = \frac{12M\left[\left(1+\frac{r}{12}\right)^{12t}-1\right]}{r}$$

where M is the regular monthly payment, r is the annual interest rate in decimal form, and t is the term of the annuity in years.

7. With a monthly payment of \$120, what would the future value be if you chose an annuity with a term of two years at 4.8% interest?

$$A = \frac{12(120)\left[\left(1+\frac{0.048}{12}\right)^{24}-1\right]}{0.048} = \$3{,}016.45$$

8. Recalculate the amount if you're willing to raise your monthly payment by \$20 per month.

$$A = \frac{12(140)\left[\left(1+\frac{0.048}{12}\right)^{24}-1\right]}{0.048} = \$3{,}519.19$$

9. Discuss the options you've seen for obtaining the furniture, focusing on which you think sounds like the best choice. You might consider combinations of these choices. Try to consider as many factors as you can.

Answers really, really vary.

Lesson 3-4 Prep Skills

This is a short review of skills that will come in handy in the next lesson. In each case, answer the question, then rate your confidence level by checking one of the boxes. If you feel like you're struggling with these skills, consult the online resources provided by your instructor for extra practice.

I KNOW IT	THINK SO	UNSURE	NO IDEA	
				SKILL 1: COMPUTE SIMPLE INTEREST
				1. Compute the simple interest on a $12,000 loan at 9% interest for 3 years.
				2. Compute the future value of a $3,700 deposit that draws 4.95% simple interest for 18 months.
				SKILL 2: PERFORM COMPLEX CALCULATIONS
				3. Use a calculator to perform each calculation: $8\left(1-\frac{.09}{12}\right)^{-24}$
				$\dfrac{17{,}000}{1-(1+0.03)^{-120}}$
				SKILL 3: EVALUATING FORMULAS
				4. Find the value of the formula below if $k = -8.1$, $m = 0.0435$, and $W = 1{,}297.3$. Round your answer to six decimal places. $\dfrac{W\left(\frac{m}{k}\right)}{k-(1+m)^{k}}$

Lesson 3-4 Investing in Yourself (Education and Home Loans)

LEARNING OBJECTIVES

- ☐ 1. Understand different student loan options.
- ☐ 2. Compute interest and monthly payments on a student loan.
- ☐ 3. Evaluate the effects of capitalizing interest.
- ☐ 4. Analyze various aspects of a mortgage.
- ☐ 5. Compare two mortgages of different lengths.
- ☐ 6. Prepare an amortization schedule.

© Digital Vision/SuperStock RF

If you think education is expensive, try ignorance.
—Derek Bok

Everyone knows that college is expensive. And some folks have opinions on whether or not it's really worth the investment. The numbers, however, say that it absolutely is, since your earning potential with a college degree increases to the point that you'll make back the cost long before retirement. When it comes to credit card debt, the best advice is often "Don't." But that's not a viable option for college costs, since the cost of NOT going to college is far more in the long run than the cost of attending. That makes student loans a necessary evil for most folks. And when you're finished with college and making a living, many of you will make the most expensive and rewarding purchase of your life: a home. If there's one common theme in this unit, it's that understanding the nuances of the financial transactions you're likely to deal with makes you far less likely to end up in a bad situation.

0. After reading the opening paragraph, what do you think the main topic of this lesson will be?

You might consider beginning this lesson by seeing how many of your students are using student loans to pay college expenses. Discuss whether or not they really understand the details of their loans.

3-4 Group

Student Loans

There are three basic types of loans that most undergraduates might consider: federal loans made by the government, federal loans made by banks or other lenders but guaranteed by the government, and private loans. Every student's first choice should be the federal loans due to stability: the interest rates are fixed for the life of the loan and are regulated by Congress, so you're unlikely to end up with a nasty surprise somewhere down the line. There are, however, limits on the amount that can be borrowed for one student through federal loan programs, and private loans might be the only alternative for an especially expensive education.

Other than the length of the loan, the big difference between student loans and other installment loans is that the payments are typically deferred until after graduation. This makes perfect sense, as in many cases students aren't working enough to make the money required to keep up with payments. So that's a pretty good deal. But it comes with a catch: in some cases, interest will accumulate on your loan while you're in school even though you're not making any payments. This can increase the principal balance considerably. The loan term can vary depending on the amount borrowed, but the standard term is 10 years, with payments deferred until 6 months after graduation.

Interest on student loans is simple interest: the interest accrues only on the principal balance, not on previously accrued interest. So while the student is still in school, the amount of interest that accrues each month remains constant. The interest is calculated using the simplified daily interest formula, which we can develop using things we know about simple interest.

1. If the principal balance of a loan is \$6,500 and the interest rate is 6.8% annually, how much simple interest will accrue in 1 year?

$I = Prt \Rightarrow I = \$6{,}500 \cdot 0.068 \cdot 1 = \$442$

2. Why does it make sense to use 365.25 days in a year for financial calculations?

Three out of every four years have 365 days, and every fourth (leap year) has 366. So, on average, the length of a year is 365.25 days. There's an expected value problem in there somewhere . . .

3. Using Questions 1 and 2, find the daily amount of interest that accrues on that \$6,500 loan. Make sure you explain how you got your answer.

$$\frac{\$442}{365.25} = \$1.21$$

We found the amount of interest per year, so we just have to divide by the number of days in a year.

4. How much interest accrues in a month? Be careful . . .

It depends on the month. For months with 31 days, it's $31 \cdot \$1.21 = \$37.51$. For months with 30 days, $30 \cdot \$1.21 = \$36.30$. For months with 29 days, \$35.09, and for months with 28 days, \$33.88.

The Simplified Daily Interest Formula

$$\text{Daily interest amount} = \frac{\text{Principal balance} \times \text{interest rate}}{365.25}$$

$$\text{Monthly interest amount} = \text{Daily interest amount} \times \text{days in month}$$

5. For that $6,500 student loan we've been working with: If you start school in August, graduate in May (3 years and 9 months later) and payments begin 6 months after graduation, how much time will pass while interest is accruing?

___4___ Years and ___3___ Months

6. Write out the months that come after the last full year ends, then the number of days in each of those months.

September: 30 days. October: 31 days. November: 30 days.

7. Find the amount of interest that accrues in the number of full years from your answer to Question 5.

$442 per year times 4 years is $1,768.

8. Write the amount of interest for each of the months listed in Question 6. Your answers to Question 4 will come in very handy.

$36.30 + $37.51 + $36.30 = $110.11

9. How much total interest accrues before payments begin?

$1,768 + $110.11 = $1,878.11

Did You Get It

Try this problem to see if you understand the concepts we just studied. The answer can be found at the bottom of the portfolio page.

1. Max borrows $11,500 for her last 2 years of college, acquiring a federal student loan at 6.2% interest.
   a. Find the monthly interest for 30- and 31-day months.
   b. Find the amount of interest that accrues if she acquires the loan in August, graduates 2 years later, and payments begin 3 months later.

When interest is covered by the government while the borrower is still in school, a student loan is called **subsidized**. If interest is added to the principal while the borrower is in school, the loan is called **unsubsidized**. For the loan in Questions 1–9, the student would have saved $1,878.11 by acquiring a subsidized loan.

When monthly payments begin on a student loan, they're based on the principal balance at that point in time. These payments can be computed using the following formula, which should look familiar: it's exactly the payment formula we studied in Lesson 3-3. We just thought we'd be nice and recopy it rather than making you page back. You should probably send us cash. Also, since most student loans have a 10-year term, we took the liberty of adapting the formula to that situation. Again, you should probably send us cash. Gift cards are appreciated, too.

Monthly Payments on a Student Loan

For a 10-year student loan with principal P and interest rate r (written as a decimal), the monthly payment M is given by

$$M = \frac{P \cdot \frac{r}{12}}{1 - \left(1 + \frac{r}{12}\right)^{-120}}$$

Reminder: Perform the calculation in one step if you can. If not, don't round too much. Always use at least four decimal places, and more than that won't kill you. I promise.

10. For the loan in Questions 1–9, what is the principal amount at the time payments begin? (This includes the original amount borrowed, plus the amount of interest that accrued until payments started.)

$\$6{,}500 + \$1{,}878.11 = \$8{,}378.11$

11. Find the monthly payment on the loan.

$$M = \frac{\$8{,}378.11 \cdot \frac{0.068}{12}}{1 - \left(1 + \frac{0.068}{12}\right)^{-120}} \approx \$96.42$$

We've seen that with a student loan, you have the option of not making any payments until after graduation. When interest on an unsubsidized student loan is not paid during college, we say that the interest is **capitalized**. When interest is capitalized, it adds to the principal balance, which of course results in larger payments. When interest is not capitalized, payments have to be made during college, but only covering the amount of interest. This keeps the principal balance the same until full monthly payments start, which makes those payments smaller.

For the student loan we've been working with, the payment while still in school would simply be the amount of interest that's accruing, which you already calculated. For reference, in the space below write the amount you'd pay in an average 30-day month (see Question 4), as well as the total amount of interest you'd pay before regular monthly payments begin (Question 9).

30-day month: $36.30

Total interest: $1,878.11

12. With interest not capitalized, the principal balance of the loan will still be $6,500 when full payments start. Find what those payments will be.

$$M = \frac{\$6{,}500 \cdot \frac{0.068}{12}}{1 - \left(1 + \frac{0.068}{12}\right)^{-120}} \approx \$74.80$$

13. You'd make 120 payments on the loan (since you have ten years to pay it back). Use your answer from Question 12 to find the total amount of money that you'd pay on the original $6,500 loan. (Hint: Don't forget the amount you would have already paid while in college!)

$120 \cdot \$74.80 + \$1{,}878.11 = \$10{,}854.11$

14. Now, the key question when studying whether or not to capitalize the interest: if interest is capitalized, after paying nothing while in school, you'll end up making 120 payments of the amount you found in Question 11. So how much money would you save by paying for the interest while still in school? More importantly, *discuss with your group whether or not you think the amount saved would make it worthwhile to scrape together those payments while in school, then write your thoughts.*

$120 \cdot \$96.42 = \$11{,}570.40$

$\$11{,}570.40 - \$10{,}854.11 = \$716.29$

Encourage students to really think about their response to this question. It's a very important part of understanding what they're getting into when acquiring student loans.

Answers vary on the value of each plan. For what it's worth, do I think it's worth making payments while in college to save an average of seventy bucks a year over ten years? Nope.

Did You Get It

Try this problem to see if you understand the concepts we just studied. The answer can be found at the bottom of the portfolio page.

2. For the loan in Did You Get It 1:
   a. Find the monthly payment after graduation if interest is not capitalized. (Note that you found how much payments would be while in school in Did You Get It 1.)
   b. Find the full monthly payment after graduation if interest is capitalized.
   c. How much money would be saved by not capitalizing interest?

Home Buying

© Brand X Pictures RF

For many people, the day they buy their first home is one of the proudest days of their life—and one of the scariest. There is nothing that compares to the feeling of looking at a house and knowing that it's all yours. But the buying process is tremendously intimidating. There are dozens of documents to sign, and the sheer numbers involved are enough to make almost everyone wonder if they're making a colossal mistake.

The most common home loans are paid over a 30-year span. That's a major commitment, and one that nobody should enter into without an understanding of the mathematics that go into the process. In the remainder of this lesson, we will study that math, hopefully helping you to become a well-informed home buyer. Maybe you're thinking, "Dude, I'm a college student. Buying a house is like a million years away." If so, you'll be surprised at how quickly that million years passes. I surely was.

A **mortgage** is a long-term loan where the lender has the right to seize the property purchased if the payments are not made. Homes are the most common items bought using mortgages. The most common mortgage term is 30 years, but they are widely available in terms from 15 to as many as 50 years.

There are several types of mortgages. With a **fixed-rate mortgage**, the rate of interest remains the same for the entire term of the loan. The payments (usually monthly) stay the same. With an **adjustable-rate mortgage**, the rate of interest may fluctuate (i.e., increase and decrease) during the period of the loan. If you opt for an adjustable-rate mortgage, it will start out with a lower interest rate than a fixed-rate loan of the same length.

15. Discuss the pros and cons of fixed and adjustable-rate loans.

Answers vary of course. Some things we'd like to see discussed are the gamble of getting a lower interest rate in the short term vs. not knowing what it might go up to in the long term. This would be a GREAT opportunity to discuss how adjustable-rate mortgages were largely responsible for the number of people getting foreclosed on when the housing market collapsed in 2006–2007.

16. Discuss the pros and cons of buying a house compared to renting a place to live.

Answers vary widely. You might suggest that students think about things like maintenance, insurance, and property taxes, which they might not have thought about.

One of the most useful (but frightening) things about home buying is calculating how much money you'll actually pay over the life of a loan. Let's see how that might work for one family.

The Petteys family plans to buy a home for $224,900, and has been offered a 30-year mortgage with a rate of 5.5% if they make a 20% down payment.

17. Calculate the down payment and the amount they'll have to borrow.

$0.20 \cdot \$224{,}900 = \$44{,}980$
$\$224{,}900 - \$44{,}980 = \$179{,}920$

Math Note

The word "mortgage" comes from a combination of Old French words "mort" (dead) and "gage" (pledge). It is believed the intent was that the debtor pledged the property to secure the loan, and if he or she failed to pay, the property was taken, and was therefore "dead" to the debtor.

18. The monthly payment on the loan will be $1,021.57. How much will the Petteys' actually pay for their $224,900 home? (Don't forget to add in the down payment.)

$\$44{,}980 + 30 \cdot 12 \cdot \$1{,}021.57 = \$412{,}745.20$

Did You Get It

Try this problem to see if you understand the concepts we just studied. The answer can be found at the bottom of the portfolio page.

3. The Trissel family agreed on a price of $229,500 for a home. Their company credit union offers a 5.0% 20-year loan with 15% down, resulting in a monthly payment of $1,287.50. Find the down payment amount, and the total amount they'll pay for the house.

3-4 Class

When you apply for a home loan, the lender will tell you what your monthly payment will be. And if you're an incredibly trusting sort, that might be good enough for you. For most of us, though, a double-check of the quoted payment would be a big comfort. In addition, knowing how the payment is calculated makes it quite a bit easier to understand the factors that determine how much you'll pay. Better still, it will allow you to compare mortgages with different rates, down payments, and lengths. So let's have a look at computing monthly payments.

1. In the space below, recopy the formula that we used for computing monthly payments on a student loan with a term of 10 years. Then describe the significance of each part of the formula requested below:

$$M = \frac{P \cdot \frac{r}{12}}{1 - \left(1 + \frac{r}{12}\right)^{-120}}$$

P: The principal amount of the loan

r: The annual interest rate

12: The number of payments per year

2. Where do you think the 120 comes from in the formula? (Hint: Think of 120 as the product of two numbers.)

It's the number of payments per year multiplied by the term of the loan in years. In other words, it's the total number of payments.

Students may need some hints here. It's really important that they have a handle on this answer . . .

3. Using your answers to Questions 1 and 2, adapt the student loan monthly payment formula into a general formula for finding the periodic payment on a loan with principal value *P*, annual interest rate *r*, term *t* (in years), and *n* payments per year. *Check with your professor to make sure that you have the right formula before moving on to the remainder of the lesson!*

$$M = \frac{P \cdot \frac{r}{n}}{1 - \left(1 + \frac{r}{n}\right)^{-nt}}$$

. . . in order to be successful here. A correct answer here is crucial, so make sure that you check all groups to make sure they have the correct answer before moving on.

Earlier we studied a loan obtained by the Petteys family with these terms: purchase price $224,900, 20% down payment, 30-year mortgage with a rate of 5.5%. The monthly payment was $1,021.57.

4. How much higher would the monthly payment be if they'd put only 10% down? This would also raise the interest rate to 5.8%.

New principal = 0.9 · $224,900 = $202,410

$$M = \frac{\$202{,}410 \cdot \frac{0.058}{12}}{1 - \left(1 + \frac{0.058}{12}\right)^{-360}} = \$1{,}187.65$$

The payment would be $166.08 more each month.

5. How much more would they end up paying over the life of the loan?

New amount = 360 · $1,187.65 + $22,490 = $450,044

Difference: $450,044 − $412,745.20 = $37,298.80

6. Now let's examine the term of the loan. If they choose a 15-year mortgage with 20% down at 5.5%, what would the monthly payment be?

$$M = \frac{\$179{,}920 \cdot \frac{0.055}{12}}{1 - \left(1 + \frac{0.055}{12}\right)^{-180}} = \$1{,}470.10$$

Math Note

A Google search for the phrase "mortgage calculator" results in over 60 million hits! There are thousands of pages available that can be used to quickly calculate monthly payments on a mortgage. But doing the calculations by hand will help you to become very familiar with the terms involved in mortgages.

7. How would their total cost compare to the other two options we studied? Which mortgage do you think is the best choice? Discuss.

New amount = 180 · $1,470.10 + $44,980 = $309,598

Difference between this and the 30-year loan with 20% down:

$412,745.20 − $309,598 = $103,147.20

Difference between this and the 30-year loan with 10% down:

$450,044 − $309,598 = $140,446

Discussions vary.

Did You Get It

Try this problem to see if you understand the concepts we just studied. The answer can be found at the bottom of the portfolio page.

4. Find the monthly payment for the Trissels (Did You Get It 3) if they choose a 15-year loan instead of 20. How much money would they save overall?

Computing an Amortization Schedule

After securing a mortgage, the lending institution will prepare an **amortization schedule**. This schedule shows what part of the monthly payment is paid on the principal and what part of the monthly payment is paid in interest. Here's the procedure for computing an amortization schedule:

Computing an Amortization Schedule

Step 1: Find the interest for the first month. Use $I = Prt$, where $t = 1/12$. Enter this value in a column labeled Interest.

Step 2: Subtract the interest from the monthly payment to get the amount paid on the principal. Enter this amount in a column labeled Payment on Principal.

Step 3: Subtract the amount of the payment on principal found in step 2 from the principal to get the balance of the loan. Enter this in a column labeled Balance of Loan.

Step 4: Repeat the steps using the amount of the balance found in step 3 for the new principal.

In the next group of questions, we'll use these steps to compute the first three months of an amortization table for the Petteys loan: principal $179,920, rate 5.5%, monthly payment $1,021.57. A template for the table is included in Question 12: fill in the information as you find it.

8. Calculate the interest for the first month on the loan, and put it into the interest column of the table.

$I = \$179{,}920 \cdot 0.055 \div 12 = \$824.63$

9. Subtract the interest from the monthly payment to get the amount paid on the principal, then put that amount in the appropriate column in the table.

$\$1{,}021.57 - \$824.63 = \$196.94$

10. Calculate the new balance of the loan by subtracting the payment on principal from the original principal, then put that in the last column.

$179,920 − $196.94 = $179,723.06

11. Repeat the previous steps for the next two months. Do your calculation in this space, then put the results in the table.

I = $179,723.06 · 0.055 ÷ 12 = $823.73
$1,021.57 − $823.73 = $197.84
$179,723.06 − $197.84 = $179,525.22
I = $179,525.22 · 0.055 ÷ 12 = $822.82
$1,021.57 − $822.82 = $198.75
$179,525.22 − $198.75 = $179,326.47

Payment Number	Interest	Payment on Principal	Balance of Loan
1	$824.63	$196.94	$179,723.06
2	$823.73	$197.84	$179,525.22
3	$822.82	$198.75	$179,326.47

12. What does this table say about what the bulk of your payment goes to when you begin making payments on a mortgage? What about near the end of the term of your loan? Explain.

With your earliest payments, most of what you pay is going toward interest, and doesn't do much to lower your balance. But by the time you get near the end of the term, the interest will be very low, and most of what you pay will go toward paying down the balance.

Did You Get It

Try this problem to see if you understand the concepts we just studied. The answer can be found at the bottom of the portfolio page.

5. The Trissel loan from Did You Get It 3 has the following parameters: $195,075 principal, 5.0% interest rate, and monthly payment $1,287.50. Complete the first two months of an amortization schedule for this loan.

Payment Number	Interest	Payment on Principal	Balance of Loan
1			
2			

After learning about amortization schedules, most people think that mortgages sound like an awful deal: your early payments go almost completely to interest, so you're paying a LOT of money, and making almost no progress on paying down your loan. But here's the benefit to you: the interest paid on a home loan can be deducted from your income when computing the amount of income tax that you owe. So at least the way mortgages work gives you a bigger break on your income taxes during the early stages of the loan.

Financial advisors have coined two terms for borrowers: "good debt" and "bad debt." Good debt is when you borrow money to buy something that is going to increase in value. Student loans and home loans are pretty much the gold standard in good debt. Houses increase in value, and the amount that you pay for a college education is more than paid back by increased earning potential.

Bad debt, on the other hand, is borrowing to buy something that loses value immediately after purchase. An obvious example is car loans, although for most people this type of bad debt is unavoidable, so we'll coin a new term: "not-so-good debt." Car loans are manageable, but it's important to avoid really long-term loans. Dealers are now offering 60, 72, and even 84 month loans. You should think carefully before taking out a 60-month loan, and you should avoid longer loans than that at all costs. If you need to finance a car for that long, you can't afford that car.

The best example of bad debt (or worst example, I guess) is credit card debt. Keeping balances on credit cards is the easiest way to put yourself into a financial hole. Pretty much everything you might buy on credit is instantly worth less than what you paid, and the high interest rates attached to most credit cards makes them a financial disaster waiting to happen. The key to managing your financial life is to be an educated borrower and educated consumer. Hopefully these lessons have given you the means to make smart financial decisions when it comes to borrowing.

3-4 Portfolio

Name ______________________________

Check each box when you've completed the task. Remember that your instructor will want you to turn in the portfolio pages you create.

Technology

1. ☐ As you probably noticed in the Class portion of this lesson, making an amortization table is a bit time-intensive. There sure are a lot of repetitive calculations. Sounds like another job for technology! Your job here is to use a spreadsheet to make an amortization table for the Petteys loan we studied in the lesson. You already have the first three months filled in (Class Questions 8–11), so you can use that to double-check your spreadsheet results. A template to help you get started can be found in the online resources for this lesson.

2. ☐ In my humble opinion, the most valuable thing you can get out of this lesson is the ability to compare different loans. The second template in the class resources is built so that you can input the principal, percent down payment, interest rate, and term of a loan, then calculate the monthly payment, total amount of interest, and total cost over the life of the loan. Use this spreadsheet to explore different options for a home that you think you might pay for some day, and report on what you found.

Online Practice

1. ☐ Include any written work from the online assignment along with any notes or questions about this lesson's content.

Applications

1. ☐ Complete the applications problems.

Reflections

Type a short answer to each question.

1. ☐ Describe some of the options you have in student loans.
2. ☐ What are the advantages and disadvantages of a mortgage with a shorter term?
3. ☐ Take another look at your answer to Question 0 at the beginning of this lesson. Would you change your answer now that you've completed the lesson? How would you summarize the topic of this lesson now?
4. ☐ What questions do you have about this lesson?

Looking Ahead

1. ☐ Complete the Lesson 3-5 Prep Skills and read the opening paragraph in Lesson 3-5 carefully, then answer Question 0 in preparation for that lesson.

Answers to "Did You Get It?"

1. a. 30 day: \$58.50; 31 day: \$60.45 **b.** \$1,603.45 **2. a.** \$128.83 **b.** \$146.79 **c.** \$551.75 **3.** Down payment: \$34,425, total amount paid: \$343,425 **4.** \$1,542.64; \$31,324.80 **5.** Payment 1: Interest = \$812.81, Payment on Principal = \$474.69, Balance = \$194,600.31; Payment 2: Interest = \$810.83, Payment on Principal = \$476.67, Balance = \$194,123.64

3-4 Applications

Name ____

Many mortgage lenders offer a way to lower your interest rate by paying some of the interest up front. This prepaid interest is known in the industry as **points**. Each point corresponds to 1% of the amount borrowed. In short, you're paying a fee to lower your interest rate. The question is whether this is a good deal for you, or for the lender. Let's look into it.

1. A lender offers you a 30-year fixed mortgage of $130,000 at 4% interest with no points. Find the monthly payment.

$$M = \frac{\$130{,}000 \cdot \frac{0.04}{12}}{1 - \left(1 + \frac{0.04}{12}\right)^{-360}} = \$620.64$$

2. You can get the same 30-year mortgage of $130,000 at 3.6% interest by paying 2 points at closing. How much would you have to come up with to cover the 2 points?

2% of $130,000 is $2,600.

3. Find the new monthly payment.

$$M = \frac{\$130{,}000 \cdot \frac{0.036}{12}}{1 - \left(1 + \frac{0.036}{12}\right)^{-360}} = \$591.04$$

4. How much are you saving each month with the new monthly payment?

$620.64 − $591.04 = $29.60

5. How long will it take to recover the amount you had to come up with to pay the 2 points? What does this mean in terms of how long you should plan to stay in that home?

$2,600/$29.60 per month = 87.8 months, or a little over seven years. If you don't stay in the home for more than seven years, then choosing the mortgage with points would have cost you money.

3-4 Applications

Name ______________________________

6. If you were to stay in that home long enough to pay off the loan, how much would the mortgage with points save you?

$620.64 per month times 360 months is $223,430.40.

$591.04 per month times 360 months is $212,774.40; adding the points, the total cost is $215,374.40.

You'd save $8,056.

7. Suppose that instead of paying the points, you chose to invest that $2,600 into a long-term fund that paid 2.7% interest compounded monthly. How much would it be worth after 30 years?

$$A = P\left(1 + \frac{r}{n}\right)^{nt} = \$2{,}600\left(1 + \frac{.027}{12}\right)^{360} = \$5{,}839.25$$

8. Discuss which option seems the best to you. Justify your conclusions.

Answers vary.

9. If you acquire a $15,000 student loan at 4.3% interest and have to begin making payments 4 years later, how much would you save in interest if the loan is subsidized?

$I = Prt = \$15{,}000(0.043)(4) = \$2{,}580.$

10. How would that subsidy affect your monthly payment?

Subsidized: $$M = \frac{\$15{,}000 \cdot \frac{0.043}{12}}{1 - \left(1 + \frac{0.043}{12}\right)^{-120}} \approx \$154.02$$

Unsubsidized: $$M = \frac{\$17{,}5800 \cdot \frac{0.043}{12}}{1 - \left(1 + \frac{0.043}{12}\right)^{-120}} \approx \$180.51$$

The subsidy would shave $26.49 from the payment.

Lesson 3-5 Prep Skills

This is a short review of skills that will come in handy in the next lesson. In each case, answer the question, then rate your confidence level by checking one of the boxes. If you feel like you're struggling with these skills, consult the online resources provided by your instructor for extra practice.

I KNOW IT | THINK SO | UNSURE | NO IDEA

SKILL 1: READING INFORMATION FROM A TABLE

1. Below is the nutrition information label for a certain brand of peanut butter.

Serving Size 32g

Amount Per Serving	
Calories 188	Calories from Fat 135
	% Daily Value*
Total Fat 16g	25%
Saturated Fat 3g	17%
Trans Fat 0g	
Cholesterol 0mg	0%
Sodium 147mg	6%
Total Carbohydrate 6g	2%
Dietary Fiber 2g	8%
Sugars 3g	
Protein 8g	
Vitamin A 0%	• Vitamin C 0%
Calcium 1%	• Iron 3%

a. How many grams of fat are there in one serving?
b. How much sodium is there in three servings?
c. What percent of the daily recommendation of fiber is contained in one serving?

SKILL 2: DO COMPUTATIONS INVOLVING PERCENTAGES

2. Based on the information in the table, what is the recommended daily value for total fat?

3. If you have 20g of this peanut butter, what percent of a full serving is that?

4. How many grams of protein would you get from that 20g of peanut butter?

SKILL 3: SOLVE A FORMULA FOR ONE VARIABLE

5. Solve the equation $\frac{k}{M} = P$ for M.

Lesson 3-5 A Walk on Wall Street (Stocks and Bonds)

© Comstock Images RF

LEARNING OBJECTIVES

- ☐ 1. Read information from a stock listing.
- ☐ 2. Calculate costs of buying stock, and profit or loss from selling.
- ☐ 3. Study the price to earnings ratio, and use it to analyze the value of a stock.
- ☐ 4. Calculate profit from a bond sale.

The real key to making money in stocks is not to get scared out of them.
—Peter Lynch

We have seen that the magic of compound interest allows your money to grow considerably over long periods of time. However, interest rates on basic savings accounts are usually quite low, so if you rely on savings alone to build a nest egg, you better hope that you live a very, very long life indeed. Most successful investors grow their money much more quickly using the stock market. In this lesson, we'll learn about the basics of stocks and bonds, and how to get information about the performance of stocks that you might be interested in. When a company files legal papers to become a corporation, it's able to issue **stock.** If you buy shares of stock, you literally become a part owner of the company; for example, if a company issues 1,000 shares of stock and you buy 250 shares, then you own one-quarter of the company. The market can be a very intimidating place, but education holds the key to overcoming that fear and getting a piece of the action.

0. After reading the opening paragraph, what do you think the main topic of this lesson will be?

3-5 Class

Basic Definitions

As a shareholder in a company, you own part of that company. When the company makes money, it can choose to distribute part of the profit to its shareholders. This money is called a **dividend.** The stockholder receives a sum of money based on the number of shares of the stock that he or she owns. Sometimes if a company doesn't make a profit or its owners or managers decide to reinvest the money into the company, no dividends are paid. So when choosing whether to buy stock in a certain company, one consideration is whether or not that company typically pays a dividend.

1. If you own stock in a certain company, why are you entitled to a periodic dividend?

Because you literally own part of the company. So if the company makes money, it's perfectly reasonable to expect that you would get a share of that profit.

2. Describe any advantages or disadvantages of buying stock in a company that pays dividends.

The advantage is that you get periodic payments, and money is good. The disadvantage is that it might help the value of the company long term if they invest that money back into growing the company rather than paying the stock holders.

Besides issuing stock, a company can also issue **bonds.** Usually bonds are issued to raise money for the company for startup costs or special projects. A person who purchases a bond is really lending money to the company. The company, in turn, repays the owner of the bond its original value plus interest. As a general rule, bonds are a safer investment than stocks, but stocks have greater growth potential.

3. Describe the difference between stocks and bonds.

If you own stock in a company, you own a part of the company. If you buy bonds issued by a company, in essence you have lent money to that company. You're entitled to getting that money back with interest, but you're not entitled to any profit made by the company.

Stocks can be bought and sold on a **stock exchange.** The price of a stock, which varies from day to day (even from minute to minute) depends completely on the amount that investors are willing to pay for it. This can be affected by the profitability of the company, the economy, scandals, even global political concerns. Investors buy or sell stock through a **stockbroker.** Traditionally, this was an individual working for a brokerage firm, but it has become common for people to use online brokers, in which the investor initiates the buying and selling of stocks online. In either case, the brokerage charges a fee, called a **commission,** for the service of having their representatives buy or sell the stock at an exchange. Bonds can also be bought and sold like stock.

4. How can an individual buy stock in a company?

By purchasing shares of stock through a stockbroker, either through a brokerage firm, or an online trading site.

5. What are some advantages of investing in the stock market?

There are two different ways that you can make money. If the company pays a dividend, you'll share in the company's profit. If the stock price increases, then you can sell your shares and make a profit.

Investing in Stocks

When all is said and done, there are two ways to earn money from stocks: buying a stock that pays a regular dividend and selling shares for more than you paid for them. This means that investors are tremendously interested in the day-to-day price fluctuations of stocks, as well as indicators of how well the company is doing. A **stock table** is a listing of key information; stock tables can be found in newspapers and online financial sites.

The stock table shown below, generated at 10:37 A.M. on September 4, 2015, is for a company called Computer Programming and Systems, Inc. Let's see if we can make sense of this mysterious mass of numbers.

Computer Programs & Systems Inc. (CPSI) - NasdaqGS ★ **Watchlist**

45.49 ↑**0.05(0.11%)** 10:37AM EDT - Nasdaq Real Time Price

Prev Close:	**45.44**	Day's Range:	**45.02 - 45.60**
Open:	**45.02**	52wk Range:	**43.24 - 64.86**
Bid:	**45.36 x 100**	Volume:	**5,119**
Ask:	**45.52 x 100**	Avg Vol (3m):	**96,911**
1y Target Est:	**46.20**	Market Cap:	**514.16M**
Beta:	**0.780949**	P/E (ttm):	**18.61**
Earnings Date:	**Oct 28 - Nov 2 (Est.)**	EPS (ttm):	**2.44**
		Div & Yield:	**2.56 (5.64%)**

Think of this material as an opportunity to run a more traditional-style class for a few minutes.

There's a whole lot of information here, and it's going to take some explanation. We suggest that you take notes as you read through the descriptions below, but you might try a quick read-through now, then refer back when you get to Questions 6–11.

At the top, we see the name of the company and the **ticker symbol** (CPSI in this case). This is a shorthand symbol assigned by the exchange that represents the stock. The "NasdaqGS" indicates that this particular stock is traded on the Nasdaq, one of the American stock exchanges. The big bold number at the top is the current price of the stock. The current price isn't like the price of milk at the grocery store: It reflects the most recent price at which the stock has been bought from someone trying to sell. This is followed by how much the price has changed during the current day's trading session.

If you can add and subtract, this makes the **Prev Close** price redundant: Note that this stock has gone up by $0.05 in this session and is currently worth $45.49, so of course the closing price on the previous day was $45.44. **Open** refers to the price at the time today's session began. This is often different from the closing price because many factors can affect what people are willing to pay from day to day. **Bid** and **Ask** are a snapshot of what people have been bidding to buy the stock, and what price current shareholders have been asking for to sell. In this case, $45.36 × 100 indicates that an offer of $45.36 for up to 100 shares has been made. $45.52 × 100 means that someone has offered up to 100 shares for sale at $45.52.

1y Target Est is what analysts are predicting the stock price will be in one year. While this is at best an educated guess, it can at least give you some guidance as to whether the stock might be a good buy. **Beta** refers to the **volatility** of the stock, which is a measure of how much the price fluctuates. If the beta is close to 1, then the stock's stability is similar to the rest of the market. Lower numbers indicate that the price is more stable than average. The **Earnings Date** reflects the next time the company is expected to release data on how well the company is doing financially.

The ranges **(Day's Range** and **52wk Range)** indicate the highest and lowest price during the current session, as well as the last full year. The latter is particularly useful, as it gives you a quick snapshot of how the stock has fluctuated, and how its current price compares. **Volume** describes how many shares have been traded in the current session, while **Avg Vol (3m)** provides a three-month average of shares traded. Lower-volume stocks can be more difficult to buy and sell due to availability.

Market Cap is the total value of the company: it's calculated by multiplying the number of shares that have been issued by the current price, so a little math allows you to calculate how many shares there are. This in turn makes the volume of sales more meaningful. **P/E (ttm)** refers to the P/E (price to earnings) ratio. This is a particularly important number, so we'll study that in detail a bit later. **EPS (ttm)** represents earnings per share, or the amount of money the company made for each share of stock that has been issued. Clearly, bigger is better here. **Div & Yield** is the dividend that the company pays for each share, and what percent of the current price that represents. If nothing else, this tells you whether or not the company pays a dividend on shares, but you can also use it for comparison purposes with other stocks.

Now let's put our new knowledge into practice by studying shares of Facebook, also on 9/4/15.

Facebook, Inc. (FB) - NasdaqGS ★ **Watchlist**

87.27 ↓**0.88(1.00%)** 1:29PM EDT - Nasdaq Real Time Price

Prev Close:	**88.15**	Day's Range:	**86.70 - 87.79**
Open:	**87.15**	52wk Range:	**70.32 - 99.24**
Bid:	**87.06 x 400**	Volume:	**16,715,420**
Ask:	**87.07 x 300**	Avg Vol (3m):	**31,684,600**
1y Target Est:	**111.75**	Market Cap:	**245.89B**
Beta:	**0.89812**	P/E (ttm):	**88.69**
Earnings Date:	**Oct 26 - Oct 30 (Est.)**	EPS (ttm):	**0.98**
		Div & Yield:	**N/A (N/A)**

6. What is the current price of the stock? What are shareholders trying to sell it currently asking for?

The current price is $87.27 per share. Shareholders are asking for $87.07.

7. What do analysts predict the price will be in one year?

The prediction is $111.75.

8. Is this stock more or less volatile than the market as a whole? How do you know?

The beta is less than 1, so the stock has been more stable than the market in general.

9. What's the highest stock price in the last 52 weeks?

$99.24

10. Does Facebook pay a dividend to shareholders? If so, how much?

The N/A in the dividend row shows that they do not pay a dividend.

11. Use the market cap to find how many shares the company issued.

245.89 billion/87.27 ≈ 2.82 billion
About 2.82 billion shares have been issued.

As previously mentioned, there are two ways to make money from stocks: buy shares of a stock that pays dividends, or buy stock at a low price and sell it at a higher price. But of course, you can't just go buy stock at the corner store—you need to use a brokerage firm, placing an order, which is then carried out by representatives at the stock exchange. In exchange for that service, the broker charges a commission, which varies among brokers. Brokers can also make recommendations concerning what stocks to buy and sell, which further justifies their commissions.

The amount that an investor receives from the sale of a stock is called the **proceeds.** The proceeds are equal to the amount of the sale minus the broker's commission. Next, we'll look at some examples of buying and selling stock.

Shares of Apple Computer (AAPL) closed at \$12.89 on April 1, 2004. Suppose that an investor bought 600 shares at that price using a broker that charged a 2% commission.

12. Find the amount of commission that was paid.

2% of \$12.89 · 600 = \$154.68

13. Find the total cost of buying the stock.

\$154.68 + 600 · \$12.89 = \$7,888.68

A **stock split** occurs when a company increases the number of outstanding shares. Most commonly, this is a **2-for-1 split,** in which the number of shares doubles. This can be done for a variety of reasons, but the simplest explanation is that when the price of one share gets particularly high, it tends to scare away smaller investors, and the split effectively halves the price of one share. In any case, the net result is that in an instant, the number of shares owned by shareholders doubles.

14. Apple did a 2-for-1 split on February 8, 2005, then a 7-for-1 split on June 9, 2014. How many shares would our investor have owned after the second split?

600 · 14 = 8,400

15. Our clever friend the Apple investor could have sold her shares for \$130.28 on June 1, 2015. If she had done so, using a brokerage with a 1.5% commission, what would her profit have been? Remember profit is the difference between cost and final proceeds from the sale, which would be selling price minus commission. Needless to say, you should show your work.

$8{,}400 \cdot \$130.28 = \$1{,}094{,}352.00$

$0.015 \cdot \$1{,}094{,}352 = \$16{,}415.28$

$\$1{,}094{,}352 - \$16{,}415.28 = \$1{,}077{,}936.72$ (Final proceeds)

$\$1{,}077{,}936.72 - \$7{,}888.68 = \$1{,}070{,}048.04$ (Profit)

16. How much money would our investor have saved by using an online trading site with a flat commission of \$19.95 per trade (for both the original purchase and the sale)?

Commission on purchase: \$154.68

Commission on sale: \$16,415.28

$\$154.68 + \$16{,}415.28 - 2 \cdot \$19.95 - = \$16{,}530.06$ saved

Good teaching moment here: Learning about investing to the point where you don't need a commissioned broker can save you a lot of money.

Did You Get It ?

Try this problem to see if you understand the concepts we just studied. The answer can be found at the bottom of the portfolio page.

1. a. When Facebook stock was first offered for public sale in May 2012, the share price was \$38. If you'd bought 500 shares at that price, using a brokerage with a 1% commission, what would your total cost have been?
   b. If you'd then sold them at the opening price on 9/4/15 (see Questions 6–11), using an online brokerage with a \$9.95 flat fee, would you have made or lost money? How much?

3-5 Group

The Price-to-Earnings Ratio For a Stock

The price-to-earnings ratio, commonly known as P/E ratio, is a key indicator of how reasonable the current price of a stock is. It compares the selling price of the stock to the annual amount of money the company has earned per share. If the P/E ratio for a certain stock is 20, that means that the current selling price is 20 times as high as the earnings per share. This tells us that for every dollar you invest in the company, that company will earn $0.05.

P/E ratio can be a little tricky to interpret. On the one hand, a lower P/E ratio indicates that a stock is priced reasonably, as you need to invest less for the company to make a certain amount of profit. But if investors believe in the growth potential of a company, they may be willing to pay a premium for the stock, which would result in a higher P/E ratio.

To help get some perspective on P/E ratio, it's useful to do some calculations. First, a formal definition:

The P/E Ratio

$$\text{P/E Ratio} = \frac{\text{Current market price}}{\text{Annual earnings per share}}$$

1. For the Facebook stock table from earlier in the lesson, calculate the P/E ratio using the current price and earnings per share. Compare to the P/E ratio listed in the table. Why do you think they're not identical?

$$\frac{\text{Current market price}}{\text{Annual earnings per share}} = \frac{\$87.27}{0.98} \approx 89.05$$

P/E ratio listed is 88.69. The discrepancy could be due to rounding in the earnings per share, as well as a recent change in market price not yet reflected in the P/E ratio.

Not every stock table you will encounter has quite the wealth of information that the ones at the beginning of this lesson have. A more abbreviated table might have the P/E ratio but not earnings per share listed, in which case you can rearrange the above formula to calculate earnings per share.

2. On September 8, 2015, Google stock closed at $600.70, and the P/E ratio was listed as 28.31. What was Google's earnings per share for the previous year? How does it compare to Facebook's? Which do you think was the better investment at that point?

Since P/E ratio is market price divided by earnings per share, earnings per share is market price divided by P/E ratio. $\$600.70/28.31 \approx \$21.22$. The earnings per share are WAY higher than Facebook. Which is the better investment is an opinion, but based on P/E ratio Google seems like a much better bet. But the high price obviously means that you could buy far fewer shares of Google.

Did You Get It

Try this problem to see if you understand the concepts we just studied. The answer can be found at the bottom of the portfolio page.

2. a. Use information in the stock table for CPSI earlier in the lesson to calculate the P/E ratio by hand, and compare to the P/E ratio in the table. If your result is different, discuss why you think that may be the case.
   b. If a competitor of CPSI has a P/E ratio of 21.4 and a market price of $38.74, calculate the earning per share.

The table below summarizes some basic rules of thumb when it comes to interpreting the P/E ratio for a stock. Remember that the best approach is to compare a stock's P/E ratio to other companies in the same industry, as current market conditions can have a pretty profound effect on P/E ratio.

P/E Ratio	Common Interpretations
N/A	Company has no earnings; it's also standard for a company with negative earnings (that is, a loss) to have an undefined P/E ratio, even though mathematically a negative P/E ratio does make sense.
0–10	The stock may be undervalued, making it a good buy; alternatively investors might feel like the earnings are likely to decline. The company may have shown short-term profit from selling assets that will hurt earnings in the long term.
10–17	P/E ratios in this range are historically considered to represent stocks being sold at fair market value.
17–25	The stock may be overvalued, so shop carefully. On the other hand, earnings may have increased since the last official earnings report was released. The company might be up and coming, with earnings expected to increase dramatically at some point.
>25	The stock might be very overpriced at the moment, or the company may be expected to have a big jump in earnings coming up. It could also be the case that the company's most recent earnings are unusually low for that company, with an expectation of recovery.

3. Come up with some ideas why the Facebook P/E ratio might have been sky-high in September of 2015. You might want to look up the history of Facebook stock on the Internet.

Answers vary. Prevailing wisdom is that this was not too long after Facebook was taken public, meaning that stock in the company was made available for the first time. Because Facebook is so ubiquitous, many investors assumed that it would be a very strong investment, leading to it being overvalued.

Investing in Bonds

When an investor buys bonds, the investor is actually loaning money to the company or government entity that issues the bonds. In exchange for lending that money, the investor will receive a fixed return on his or her investment for a given period of time. The ending value of the bond is called the **face value;** this is typically $1,000, but can vary. Bonds also have a **maturity date,** which is the date that the face value is returned to the investor, as well as interest. In some cases, no interest is paid until the maturity date, at which time the full amount of interest is paid. **Coupon bonds,** on the other hand, pay an annual or semi-annual interest payment, known as a **coupon.**

You can think of it this way: If someone offers to sell you a $20 bill for $15, that's a pretty darn good deal. But the catch is that you don't get the $20 until some predetermined time later down the road. That's what a bond is; it's actually pretty simple.

But here's where it can get more interesting. Suppose that you get tired of waiting for the time when you get that $20 bill, and you sell the bond to your neighbor for $18. You've made $3 on your investment and you're done. Your neighbor now waits until the maturity date, cashes in the bond, and made the other $2 of the original $5 in profit. This is how bonds are bought and sold on the open market. Of course, your neighbor wouldn't pay more than $20 for a bond with a maturity value of $20. But if you were desperate for cash, you might be compelled to unload the bond for less than the $15 you originally paid. You've taken a loss, but at least you got some money out at the time you needed it.

Investors buy bonds for two main reasons. First, they're much safer investments than stocks, even though they have much less opportunity for a big profit. The interest is guaranteed, as long as the company that issued the bond stays in business and has the money to pay its debts. This makes bonds a safe, but not guaranteed, investment. Second, bonds are used to offset losses on stocks when the economy is in a down cycle, because the value of bonds tends to go up when the value of stocks goes down.

Here's a short version of why that makes sense: In our earlier hypothetical example where you're making a $5 profit for waiting to cash in your bond, if the economy is booming in the meantime, prices in general will tend to go up, so the $5 you're getting at some point in the future has less buying power than the $5 you were expecting when you made the investment. This makes the bond in effect worth less. The opposite is also true: if the economy fizzles after you buy that bond, the $5 you eventually get will in effect be worth more than the $5 you thought you were getting, so the bond is worth more. Stocks tend to rise when the economy is doing well and fall when it's not, so the value of bonds usually goes in the opposite direction of the stock market.

Suppose that you buy a bond with face value $1,000 that was originally issued 18 months ago. The maturity date is 4 years from the time it was issued, and the interest rate is 4% simple interest per year. If you pay $820 for the bond and keep it until the maturity date, what is your profit? What percent return would you get per year? Let's do some calculating.

4. Find the amount of interest that will be paid on the bond at maturity, and the total amount of money you'll get from the issuer at that point. You'll want to reach back into the archives and use the simple interest formula.

$I = Prt \Rightarrow I = \$1{,}000 \cdot 0.04 \cdot 4 = \$160$ in interest

Total amount received: $1,160.

5. Find the profit you'd make on this transaction.

Total amount received − Amount paid = \$1,160 − \$820 = \$340

Math Note

There are fees and/or commissions for buying bonds as well as stocks. We're omitting fees from this discussion to focus on main issues involved in buying and selling bonds.

6. What percent of your investment is the profit made?

\$340/\$820 ≈ 41.5%

7. If you divide your total percent return from Question 6 by the amount of time in years that you hold the bond, you'll get your percent return per year, or **annual** return, which is a key indicator of how well an investment performs. A long-standing rule of thumb on long-term investments in the stock market is that you can expect annual returns in the 6–7% range. How would this bond investment compare?

41.5%/2.5 years = 16.6% per year. By comparison, this is a very good investment.

Did You Get It

Try this problem to see if you understand the concepts we just studied. The answer can be found at the bottom of the portfolio page.

3. Rashard bought four bonds with face values of \$1,000 each, a simple interest rate of 5.8% per year, and a maturity date 10 years after they were issued. He paid \$4,725 three years after the bonds were issued. If he keeps the bonds until maturity, find his total profit, and his percent return per year.

3-5 Portfolio

Name ______________________________

Check each box when you've completed the task. Remember that your instructor will want you to turn in the portfolio pages you create.

Technology

1. ☐ This assignment will last longer than most. The plan is for you to pick three stocks to buy with a hypothetical investment of $5,000. Find a financial website that lists current stock prices, find three stocks you're interested in, and calculate how many shares you can buy with your $5,000. Keep in mind that you have to buy three different stocks, and assume a fixed commission of $9.95 for each purchase. Then track the closing price of that stock each day for the time period chosen by your professor, and input the closing price into a spreadsheet. Finally use formulas to calculate the amount of gain or loss on your investment. A template to help you set up the calculations can be found in the online resources for this lesson.

Online Practice

1. ☐ Include any written work from the online assignment along with any notes or questions about this lesson's content.

Applications

1. ☐ Complete the applications problems.

Reflections

Type a short answer to each question.

1. ☐ Describe the difference between stocks and bonds.
2. ☐ What are some things you should consider when deciding whether or not to buy stock in a particular company?
3. ☐ What is the advantage of having stock in a company that pays dividends?
4. ☐ Take another look at your answer to Question 0 at the beginning of this lesson. Would you change your answer now that you've completed the lesson? How would you summarize the topic of this lesson now?
5. ☐ What questions do you have about this lesson?

Looking Ahead

1. ☐ Complete the Lesson 3-6 Prep Skills and read the opening paragraph in Lesson 3–6 carefully, then answer Question 0 in preparation for that lesson.

Answers to "Did You Get It?"

1. a. $19,190 **b.** You would have made $24,375.05

2. a. 18.64; answers vary on reasoning **b.** $1.81 **3.** $1,595; 4.8% per year

3-5 Applications

Name ______________________________

Use the stock listing for Wal-Mart to answer Questions 1–10.

Wal-Mart Stores Inc. (WMT) - NYSE

60.75 ↓0.08(0.13%) 4:01PM EST

After Hours : **60.75** 0.00 (0.00%) 4:12PM EST

Prev Close:	60.83	Day's Range:	60.50 - 60.97
Open:	60.58	52wk Range:	56.30 - 90.97
Bid:	60.73 x 4300	Volume:	5,296,196
Ask:	60.74 x 2900	Avg Vol (3m):	12,578,300
1y Target Est:	63.63	Market Cap:	194.51B
Beta:	0.444396	P/E (ttm):	13.01
Next Earnings Date:	N/A	EPS (ttm):	4.67
		Div & Yield:	1.96 (3.24%)

1. What is the ticker symbol you'd use to look up stock quotes for Wal-Mart?

   WMT

2. What was the highest and lowest price that the stock sold for during the last 52 weeks?

   $90.97 and $56.30

3. What was the amount of the dividend per share that the company paid last year?

   $1.96

4. If you own 682 shares, how much in dividends did you make last year?

   682 times $1.96 is $1,336.72.

5. How many shares were traded yesterday?

   5,296,196

6. What was the closing price of the stock yesterday?

   $60.83

3-5 Applications

Name ____________________

7. Find the annual earnings per share.

$4.67

8. If you purchase 842 shares of Wal-Mart stock at $60.67 per share and the broker's commission is 2%, find the total cost of the purchase.

842 · $60.67 = $51,084.14
1.02 · $51,084.14 = $52,105.82

9. Using information in the table, do a calculation to confirm that the P/E ratio is correct as listed.

Current price is $60.75, and earnings per share are $4.67. $60.75/$4.67 = 13.01

10. Based on the information in the table below, do you think Home Depot stock was a better buy than Wal-Mart at that point? Justify your response.

This is an opinion question, but Home Depot had higher earnings per share and a much higher P/E ratio at that point.

The Home Depot, Inc. (HD) - NYSE

132.68 ↓0.22(0.17%) 4:00PM EST

After Hours : **132.43** ↓0.25 (0.19%) 5:19PM EST

Prev Close:	132.90	Day's Range:	131.89 - 132.90
Open:	132.25	52wk Range:	92.17 - 135.47
Bid:	132.40 x 300	Volume:	2,607,226
Ask:	132.88 x 100	Avg Vol (3m):	5,027,970
1y Target Est:	141.00	Market Cap:	168.22B
Beta:	0.965301	P/E (ttm):	24.89
Next Earnings Date:	N/A	EPS (ttm):	5.33
		Div & Yield:	2.36 (1.79%)

3-5 Applications

Name ______________________________

11. Compare the two investments below and decide which would have been the better choice, and why.

Investment 1: \$10,000 was invested in a 24-month CD that earned 5.1% annual interest compounded monthly.

Investment 2: 1,400 shares of stock in the Lybarger Aviation Company were bought at \$7.11 per share using a brokerage with a 0.75% commission rate on both buying and selling stock. Over the 2 years the stock was held, it paid a dividend of \$0.48 per share in the first year and \$0.36 per share in the second year. The stock was sold through the same brokerage for \$7.95 per share.

Investment 1: $A = \$10{,}000\left(1 + \frac{.051}{12}\right)^{24} = \$11{,}071.44$; Profit $= \$1{,}071.44$

Investment 2: Cost $= 1{,}400 \cdot \$7.11 = \$9{,}954$; commission $= 0.0075 \cdot \$9{,}954 = \$74.66$

Total investment: $\$9{,}954 + \$74.66 = \$10{,}028.66$

Dividends: $1{,}400 \cdot \$0.48 + 1{,}400 \cdot \$0.36 = \$1{,}176$

Proceeds from sale: $1{,}400 \cdot \$7.95 = \$11{,}130$; commission $= 0.0075 \cdot \$11{,}130 = \$83.48$

Profit $= \$11{,}130 + \$1{,}176 - \$83.48 - \$10{,}028.66 = \$2{,}193.86$

Investment 2 made more than twice as much profit.

3-5 Applications

Name ______________________________

12. A bond with face value \$10,000, simple interest 6.45%, and term 12 years is originally bought by Larry. After 33 months, he sells it to Curly for \$11,400. Curly then holds on to it for 7 years, eventually selling it to Moe for \$14,950. Moe keeps the bond until it matures and cashes it in.

a. Which investor made the greatest profit?

b. Which got the greatest percent return on his investment?

c. Which got the greatest percent return per year?

Larry's profit: \$1,400

Curly's profit: \$3,550

Final value of bond: $\$10{,}000(1 + 0.0645 \cdot 12) = \$17{,}740$

Moe's profit: $\$17{,}740 - \$14{,}950 = \$2{,}790$

Greatest profit was made by Curly.

Larry's percent profit: $\dfrac{\$1{,}400}{\$10{,}000} = 0.14$, *or 14%, which is* $\dfrac{14\%}{33 \div 12 \text{ yrs}} = 5.09\%$ *per year.*

Curly's percent profit: $\dfrac{\$3{,}550}{\$11{,}400} \approx 0.311$, *or 31.1%, which is* $\dfrac{31.1\%}{7 \text{ yrs}} = 4.44\%$ *per year.*

Moe's percent profit: $\dfrac{\$2{,}790}{\$14{,}950} \approx 0.187$, *or 18.7%, which is* $\dfrac{18.7\%}{27 \div 12 \text{ yrs}} = 8.31\%$ *per year.*

Greatest percent return was Curly, but greatest percent per year return was Moe.

Lesson 3-6 Prep Skills

This is a short review of skills that will come in handy in the next lesson. In each case, answer the question, then rate your confidence level by checking one of the boxes. If you feel like you're struggling with these skills, consult the online resources provided by your instructor for extra practice.

I KNOW IT	THINK SO	UNSURE	NO IDEA	
				SKILL 1: CONVERT DECIMALS TO PERCENT FORM
				1. Write each decimal in percent form. 0.23
				0.794
				0.035
				SKILL 2: COMPUTE PERCENTAGES OF AN AMOUNT
				2. What is 10% of $8,603?
				3. What is 15% of $23,462.12?
				SKILL 3: COMPUTE PERCENTAGE OF A DIFFERENCE
				4. Find 20% of the difference between $83,402 and $36,901.

Lesson 3-6 A Taxing Situation (Income Taxes)

LEARNING OBJECTIVES

- ☐ 1. Understand why we pay taxes.
- ☐ 2. Explain the basic process of paying taxes.
- ☐ 3. Determine the amount of tax due based on taxable income.
- ☐ 4. Complete a 1040EZ form.

© Steven Cole/Getty Images RF

In this world nothing can be said to be certain, except death and taxes.
—Benjamin Franklin

"I really enjoy paying taxes," said no one. Ever. Of course, if we didn't have to pay taxes, we'd be missing out on some pretty cool stuff, like roads, police, trash pickup, fire fighters, schools, parks. . . . Okay, so maybe taxes aren't THAT bad after all. And as our friend Mr. Franklin pointed out, it's not like you get a choice. Taxes are necessary to fund governments at three levels: local, state, and federal. Property taxes, local sales tax, and local income taxes are kept locally to fund services within your community. State income taxes, which are used to fund state governments, vary from state to state depending on revenue a state can generate from other sources, like tourism, oil, or sales tax. Some states have no income tax at all, and are funded entirely through other sources. Federal income tax is something that everyone who earns income has to pay. Based on your income and **deductions** (which are characteristics of your financial situation that allow you to subtract from the tax you have to pay), the percentage of your income that you pay can vary considerably. In this lesson we'll study some common aspects of taxation. These are things that every American *should* be aware of, but far too few actually are.

0. After reading the opening paragraph, what do you think the main topic of this lesson will be?

3-6 Class

How Does the Tax Process Work?

If you earn a paycheck, through a process called **withholding,** taxes are automatically taken from the amount you've earned. The amount of tax that is withheld is based on two main factors: how much you earn, and paperwork you fill out through your employer describing how many other people depend on you financially. The intent is to estimate the amount of tax that you'll owe at the end of the year and pay some of it during each pay period so that you don't end up with a big bill that you can't afford to pay at the end of the year. When it's time to file a tax return, you'll find out if you paid too much, which entitles you to a refund, or too little, which means you need to pay more.

Our tax system is designed so that people that make more money pay a higher percentage of their income in taxes than do those who make less money. The range in which your income falls is known as a **tax bracket,** and the bracket you're in will determine how much of your income you will owe. It's essentially impossible to understand taxes without understanding tax brackets, so that's where we'll begin.

There are four different filing statuses: Single, which is self-explanatory; married filing jointly, which is when a married couple files a single return for both; married filing separately, which is when married individuals decide to file separate returns; and head of household, which is for unmarried individuals that financially support others. There are different income levels that qualify you for certain brackets depending on which filing status applies to you. Below is a table of tax brackets for 2014.

Tax Rate	Single	Married Filing Jointly and Widow(er)	Married Filing Separately	Head of Household
10%	$1–$9,075	$1–$18,150	$1–$9,075	$1–$12,950
15%	$9,076–$36,900	$18,151–$73,800	$9,076–$36,900	$12,951–$49,400
25%	$36,901–$89,350	$73,801–$148,850	$36,901–$74,425	$49,401–$127,550
28%	$89,351–$186,350	$148,851–$226,850	$74,426–$113,425	$127,551–$206,600
33%	$186,351–$405,100	$226,851–$405,100	$113,426–$202,500	$206,601–$405,100
35%	$405,101–$406,750	$405,101–$457,600	$202,551–$228,800	$405,101–$432,200
39.6%	over $406,750	over $457,600	over $228,800	over $432,200

For each filing status, the range of incomes listed in that column determines which tax bracket an individual falls into. For example, notice that in the second row of the Married Filing Separately column, the range is $9,076–$36,900, and that row has 15% in the Tax Rate column. This tells us that a married taxpayer filing separately with taxable income between $9,076 and $36,900 is in the 15% tax bracket. (In a while, we'll study exactly what that tells us about the amount of tax he or she would pay.)

1. If a person filing as single has taxable income of $42,000, what tax bracket is she in? What about a married person filing jointly?

   Single: 25% bracket
   Married filing jointly: 15%

Now for the big question: How much would that single person pay in taxes? If you said 25% of $42,000, that's a fine guess. An incorrect one, mind you, but a fine guess nonetheless. So what's the deal? What does the 25% tax bracket mean?

In short, all taxpayers that are single pay 10% on the first $9,075. They then pay 15% on the next $27,825, which is the amount that falls into the range $9,076–$36,900. Finally, they pay 25% of the amount over $36,900.

2. Using the information just provided, calculate the amount of tax that our single person would pay on income of $42,000.

10% of $9,075 = $907.50

15% of ($36,900 − $9,075) = $4,173.75

25% of ($42,000 − $36,900) = $1,275

Total tax = $907.50 + $4,173.75 + $1,275 = $6,356.25

Math Note

The taxes here are federal only: the amount of tax due at the state and local level vary widely depending on where you live.

3. How much would she have overpaid had she simply paid 25% of her taxable income?

25% of $42,000 = $10,500

$10,500 − $6,356.25 = $4,143.75

She would have overpaid by $4,143.75.

Since many Americans are, shall we say, somewhat math-challenged, the Internal Revenue Service (IRS), the government agency responsible for collecting taxes, publishes tables that let filers look up the amount of tax that they should pay. Their documentation even includes an example with the instructions, in this case for form 1040EZ, which we'll study in a bit.

2014 Tax Table

Example. Mr. Brown is single. His **taxable income** on line 6 of Form 1040EZ is $26,250. He follows two easy steps to figure his tax: **1.** He finds the $26,250-26,300 taxable income line. **2.** He finds the Single filing status column and reads down the column. The **tax** amount shown where the taxable income line and the filing status line meet is $3,488. He enters this amount on line 10 of Form 1040EZ.

At least	But less than	Single	Married filing jointly
		Your tax is—	
26,200	**26,250**	3,480	3,026
→ **26,250**	**26,300**	(3,488)	3,034
26,300	**26,350**	3,495	3,041
26,350	**26,400**	3,503	3,049

4. Using the tax bracket info on the previous page, if Mr. Brown believes the tax table, will he pay too much or too little? By how much?

Mr. Brown's income puts him in the 15% bracket, so his tax computation is:

10% of $9,075 + 15% of ($26,250 − $9,075) = $907.50 + (0.15)($17,175) = $3,483.75

The amount shown in the table is $3,488, so Mr. Brown will pay $4.25 too much.

5. What would happen to you if you fail to pay your taxes? Discuss, and if you're able, maybe do a little bit of Internet research.

First, you'll get fined and owe more. Other consequences range from your credit rating taking a hit, to liens placed on your belongings, your wages being garnished, and in some extreme cases even prison time.

Where Do Your Tax Dollars Go?

While paying taxes really stings, it can help at least some to know exactly what those dollars are accomplishing. The government maintains a website (the link is listed under the graphic below) that illustrates where federal tax dollars go. And for people that stink at percents, which of course does not include you, it's possible to just enter a taxable income and see how that money gets divided up. But we don't need that . . . we can just use the percents.

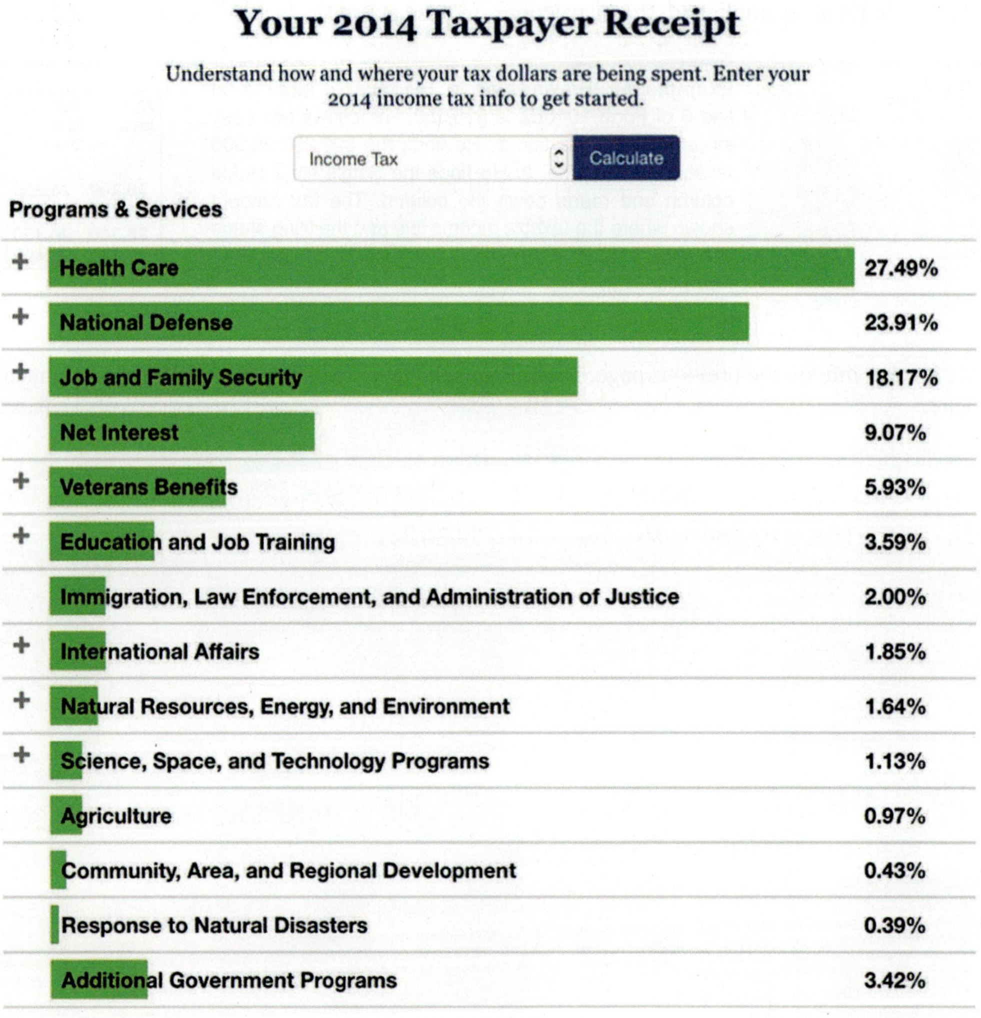

Source: https://www.whitehouse.gov/2014-taxreceipt

6. If you paid $6,500 in federal taxes in 2014, how much of your money went to education and job training?

3.59% of $6,500 is $233.35.

7. In 2014, the President paid $93,362 in federal taxes. How much of his money went to national defense?

23.91% of $93,362 is $22,322.85.

Did You Get It

Try these problems to see if you understand the concepts we just studied. The answer can be found at the bottom of the portfolio page.

1. Find the tax bracket for a person with status married filing jointly if she has a taxable income of $97,800.
2. Calculate the amount of federal tax that person would owe. What percentage of her taxable income would that be?

3-6 Group

Tax Forms: A Lesson in Following Directions

Many people are intimidated into submission by filling out tax forms. To be fair, you don't want to be making mistakes on forms this important, but with a bit of education and a solid ability to follow written instructions, completing basic tax forms is very doable. In this case, we're going to study federal form 1040EZ, which is a short form that can be filled out by taxpayers that meet the following qualifications:

- Filing status is Single or Married Filing Jointly.
- Taxpayer (and spouse if filing jointly) were under age 65 and not blind at the end of the tax year.
- Taxpayer does not claim any dependents.
- Taxable income is less than $100,000.
- Income came from only wages, salaries, tips, taxable scholarship or fellowship grants, or unemployment compensation, or Alaska Permanent Fund dividends, and taxable interest was not over $1,500.

(There are a couple of other technical provisions, but these are the main qualifications.)

The next page is a copy of the 1040EZ form for 2014, and we'll practice filling out the form. But first, we should discuss how you get information about your income. For many folks, this will come from two sources: a W-2 form, which reports your wages, and a 1099-INT, which reports interest income.

a Employee's social security number XXX-XX-XXXX	OMB No. 1545-0008	Safe, accurate, FAST! Use IRS e-file	Visit the IRS website at www.irs.gov/efile
b Employer identification number (EIN)		**1** Wages, tips, other compensation 58,584.78	**2** Federal income tax withheld 8,941.46
c Employer's name, address, and ZIP code Kramerica Industries 129 West 81st Street, Apartment 5B New York, NY 10029		**3** Social security wages 66,482.21	**4** Social security tax withheld 2,132.11
		5 Medicare wages and tips 66,482.21	**6** Medicare tax withheld 963.99
		7 Social security tips	**8** Allocated tips
d Control number		**9**	**10** Dependent care benefits
e Employee's first name and initial Last name Suff. Darin Roberts 18 Newman Ave, Apt. 4B New York, NY 10012		**11** Nonqualified plans	**12a** See instructions for box 12
		13 Statutory employee [X] Retirement plan [ ] Third-party sick pay [ ]	**12b**
		14 Other	**12c**
f Employee's address and ZIP code			**12d**

15 State	Employer's state ID number	16 State wages, tips, etc.	17 State income tax	18 Local wages, tips, etc.	19 Local income tax	20 Locality name
NY	47-325941	58,584.78	1,762.12	66,482.21	1,329.66	NYC

Form **W-2** **Wage and Tax Statement**

Department of the Treasury—Internal Revenue Service

Copy B—To Be Filed With Employee's FEDERAL Tax Return.
This information is being furnished to the Internal Revenue Service.

☐ VOID ☐ CORRECTED

PAYER'S name, street address, city or town, state or province, country, ZIP or foreign postal code, and telephone no. Great Eastern Savings and Loan 400 West 79th Street New York, NY 10029	Payer's RTN (optional) 1 Interest income $ 749.36	OMB No. 1545-0112 Form **1099-INT**	**Interest Income**
PAYER'S federal identification number / RECIPIENT'S identification number XXX-XX-XXXX	2 Early withdrawal penalty $		**Copy 1**
	3 Interest on U.S. Savings Bonds and Treas. obligations $		**For State Tax Department**
RECIPIENT'S name Darin Roberts	4 Federal income tax withheld $	5 Investment expenses $	
Street address (including apt. no.) 18 Newman Ave, Apt. 4B	6 Foreign tax paid $	7 Foreign country or U.S. possession	
City or town, state or province, country, and ZIP or foreign postal code New York, NY 10012	8 Tax-exempt interest $	9 Specified private activity bond interest $	
FATCA filing requirement ☐	10 Market discount $	11 Bond premium $	
	12	13 Bond premium on tax-exempt bond $	
Account number (see instructions)	14 Tax-exempt and tax credit bond CUSIP no.	15 State / 16 State identification no.	17 State tax withheld $ $

Form **1099-INT** www.irs.gov/form1099int Department of the Treasury - Internal Revenue Service

Form **1040EZ**

Department of the Treasury—Internal Revenue Service

Income Tax Return for Single and Joint Filers With No Dependents (99) 2014

OMB No. 1545-0074

Your first name and initial: A = Archibald, D = Darin	Last name	**Your social security number**
If a joint return, spouse's first name and initial	Last name	**Spouse's social security number**
Home address (number and street). If you have a P.O. box, see instructions.	Apt. no.	▲ Make sure the SSN(s) above are correct.
City, town or post office, state, and ZIP code. If you have a foreign address, also complete spaces below (see instructions).		**Presidential Election Campaign** Check here if you, or your spouse if filing jointly, want $3 to go to this fund. Checking a box below will not change your tax or refund. ☐ **You** ☐ **Spouse**
Foreign country name	Foreign province/state/county	Foreign postal code

Income

Attach Form(s) W-2 here.

Enclose, but do not attach, any payment.

1	Wages, salaries, and tips. This should be shown in box 1 of your Form(s) W-2. Attach your Form(s) W-2.	1	A: 26,667 D: 58,584.78
2	Taxable interest. If the total is over $1,500, you cannot use Form 1040EZ.	2	A: 0 D: 749.36
3	Unemployment compensation and Alaska Permanent Fund dividends (see instructions).	3	
4	Add lines 1, 2, and 3. This is your **adjusted gross income.**	4	A: 26,667 D: 59,334.14
5	If someone can claim you (or your spouse if a joint return) as a dependent, check the applicable box(es) below and enter the amount from the worksheet on back. ☐ **You** ☐ **Spouse** If no one can claim you (or your spouse if a joint return), enter $10,150 if **single;** $20,300 if **married filing jointly.** See back for explanation.	5	A: 10,150 D: 10,150
6	Subtract line 5 from line 4. If line 5 is larger than line 4, enter -0-. This is your **taxable income.** ▶	6	A: 16,517 D: 49,184.14

These lines don't apply to Darin or Archibald.

Payments, Credits, and Tax

Both Darin and Archibald have health insurance, so check this box.

7	Federal income tax withheld from Form(s) W-2 and 1099. A: 3,518	7	D: 8,941.46
8a	**Earned income credit (EIC)** (see instructions)	8a	
b	Nontaxable combat pay election. 8b		
9	Add lines 7 and 8a. These are your **total payments and credits.** ▶	9	A: 3,518 D: 8,941.46
10	**Tax.** Use the amount on **line 6 above** to find your tax in the tax table in the instructions. Then, enter the tax from the table on this line.	10	A: 2,023.80 D: 8,152.29
11	Health care: individual responsibility (see instructions) Full-year coverage ☐	11	
12	Add lines 10 and 11. This is your **total tax.** A: 2,023.80	12	D: 8,152.29

Refund

Have it directly deposited! See instructions and fill in 13b, 13c, and 13d, or Form 8888.

13a	If line 9 is larger than line 12, subtract line 12 from line 9. This is your **refund.** If Form 8888 is attached, check here ▶ ☐	13a	A: 1,494.20 D: 789.17
▶ b	Routing number ☐☐☐☐☐☐☐☐☐ ▶ c Type: ☐ Checking ☐ Savings		
▶ d	Account number ☐☐☐☐☐☐☐☐☐☐☐☐☐☐☐☐☐		

Amount You Owe

14	If line 12 is larger than line 9, subtract line 9 from line 12. This is the **amount you owe.** For details on how to pay, see instructions. ▶	14	

Third Party Designee

Do you want to allow another person to discuss this return with the IRS (see instructions)? ☐ **Yes.** Complete below. ☐ **No**

Designee's name ▶ Phone no. ▶ Personal identification number (PIN) ▶

Sign Here

Joint return? See instructions.

Keep a copy for your records.

Under penalties of perjury, I declare that I have examined this return and, to the best of my knowledge and belief, it is true, correct, and accurately lists all amounts and sources of income I received during the tax year. Declaration of preparer (other than the taxpayer) is based on all information of which the preparer has any knowledge.

Your signature	Date	Your occupation	Daytime phone number
Spouse's signature. If a joint return, **both** must sign.	Date	Spouse's occupation	If the IRS sent you an Identity Protection PIN, enter it here (see inst.)

Paid Preparer Use Only

Print/Type preparer's name	Preparer's signature	Date	Check ☐ if self-employed	PTIN
Firm's name ▶		Firm's EIN ▶		
Firm's address ▶		Phone no.		

For Disclosure, Privacy Act, and Paperwork Reduction Act Notice, see instructions. Cat. No. 11329W Form **1040EZ** (2014)

The goal of the group activity is to fill in the 1040EZ form on the previous page for one of two individuals: Darin, whose W-2 and 1099-INT were displayed earlier, and Archibald, whose W-2 is below. Each is single and has no dependents. Archibald works part time while still in college, and received $3,000 in scholarship money. Because he's not currently in a Bachelor's degree program or higher, that scholarship money is considered taxable income.

Choose which individual you want to help with their 1040EZ form. After filling it out, have another group perform an audit to check your work. In each case, you'll need to use the tax bracket information from earlier to calculate the amount of tax that should be paid.

a Employee's social security number: XXX-XX-XXXX
OMB No. 1545-0008
Safe, accurate, FAST! Use IRS e-file
Visit the IRS website at www.irs.gov/efile

b Employer identification number (EIN)

c Employer's name, address, and ZIP code
Heisman Marketing
1975 Lane Ave.
Columbus, OH 43018

d Control number

e Employee's first name and initial | Last name | Suff.
Archibald Griffen
45 Buckeye Way
Columbus, OH 43085

f Employee's address and ZIP code

1 Wages, tips, other compensation: 23,667	2 Federal income tax withheld: 3,518
3 Social security wages: 23,667	4 Social security tax withheld: 943.12
5 Medicare wages and tips: 23,667	6 Medicare tax withheld: 247.01
7 Social security tips	8 Allocated tips
9	10 Dependent care benefits
11 Nonqualified plans	12a See instructions for box 12
13 Statutory employee [X] Retirement plan [ ] Third-party sick pay [ ]	12b
14 Other	12c
	12d

15 State	Employer's state ID number	16 State wages, tips, etc.	17 State income tax	18 Local wages, tips, etc.	19 Local income tax	20 Locality name
OH	11-004141	23,667	852.19	23,667	356.97	Colum

Form **W-2** **Wage and Tax Statement**

Department of the Treasury—Internal Revenue Service

Copy B—To Be Filed With Employee's FEDERAL Tax Return.
This information is being furnished to the Internal Revenue Service.

Did You Get It

Try this problem to see if you understand the concepts we just studied. The answer can be found at the bottom of the portfolio page.

3. A taxpayer with status married filing jointly has no dependents other than her spouse. Her federal wages are $81,112, and she also had $1,462.12 in interest income. What would you write in the following lines on her 1040EZ form?

   a. 4 b. 5 c. 6 d. 10

3-6 Portfolio

Name ______________________________

Check each box when you've completed the task. Remember that your instructor will want you to turn in the portfolio pages you create.

Technology

1. ☐ There are TONS of free online tools to help you with tax forms and calculations. Find one you like and use it to calculate the amount of tax that Darin Roberts (tax forms displayed within this lesson) owes in federal taxes. Then decide if he would be due a refund or have to pay more, and describe how you decided. Make sure that you include a reference to the online tool you used. If you could include a screen shot of your work, your professor would be most appreciative.

Online Practice

1. ☐ Include any written work from the online assignment along with any notes or questions about this lesson's content.

Applications

1. ☐ Complete the applications problems.

Reflections

Type a short answer to each question.

1. ☐ Why do we have to pay taxes?
2. ☐ What is a tax bracket? What is the point of tax brackets?
3. ☐ What is a 1040 form used for?
4. ☐ Take another look at your answer to Question 0 at the beginning of this lesson. Would you change your answer now that you've completed the lesson? How would you summarize the topic of this lesson now?
5. ☐ What questions do you have about this lesson?

Looking Ahead

1. ☐ Complete the Lesson 4-1 Prep Skills and read the opening paragraph in Lesson 4-1 carefully, then answer Question 0 in preparation for that lesson.

Answers to "Did You Get It?"

1. 25% **2.** $16,162.50; 16.5% **3.** **a.** Line 4: $82,574.12; **b.** Line 5: $20,300; **c.** Line 6: $62,274.12; **d.** Line 10: $8,433.62

3-6 Applications

Name ______________________________

Everyone fears the tax auditor—that's a government employee who checks over tax returns for accuracy. This is your chance to play auditor. Check the 1040-EZ filed by the deputy sheriff whose W-2 and 1099 forms are below. The completed 1040 is on page 295. Rick is filing jointly for himself and his wife.

a Employee's social security number XXX-XX-XXXX	OMB No. 1545-0008	Safe, accurate, FAST! Use IRS e-file	Visit the IRS website at www.irs.gov/efile
b Employer identification number (EIN)		1 Wages, tips, other compensation 92,513	2 Federal income tax withheld 14,319
c Employer's name, address, and ZIP code King County Sherriff 411 W. Main Ave Cynthiana, KY 41031		3 Social security wages 96,303	4 Social security tax withheld 3,601.11
		5 Medicare wages and tips 96,303	6 Medicare tax withheld 952.00
		7 Social security tips	8 Allocated tips
d Control number		9	10 Dependent care benefits
e Employee's first name and initial, Last name, Suff. Richard Grimes 11 Apocalypse Circle Cynthiana, KY 41031		11 Nonqualified plans	12a See instructions for box 12
		13 Statutory employee [x] Retirement plan [ ] Third-party sick pay [ ]	12b
		14 Other	12c
f Employee's address and ZIP code			12d

15 State	Employer's state ID number	16 State wages, tips, etc.	17 State income tax	18 Local wages, tips, etc.	19 Local income tax	20 Locality name
KY	21-006390	96,303	3,774	96,303	917	Cynthi

Form **W-2** **Wage and Tax Statement**

Department of the Treasury—Internal Revenue Service

Copy B—To Be Filed With Employee's FEDERAL Tax Return.
This information is being furnished to the Internal Revenue Service.

[ ] VOID [ ] CORRECTED

PAYER'S name, street address, city or town, state or province, country, ZIP or foreign postal code, and telephone no. Kentucky State Employee Credit Union 142 Poplar St Cynthiana, KY 41031	Payer's RTN (optional)	OMB No. 1545-0112	**Interest Income**
	1 Interest income $ 1,104.61	Form **1099-INT**	
	2 Early withdrawal penalty $		**Copy 1**
PAYER'S federal identification number / RECIPIENT'S identification number XXX-XX-XXXX	3 Interest on U.S. Savings Bonds and Treas. obligations $		**For State Tax Department**
RECIPIENT'S name Richard Grimes	4 Federal income tax withheld $	5 Investment expenses $	
Street address (including apt. no.) 11 Apocalypse Circle	6 Foreign tax paid $	7 Foreign country or U.S. possession	
City or town, state or province, country, and ZIP or foreign postal code Cynthiana, KY 41031	8 Tax-exempt interest $	9 Specified private activity bond interest $	
	10 Market discount $	11 Bond premium $	
FATCA filing requirement [ ]	12	13 Bond premium on tax-exempt bond $	
Account number (see instructions)	14 Tax-exempt and tax credit bond CUSIP no.	15 State / 16 State identification no.	17 State tax withheld $ $

Form **1099-INT** www.irs.gov/form1099int Department of the Treasury - Internal Revenue Service

3-6 Applications

Name ______________________________

1. Using the table on page 284, calculate the tax due on the taxable income shown on Rick's 1040. Then compare to the amount shown in the IRS tax table to the right. Is Rick better off using his amount, or the IRS amount?

0.10($18,150) + 0.15($73,800 − $18,150) + 0.25($76,003 − $73,800) = $10,713.25

The table shows $10,719, so Rick is $5.75 ahead if he calculates the tax himself.

If line 43 (taxable income) is—		And you are—			
At least	But less than	Single	Married filing jointly *	Married filing sepa-rately	Head of a house-hold
		Your tax is—			
76,000					
76,000	76,050	14,863	10,719	14,911	13,419
76,050	76,100	14,875	10,731	14,925	13,431
76,100	76,150	14,888	10,744	14,939	13,444
76,150	76,200	14,900	10,756	14,953	13,456
76,200	76,250	14,913	10,769	14,967	13,469

2. List any mistakes you found on the 1040, and write how they should be corrected. If you need extra space for calculations, use the blank space on the next page.

1. He used the wrong income on line 1: the federally taxable income is in box 1 of his W2, $92,513.
2. He failed to report $1,104.61 in interest income on line 2.
3. This makes his corrected taxable income $92,513 + $1,104.61 − $20,300 = $73,317.61. This should go on line 6.
4. He put the amount of state tax he paid on line 7, rather than the federal amount, which is $14,319. This goes on lines 7 and 9.
5. He calculated the tax correctly, but since it was on the wrong amount, that needs to be redone. The correct tax is 0.10($18,150) + 0.15($73,317.61 − $18,150) = $10,090.14, which goes on lines 10 and 12.
6. The amount of tax paid and owed are both wrong, so where he calculated what he owed, he was way off. See below.

3. Would our friend Rick have paid too much or too little in taxes without your corrections? By how much?

WAY too much. He is actually due a refund of $14,319 − $10,090.14 = $4,228.86. So he cheated himself out of $7,112.14 + $4,228.86, which is $11,341.

3-6 Applications

Name ______________________________

4. Using the corrected tax that Rick paid after you finished your audit, how much of Rick's taxes will go to paying interest on the national debt?

The amount he pays in taxes is $10,090.14. The bar graph on Page 286 shows that 9.07% goes to interest, so Rick is paying $0.0907 \cdot \$10{,}090.14 = \$915.18$ toward interest on the national debt. Ugh.

3-6 Applications

Name ______________________________

Form **1040EZ**

Department of the Treasury—Internal Revenue Service

Income Tax Return for Single and Joint Filers With No Dependents (99) **2014**

OMB No. 1545-0074

Your first name and initial	Last name	Your social security number
Richard	Grimes	xxx | xx | xxxx
If a joint return, spouse's first name and initial	Last name	Spouse's social security number
Lori	Grimes	xxx | xx | xxxx

Home address (number and street). If you have a P.O. box, see instructions.	Apt. no.	▲ Make sure the SSN(s) above are correct.
11 Apocalypse Circle		

City, town or post office, state, and ZIP code. If you have a foreign address, also complete spaces below (see instructions).
Cynthiana, KY 41031

Foreign country name	Foreign province/state/county	Foreign postal code

Presidential Election Campaign
Check here if you, or your spouse if filing jointly, want $3 to go to this fund. Checking a box below will not change your tax or refund. ☐ You ☐ Spouse

Income

Attach Form(s) W-2 here.

Enclose, but do not attach, any payment.

1	Wages, salaries, and tips. This should be shown in box 1 of your Form(s) W-2. Attach your Form(s) W-2.	1	96.303
2	Taxable interest. If the total is over $1,500, you cannot use Form 1040EZ.	2	
3	Unemployment compensation and Alaska Permanent Fund dividends (see instructions).	3	
4	Add lines 1, 2, and 3. This is your **adjusted gross income.**	4	96,303
5	If someone can claim you (or your spouse if a joint return) as a dependent, check the applicable box(es) below and enter the amount from the worksheet on back. ☐ **You** ☐ **Spouse** If no one can claim you (or your spouse if a joint return), enter $10,150 if **single;** $20,300 if **married filing jointly.** See back for explanation.	5	20,300
6	Subtract line 5 from line 4. If line 5 is larger than line 4, enter -0-. This is your **taxable income.** ▶	6	76,003

Payments, Credits, and Tax

7	Federal income tax withheld from Form(s) W-2 and 1099.	7	3,601.11
8a	**Earned income credit (EIC)** (see instructions)	8a	
b	Nontaxable combat pay election. 8b		
9	Add lines 7 and 8a. These are your **total payments and credits.** ▶	9	3,601.11
10	**Tax.** Use the amount on **line 6 above** to find your tax in the tax table in the instructions. Then, enter the tax from the table on this line.	10	10,713.25
11	Health care: individual responsibility (see instructions) Full-year coverage ☐	11	
12	Add lines 10 and 11. This is your **total tax.**	12	10,713.25

Refund

Have it directly deposited! See instructions and fill in 13b, 13c, and 13d, or Form 8888.

13a	If line 9 is larger than line 12, subtract line 12 from line 9. This is your **refund.** If Form 8888 is attached, check here ▶ ☐	13a	
▶ b	Routing number ▶ c Type: ☐ Checking ☐ Savings		
▶ d	Account number		

Amount You Owe

14	If line 12 is larger than line 9, subtract line 9 from line 12. This is the **amount you owe.** For details on how to pay, see instructions. ▶	14	7,112.14

Third Party Designee

Do you want to allow another person to discuss this return with the IRS (see instructions)? ☐ **Yes.** Complete below. ☐ **No**

Designee's name ▶ Phone no. ▶ Personal identification number (PIN) ▶

Sign Here

Under penalties of perjury, I declare that I have examined this return and, to the best of my knowledge and belief, it is true, correct, and accurately lists all amounts and sources of income I received during the tax year. Declaration of preparer (other than the taxpayer) is based on all information of which the preparer has any knowledge.

Joint return? See instructions.

Keep a copy for your records.

Your signature	Date	Your occupation	Daytime phone number
Spouse's signature. If a joint return, **both** must sign.	Date	Spouse's occupation	If the IRS sent you an Identity Protection PIN, enter it here (see inst.)

Paid Preparer Use Only

Print/Type preparer's name	Preparer's signature	Date	Check ☐ if self-employed	PTIN
Firm's name ▶		Firm's EIN ▶		
Firm's address ▶		Phone no.		

For Disclosure, Privacy Act, and Paperwork Reduction Act Notice, see instructions. Cat. No. 11329W Form **1040EZ** (2014)

Unit 4
Statistically Speaking

Outline

Lesson 4-1 Prep Skills

This is a short review of skills that will come in handy in the next lesson. In each case, answer the question, then rate your confidence level by checking one of the boxes. If you feel like you're struggling with these skills, consult the online resources provided by your instructor for extra practice.

I KNOW IT	THINK SO	UNSURE	NO IDEA	
				SKILL 1: REDUCING FRACTIONS
				1. Write each fraction in lowest terms: $\frac{12}{16}$ $\frac{65}{80}$
				SKILL 2: CONVERTING BETWEEN FRACTIONS, DECIMALS, AND PERCENTS
				2. Write each percent in decimal form: 60%
				12.5%
				3. Write each fraction in percent form: $\frac{3}{5}$
				$\frac{12}{16}$
				4. Write each decimal in percent form. 0.13
				0.058
				SKILL 3: INTERPRETING "AT MOST" AND "AT LEAST"
				5. Write the whole numbers between 1 and 20 that are at least 15, and those that are at most 8.

Lesson 4-1 A Likely Story (Basic Probability)

LEARNING OBJECTIVES

- ☐ 1. Understand key terminology in the study of probability.
- ☐ 2. Compute and interpret theoretical and empirical probabilities.
- ☐ 3. Compare theoretical and empirical probability.

© Brand X Pictures/Jupiterimages RF

A scientist worthy of a lab coat should be able to make original discoveries while wearing a clown suit, or give a lecture in a high squeaky voice from inhaling helium. It is written nowhere in the math of probability theory that one may have no fun.

—Eliezer Yudkowsky

Walking into a casino without knowing anything about probability is kind of like going to a stick fight without a stick—you're likely to take a beating. Casinos aren't in the business of losing money, and the games are designed so that most people lose more than they win. But an understanding of what is likely to happen in a given situation can give you an advantage over other players, giving you a better chance of walking out the door with some cash in your pockets. (Although if you learn enough about probability, you might decide that staying away from casinos is your best approach.)

The study of probability originated in an effort to understand games of chance, like those that use coins, dice, and playing cards. Generally speaking, probability is simply a number that describes how likely an event is to occur. We will use games of chance to illustrate the ideas, but will eventually see that probability has many applications beyond simple games.

0. After reading the opening paragraphs, what do you think the main topic of this lesson will be?

4-1 Class

In this lesson, we'll examine the basic concepts involved in studying **probability**. In short, the probability of something occurring is a number that represents how likely it is to occur. You can think of it as the percent chance of something happening. The major difference between percent chance and probability is that we write probability as a number between zero and one. So in that regard, it's like percent chance with the percentage written in decimal or fractional form.

1. A forecaster says that there's a 70% chance of rain today. What is the probability of rain?

The probability is 0.7, or 7/10.

Sample Spaces

Processes such as flipping a coin, rolling a die, or drawing a card from a deck are called probability experiments.

A **probability experiment** is a process that leads to well-defined results called **outcomes**. An outcome is the result of a single trial of a probability experiment.

Some examples of a trial are flipping a coin once, rolling a single die, and drawing one card from a deck.

2. When a coin is tossed, there are ___2___ possible outcomes. List them.

heads, tails

3. When rolling a single die, there are ___6___ possible outcomes. List them.

1, 2, 3, 4, 5, 6

In a probability experiment, we can predict what outcomes are possible, but we can't predict with certainty which one will occur. We say that the outcomes occur at **random**. In any experiment, the set of all possible outcomes is called the **sample space**. What you did in Questions 2 and 3 was find the sample space for two different probability experiments.

4. For each experiment described below, write the sample space. The first is filled in for you as a reminder that a sample space is a SET, and we write sets inside braces with commas separating the elements.

Experiment	Sample Space
Flip one coin	{head, tail}
Roll one die	{1, 2, 3, 4, 5, 6}
Answer a true-false question	{Right, wrong} OR {True, False}
Flip two coins	{Tails tails, tails heads, heads heads, tails tails}

In finding probabilities, it's sometimes necessary to consider several outcomes of a probability experiment.

5. For example, when a die is rolled, we may want to consider obtaining an odd number; there are ___3___ outcomes that satisfy this criterion. List them.

1, 3, 5

Getting an odd number when rolling a die is an example of an *event.*

An **event** is any subset of the sample space for a probability experiment. This means that an event could be a single outcome, or a set of several outcomes.

There's a subtle distinction between an outcome and an event: an outcome is a single occurrence, while an event can contain a number of outcomes. So rolling 2 with a single die is an outcome, but can also be considered an event. Rolling an odd number, on the other hand, is an event made up of three outcomes, but is not itself an outcome.

Let's look at that distinction more closely. You already listed out the sample space for flipping two coins. Look back at the table on the previous page, and make sure that you have FOUR outcomes in your sample space.

© Photodisc/Getty Images RF

Experiment: draw a card. Sample space: 52 cards. Outcome: draw the ace of spades. Event: draw an ace.

6. How many results match getting heads on the first flip and tails on the second?

Just one

That's an outcome.

7. How many results match getting one head and one tail?

Two

That's an event.

Identify each scenario in Questions 8–10 as an event, an outcome, or both.

8. When rolling two dice, getting 3 on the first die and 5 on the second.

This is an outcome. It's a specific single occurrence. That makes it an event as well.

9. When rolling two dice, getting a total of 10.

This is an event. There are several different ways to get a total of 10.

10. When flipping a coin ten times consecutively, getting tails eight times.

This is an event. You could get tails on the first eight trials, or the last eight, or . . . well, you get the picture.

Some students may be unfamiliar with a standard deck of cards. Not surprisingly, we'll be using cards quite a bit in the next few lessons, so you might want to bring in and describe a deck to be on the safe side.

Did You Get It

Try this problem to see if you understand the concepts we just studied. The answer can be found at the bottom of the portfolio page.

1. A baseball team has 12 pitchers on the roster, and the manager, who evidently is intent on losing his job, decides to pick the starting pitcher for tonight's game randomly. What is the sample space? Write an example of an outcome, and an example of an event.

Theoretical Probability

Now we're ready to specifically define what is meant by probability. The first type we'll study is called **theoretical probability**. The goal is to determine all of the possible outcomes in a sample space and determine the probability, or likelihood, of an event occurring without actually performing experiments. There is one key assumption we make in theoretical probability: that every outcome in a sample space is equally likely. For example, when a single die is rolled, we assume that each number is equally likely to come up. When a card is chosen from a deck of 52 cards, we assume that each card has the same probability of being drawn.

11. You'd have to be awfully bored to flip a coin 100 times, but if you did, about how many times do you think it would land tails side up?

Answers can vary, but a reasonable answer would have to be something close to 50.

12. What percentage of the time would you expect the coin to land tails up?

50%

13. Write a fraction with the number of flips in the denominator and the number of times you're expecting it to land tails up in the numerator. What does this have to do with your answer to Question 12?

50/100 This provides the percentage from Question 12.

Congratulations: you've just computed your first theoretical probability. You didn't actually flip a coin 100 times: you just used reasoning ability to decide what the likelihood of getting tails is. That's what makes it a theoretical probability. You developed a theory by thinking, as opposed to gathering data experimentally. (We'll get to that in a bit.)

Calculating a Theoretical Probability

The probability of an event E in a sample space S occurring is given by

$$P(E) = \frac{\text{The number of outcomes in } E}{\text{The number of outcomes in } S} = \frac{n(E)}{n(S)}$$

Take a minute to describe how students really used this formula mentally when they answered Question 12.

For Questions 14–16, a single six-sided die is rolled. Find the probability of each event. If you need to, look back at your sample space on page 300 for reference.

14. Rolling a 2.

1/6

© Jules Frazier/Getty Images RF

15. Rolling a number less than 5.

4/6, or 2/3

16. Rolling an odd number.

3/6, or 1/2

17. If THREE coins are flipped, write the sample space. Remember that Tails, Heads, Heads is a different outcome than Heads, Heads, Tails even though both result in one tail and two heads (which, by the way, sounds like a bunny rabbit from a horror story).

{TTT, TTH, THH, THT, HTT, HTH, HHT, HHH}

Now is a good time to point out that we often use abbreviations when listing sample spaces, like T and H for tails and heads.

For Questions 18–20, three coins are flipped. Find the probability of each event.

18. Two tails

3/8

19. At least one tail

7/8

20. At most one tail

1/2

Did You Get It ?

Try these problems to see if you understand the concepts we just studied. The answers can be found at the bottom of the portfolio page.

2. There are 75 balls in a bingo hopper. The ones numbered 1–15 are labeled with B, the ones from 16–30 with I, from 31–45 N, from 46–60 G, and from 61–75 O. Find the probability of these events:

   a. Drawing N42
   b. Drawing N11
   c. Drawing a number less than 50
   d. Drawing an odd number

3. In the game of rock, paper, scissors, two people play, and it's equally likely that you will win, lose, or tie. Find the probability of

   a. Losing twice in a row.
   b. Winning at least once in two tries.
   c. Having the same outcome twice in a row.

Now that we know a little bit about probability, we can make a series of simple but important observations.

Observation 1: Probability is never negative.

21. Explain why a probability can't be negative.

Because the formula divides two numbers, both of which are the number of ways something can occur. You can't have a negative number of possibilities.

Observation 2: Probability is never greater than one.

22. Explain why probabilities are never greater than one.

The denominator of the fraction is the total number of possible ways that something can occur. The numerator consists of some specific possibilities from that set, so it can't be more than the number in the denominator. To be greater than one, the numerator would have to be bigger.

Observation 3: When an event can't possibly occur, its probability is zero. When an event is certain to occur, its probability is one.

23. Explain why impossible events have probability zero and certain events have probability one.

If an event is impossible, the number of ways it can happen is zero. The only way we can consider an event to be "certain" is if it contains every possible outcome. In that case, it's obvious that the number of outcomes in the event is the same as the number of total outcomes, so the fraction has value one.

Observation 4: If you add the probabilities for every outcome in the sample space, the result is always one.

24. Explain why the probabilities of all outcomes have to add up to one. Thinking of rolling a die might be helpful.

If you add the fractions for all possible outcomes, they'll all have the same denominator. So the sum would come from adding all the numerators. But the numerators will consist of all possible outcomes, which gets us back to a fraction with the same numerator and denominator again.

4-1 Group

If it doesn't make you uncomfortable, exchange the following information with the classmates in your Unit 4 group. This will be your small group for the fourth unit. It would be a good idea to schedule a time for the group to meet to go over homework, ask/answer questions, or prepare for exams. You can use this table to help schedule a mutually agreeable time.

Name	Phone Number	Email	Available times

Probabilities are usually expressed as fractions or decimals between (and including) zero and one. But occasionally we'll express probabilities as percents. For example, when the probability of an event is 1/2, there's a 50% chance that it'll occur. If an event has probability close to zero, it's very unlikely to occur, and if the probability is close to one, it's very likely to occur.

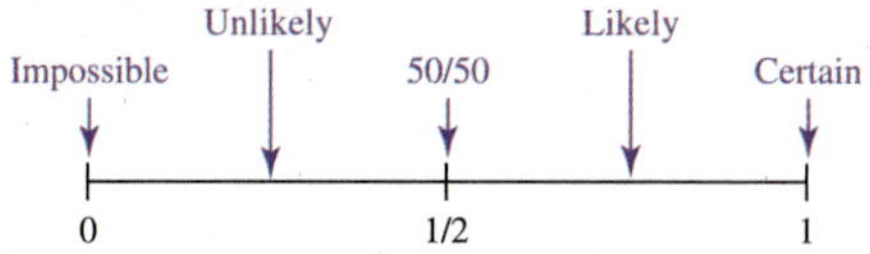

© Stockbyte/Alamy RF

1. Describe exactly what you think it means when a weather forecaster says that there's a 10% chance of rain today.

Answers vary.

A lot of people think that a 10% chance of rain means that it will rain for 10% of the day, but that's not correct. The forecaster is expressing a probability—how likely he or she thinks it is that there will be SOME measurable precipitation in that area today. So in essence, it means that if you keep track of ten days with a 10% chance of rain, you would expect rain on one of those ten days.

In Questions 2–5, estimate the probability of each event. You don't have to do any calculations: The idea is just to think about the size of a probability compared to how likely something is to occur.

2. Passing this course

Answers vary for each of these.

3. Living past the age of 100

4. Winning a million-dollar lottery

5. Doing tonight's homework

Sometimes it's useful to find the probability that a given event will NOT occur. Here's an example.

6. What's the probability of rolling a 2 with one die?

1/6

7. How many outcomes are in the event of NOT rolling a 2?

Five

8. Use your answer to Question 7 to find the probability of NOT rolling a 2.

5/6

When you roll one die, you will either get 2, or you will not. If we call the event of rolling a 2 *E*, then we call the event of not rolling a 2 *E*′, and read it as "*E* complement" or "the complement of *E*."

For an event *E* in a sample space, the set of all outcomes that are not in *E* is called the **complement** of *E*.

9. Describe the complement of each event in detail. Don't just put "not" in front. Think hard! The full answer may not be the obvious one.

a. Passing this course

Sample answer: failing the course, dropping the course, or taking an incomplete in the course.

b. Winning $1,000,000 when you buy a lotto ticket

Sample answer: winning nothing, winning some amount less than a million dollars, or winning some amount more than a million dollars.

Given the fact that all probabilities for an experiment have to add up to 1, it makes perfect sense that the probability of rolling 2 plus the probability of not rolling 2 is 1. This line of reasoning gives us a general rule:

The Probability of a Complement

For any event *E*, if *E*′ is the event "*E* does not occur," then

$P(E') = 1 - P(E)$

10. Of the next 32 trials on the docket in a county court, 5 are homicides, 12 are drug offenses, 6 are assaults, and 9 are property crimes. If jurors are assigned to trials randomly, what's the probability that a given juror won't get a homicide case? Describe two different ways you could calculate this probability.

Brute force method: $12 + 6 + 9 = 27$ are not homicides, so the probability of not getting a homicide is 27/32.

Clever way: The probability of getting a homicide is 5/32, so the probability of not getting a homicide is $1 - (5/32)$, which is 27/32.

Did You Get It

Try this problem to see if you understand the concepts we just studied. The answer can be found at the bottom of the portfolio page.

4. For the docket in Question 10, what's the probability that a juror gets assigned to a case that isn't a drug offense?

Empirical Probability

The second approach to probability we will study is computed using experimental data, rather than counting equally likely outcomes. For example, suppose that out of the last 100 days, you've gone to the gym 60 times. It would be reasonable to guess that the probability of you going to the gym on any given day is about 60/100, or 0.6.

This type of probability is called **empirical probability**, and is based on *observed frequencies*—that is, the number of times a particular event has occurred out of a certain number of trials. In this case, the observed frequency of days going to the gym is 60, the observed frequency of not going is 40, and the total number of trials is $60 + 40 = 100$.

11. On any given day, there are exactly two possibilities: you can go to the gym, or not go to the gym. Why would this fact and theoretical probability not be useful in deciding the probability that you'll go to the gym?

Because whether or not you go to the gym isn't a random occurrence. Although there are two outcomes, there's no reason to believe that they're equally likely.

This is probably THE key idea in understanding empirical vs. theoretical probability. Have students present their answers to this question and compare.

When calculating theoretical probabilities, we assumed that all outcomes were equally likely. But there are plenty of situations where that assumption would be faulty. In that case, we often perform an experiment to estimate probabilities.

12. If you flip a paper or plastic cup, it can land on its top, its bottom, or its side. Are these outcomes equally likely? Explain.

They are not. The cup is by far most likely to land on its side.

13. Estimate the probability of each outcome.

It would be nice if students point out that it depends on the cup. Weight, shape, and width of the base play a role. Make sure that their probabilities add up to one!

The way that we estimated the probability that you go to the gym on any given day is a good model for how to compute empirical probabilities in general:

Calculating an Empirical Probability

The empirical probability of an event E is given by

$$P(E) = \frac{\text{Observed frequency of the event}}{\text{The number of trials}} = \frac{f}{n}$$

14. Describe how you could design an experiment to develop empirical probabilities for flipping a cup.

You could flip the cup a bunch of times, keeping track of how many times it lands on its side, top, and bottom. Then divide each frequency by the total number of trials.

15. Now let's perform that experiment. Your professor will supply your group with a cup. Flip it from about waist level and record how it lands in the table below. Complete as many trials as you feel like you need in order to get accurate empirical probabilities, and then compute those probabilities.

Position	Put a mark for each occurrence	Add the marks
Land on top		
Land on bottom		
Land on side		

Answers vary.

If possible, choose cups that are at least reasonably likely to land on their top or bottom. The experiment isn't particularly interesting if the cup always ends up on its side. Avoid glass. :)

The table you used to organize the results of your experiment is an example of a **frequency distribution**. This is a table that list categories that we call **classes** (in this case the outcomes of our experiment) and frequencies for each class.

Earlier, we talked about the number of times you might go to the gym in 100 days.

16. Complete a frequency distribution for going to the gym based on the data provided earlier.

Result (Class)	Observed Frequency
Gym	60
No gym	40
Total	100

Math Note

We'll study frequency distributions in much more depth a bit later in this course.

17. In a random sample of 500 people, 210 had type O blood, 223 had type A, 51 had type B, and 16 had type AB. Set up a frequency distribution for the data.

Result (Class)	Observed Frequency
O	210
A	223
B	51
AB	16

For Questions 18–21, use your frequency distribution to find the probability that a randomly selected person from the population has the given blood type.

18. Type O

$210/500 = 0.42$

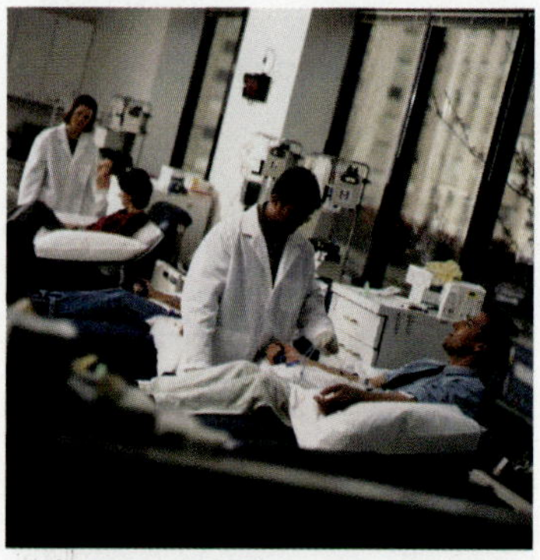

© Keith Brofsky/Getty Images RF

19. Type A or B

$(223 + 51)/500 = 0.548$

20. Neither type A nor type O

$(51 + 16)/500 = 0.134$

21. A blood type other than AB

$(210 + 223 + 51)/500 = 0.968$

OR

$1 - (16/500) = 0.968$

Did You Get It

Try this problem to see if you understand the concepts we just studied. The answer can be found at the bottom of the portfolio page.

5. A 2-lb bag of Hershey's miniatures contains 33 milk chocolate bars, 27 Krackel bars, 26 Mr. Goodbars, and 19 Special Dark bars. Set up a frequency distribution and find the probability that a bar chosen from a randomly selected bag is

   a. A Mr. Goodbar.
   b. A Krackel or Special Dark.
   c. Not a milk chocolate bar.

We'll close the lesson with a comparison between the two types of probability we've studied, theoretical and empirical. The key thing to remember is that in theoretical probability, we use a sample space with a definite number of outcomes, and assume that all of them are equally likely. This makes theoretical probability a bad option for something like the probability of a team winning a given game, because in most cases it would be silly to assume that who wins a game is random.

Empirical probability, on the other hand, uses observed frequencies and a total number of trials to compute a probability. A perfectly reasonable question is, "How many trials are needed to get an accurate probability?" The answer, of course, is seven. Just kidding. There IS no answer to that question. All we can say is that the more trials you do in an experiment, the more likely it is that the probabilities you calculate are true representatives of the situation.

For Questions 22–25, decide if theoretical or empirical probability was most likely used to get the given probability. Justify your answer, of course.

22. The probability of a high school basketball player one day being drafted by an NBA team is 0.000408.

Empirical. This would be based on gathering data about the number of high school players compared to the number of players that get drafted.

23. The probability of being dealt a one pair in five-card poker is 0.422569.

Theoretical. Every hand is equally likely, so you can count the number of possible outcomes and the number that result in one pair.

> **Math Note**
>
> Each of these probabilities was found by doing a Google search for the string "The probability of."

24. The probability of being in a fatal accident if you make two car trips a day for one year is $\frac{1}{13,750}$.

Empirical. This is based on the number of fatal accidents and the total number of driving trips.

25. The probability of winning some prize when buying one New York State Lottery scratch-off ticket is 0.216.

Theoretical. The games are designed to have a certain number of winning tickets out of the total number that are sold.

4-1 Portfolio

Name ____

Check each box when you've completed the task. Remember that your instructor will want you to turn in the portfolio pages you create.

Technology

1. ☐ Have you ever seen a fifteen-sided die? I know I haven't. But you can roll one virtually in a spreadsheet. Actually, you can pick as many sides as you want using the **randbetween** command. Entering "=randbetween(1,15)" will generate a randomly selected number between 1 and 15, which of course is exactly the same thing as rolling a 15-sided die. In the tech template for this lesson (see online resources), we've set up a die rolling experiment. You can enter the number of sides you want, then copy down the formula in cell B2 as many times as you want. Each represents a single roll of the die. You decide on a number of trials, then compute empirical probabilities for each outcome. Finally, double the number of trials and repeat, comparing the second result to the theoretical probability of each number.

Online Practice

1. ☐ Include any written work from the online assignment along with any notes or questions about this lesson's content.

Applications

1. ☐ Complete the applications problems.

Reflections

Type a short answer to each question.

1. ☐ Describe the connection between the probability of an event occurring and the percent chance that it will occur.
2. ☐ Explain how theoretical probability differs from empirical probability.
3. ☐ When would you be likely to use empirical methods to calculate a probability rather than theoretical methods?
4. ☐ Take another look at your answer to Question 0 at the beginning of this lesson. Would you change your answer now that you've finished the lesson? How would you summarize the topic of this lesson now?
5. ☐ What questions do you have about this lesson?

Looking Ahead

1. ☐ Complete the Lesson 4-2 Prep Skills and read the opening paragraph in Lesson 4-2 carefully, then answer Question 0 in preparation for that lesson.

Answers to "Did You Get It?"

1. Sample space: all 12 pitchers. An outcome would be choosing one specific pitcher. An event would be something like choosing a left-handed pitcher.

2. a. $\frac{1}{75}$ **b.** 0 **c.** $\frac{49}{75}$ **d.** $\frac{38}{75}$ **3. a.** $\frac{1}{9}$ **b.** $\frac{5}{9}$ **c.** $\frac{1}{3}$ **4.** $\frac{5}{8}$

5. a. 0.248 **b.** 0.438 **c.** 0.686

4-1 Applications

Name ______________________________

1. On *The Price Is Right* game show, to get into the showcase at the end, contestants spin a wheel with 20 equally sized spaces, numbered 5 cents, 10 cents, 15 cents, and so on, up to $1.00. If the contestant spins once, find the probability of each event below.

a. $1.00

1/20

b. An amount less than 50 cents.

9/20

c. An amount greater than 75 cents

1/4

d. An amount that is divisible by five.

1

e. Which type of probability (empirical or theoretical) are you using to answer these questions? Explain.

Theoretical. We know for sure how many total outcomes there are, and we count up how many outcomes satisfy the given event.

2. If there are only 50 lottery tickets for the Big Game, one of which is a winning ticket, and you buy 7 of those tickets at random, what is the probability that you'll win the super jackpot? Which type of probability (empirical or theoretical) did you use to answer these questions? Explain.

7/50. This is also theoretical, since we know the total number of outcomes, and the number that satisfy the event.

4-1 Applications

Name ______________________________

3. A recent survey reported that 67% of Americans approve of human embryonic stem cell research. If an American is selected at random, find the probability that he or she will disapprove or have no opinion on the issue. Which type of probability (empirical or theoretical) did you use to answer this question? Explain.

If 67% approve, then either 33% either disapprove or have no opinion. This corresponds to a probability of 0.33. This is empirical probability, since we're basing our calculation on numbers that were experimentally determined.

4. There were 7,690 single-bias hate crime offenses reported in the United States in 2010. These crimes were divided into categories, with the following breakdown: 3,725 were based on racial bias, 1,409 on religious bias, 1,470 on sexual orientation bias, 1,040 on ethnic bias, and 46 on bias against those with disabilities.

a. Make a frequency distribution for this data.

Class	Frequency
Racial	3,725
Religious	1,409
Sexual Orientation	1,470
Ethnic	1,040
Disabilities	46

Now use your frequency distribution to find the probability of each event in parts b–d.

b. Crime was motivated by sexual orientation bias.

$1{,}470/7{,}690 \approx 0.19$

c. Crime was motivated by bias against race or ethnicity.

$(3{,}725 + 1{,}040)/7{,}690 \approx 0.62$

d. Crime was not motivated by bias against religion or those with disabilities.

$1 - (1{,}409 + 46)/7{,}690 \approx 0.81$

e. Which type of probability (empirical or theoretical) are you using to answer these questions? Explain.

Empirical. All are based on actual gathered data. Also, there's no assumption about outcomes being equally likely.

4-1 Applications

Name ______________________________

5. In December 2008, 500 men and 500 women were surveyed by Omnitel about their opinion on whether federal government bailout money should be used to help homeowners in default. The number giving each response is shown below.

Response	Men	Women
Yes	195	235
No	290	220
Not sure	15	45

a. If a person who participated in the survey is selected at random, what is the probability that he or she answered no?

(290 + 220)/1,000 = 0.51

b. What is the probability that the person selected is a man who answered either yes or no?

(195 + 290)/1,000 = 0.485

c. Based on the data from the survey, if you had stopped a random woman on the street in December 2008 and asked her opinion, what is the probability that she would have said that bailout money should not be used to help homeowners in default?

220/500 = 0.44

d. Which type of probability (empirical or theoretical) are you using to answer these questions? Explain.

Empirical. All are based on actual gathered data. Also, there's no assumption about outcomes being equally likely.

Lesson 4-2 Prep Skills

This is a short review of skills that will come in handy in the next lesson. In each case, answer the question, then rate your confidence level by checking one of the boxes. If you feel like you're struggling with these skills, consult the online resources provided by your instructor for extra practice.

I KNOW IT	THINK SO	UNSURE	NO IDEA	
				SKILL 1: MULTIPLY STRINGS OF INTEGERS
				1. Perform each multiplication without a calculator. $3 \cdot 2 \cdot 5$
				$4 \cdot 3 \cdot 5 \cdot 2$
				SKILL 2: COMPUTE PROBABILITIES
				2. The math department at my school offers a total of 27 different classes, three of which are calculus classes. If someone is goofy enough to pick a math class randomly, what's the probability that they'll end up in a calculus class?
				3. If you flip a coin twice, what is the probability of it landing heads up both times?
				4. What's the probability of rolling a 5 with one die?
				5. What's the probability of rolling a number less than 5 with one die?

Lesson 4-2 Make It Count (Sample Spaces and Counting Techniques)

LEARNING OBJECTIVES

- ☐ 1. Describe how counting techniques are useful in probability theory.
- ☐ 2. Use tree diagrams and tables to determine sample spaces and compute probabilities.
- ☐ 3. Develop and use the fundamental counting principle.

© MedioImages RF

Why does my brain insist on counting the steps every time I walk up a flight of stairs? I just can't help myself. There's something about my mind that always wants to keep counting.

—Rachel Nichols

For centuries, people have tried a wide variety of techniques, some of them pretty bizarre, to try and influence the gender of their children. (Look it up sometime—it's pretty fascinating.) The truth is, without the aid of cutting-edge science, you don't get to choose. But that doesn't stop many young couples from planning the type of family they hope to have. Suppose that one couple would like to have three children, but they definitely want to have at least one boy and one girl. What is the probability that they'll get their wish without having to go beyond three kids?

Believe it or not, the subject of this lesson is counting. Seriously. When working with theoretical probabilities, we know that we need to decide on the sample space for an event, and then find how many individual outcomes are in that event. When situations start to get complicated, it might not always be apparent how to do so. That's where developing counting techniques will be a big help.

0. After reading the opening paragraphs, what do you think the main topic of this lesson will be?

4-2 Class

Alright, let's start counting. One, two, three, four . . . just kidding. That's not the kind of counting we're talking about. The first technique we will study involves diagramming out all of the possible outcomes for probability experiments, like the one referenced in the opening paragraph above. The specific type of diagram we'll study is called a *tree diagram.*

A **tree diagram** is a diagram where we use branches, similar to a flow chart, to list the outcomes of two or more probability experiments that are done in sequence.

While it's not the most sensitive thing we can do, we can regard the birth of each child for the couple above as a probability experiment: each child is either a boy or a girl. When you start to look at three such experiments in succession, it's helpful to diagram out the possible outcomes.

To draw a tree diagram, we start with branches emanating, or starting from, a single point to show the possible outcomes for the first experiment. Next, we show the outcomes for the second experiment using branches emanating from the end of each branch that was used for the first experiment, and so on.

1. Complete the tree to find the sample space for the genders of three children in a family. Use B to indicate the outcome of having a boy and G to indicate the outcome of having a girl. One outcome is completed for you.

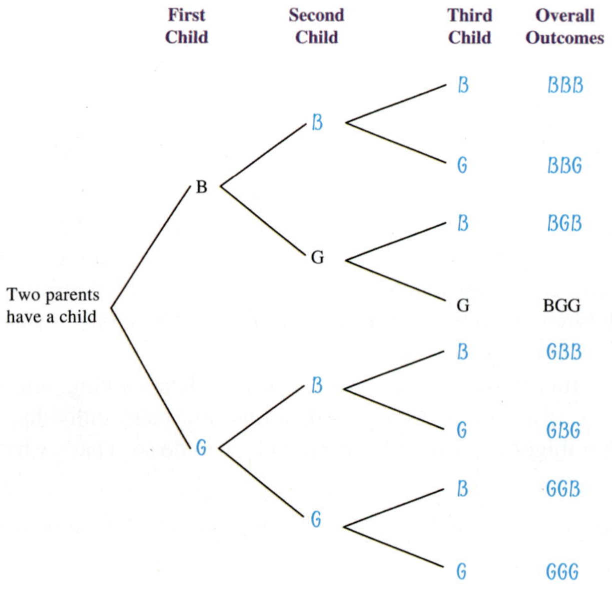

2. Now that a tree diagram is drawn, you can find the outcomes by tracing through all of the branches. List the sample space.

{BBB, BBG, BGB, BGG, GBB, GBG, GGB, GGG}

Make sure that the students have eight outcomes in their sample space. They may try to list the genders of individual children rather than the genders of all three.

3. Find the probability that a family with three children will have exactly two boys.

3/8

Math Note

Computing probability here requires assuming that there's a 50-50 chance of getting a boy every time. That's not exactly true, but it's pretty close.

4. Find the probability that a family with three children will have a least two girls.

4/8 or 1/2

5. Look back at the opening paragraph. What is the percent chance that the couple will get their wish?

6/8 = 75%

6. Describe how being able to count different possibilities helps us in computing probability.

Answers vary. Look for references to the formula for probability, in which we need to know the total number of outcomes in a sample space, and the number that make up a given event.

Did You Get It

Try these problems to see if you understand the concepts we just studied. The answers can be found at the bottom of the portfolio page.

1. A soda machine dispenses both Coke and Pepsi products, in both 12-ounce cans and 20-ounce bottles. For each brand, it has a regular cola, diet cola, and lemon-lime drink. Use a tree diagram to find the sample space for the experiment of choosing one drink at random from this machine.
2. Suppose the soda machine goes berserk and starts dispensing drinks randomly. If you want a diet cola, what is the probability that you'll get one?

In order to collect information for a student survey, a researcher classifies students according to eye color (blue, brown, green), gender (male, female), and class rank (freshman, sophomore). A folder for each classification is then made up (e.g., freshman/female/green eyes).

7. How many different outcomes do you think make up the sample space here? Discuss and see if you can come up with an answer without using a tree diagram.

We kind of want answers to vary here, although if they guess the fundamental counting principle before we develop it that would be pretty cool. For what it's worth, the answer is 12.

8. Find the sample space for the folders using a tree diagram.

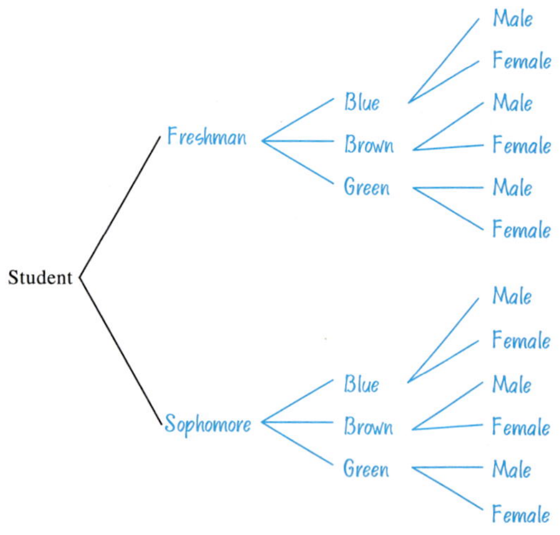

Write sample space here

Freshman, Blue, Male
Freshman, Blue, Female
Freshman, Brown, Male
Freshman, Brown, Female
Freshman, Green, Male
Freshman, Green, Female
Sophomore, Blue, Male
Sophomore, Blue, Female
Sophomore, Brown, Male
Sophomore, Brown, Female
Sophomore, Green, Male
Sophomore, Green, Female

9. If a folder is selected at random, find the probability that it includes students with blue eyes.

4/12 or 1/3

10. If a folder is selected at random, find the probability that it includes students who are female and don't have brown eyes.

4/12 or 1/3

11. If a folder is selected at random, find the probability that it includes students who are male freshmen.

3/12 or 1/4

Did You Get It ?

Try this problem to see if you understand the concepts we just studied. The answer can be found at the bottom of the portfolio page.

3. A coin is flipped, and then a die is rolled. Use a tree diagram to find the probability of getting heads on the coin and an even number on the die.

Tables

Another way of determining a sample space is by making a table. Consider the sample space of selecting a card from a standard deck of 52 cards. (The cards are assumed to be shuffled to make sure that the selection occurs at random.) There are four suits—hearts, diamonds, spades, and clubs, and 13 cards of each suit consisting of the denominations ace (A), 2, 3, 4, 5, 6, 7, 8, 9, 10, and 3 picture or face cards—jack (J), queen (Q), and king (K). The sample space is shown in the table below.

	A	2	3	4	5	6	7	8	9	10	J	Q	K
♥	A♥	2♥	3♥	4♥	5♥	6♥	7♥	8♥	9♥	10♥	J♥	Q♥	K♥
♦	A♦	2♦	3♦	4♦	5♦	6♦	7♦	8♦	9♦	10♦	J♦	Q♦	K♦
♠	A♠	2♠	3♠	4♠	5♠	6♠	7♠	8♠	9♠	10♠	J♠	Q♠	K♠
♣	A♣	2♣	3♣	4♣	5♣	6♣	7♣	8♣	9♣	10♣	J♣	Q♣	K♣

A card is drawn from an ordinary deck. In Questions 12–14, use the sample space shown above to find the probabilities of getting

12. A jack.

$4/52 = 1/13$

13. A black card with a number between 2 and 7.

$8/52 = 2/13$

14. A 3 or a diamond.

$16/52 = 4/13$

15. Describe why having the sample space listed out in table form is so useful for probability questions like this one.

You don't really have to think that much-you just have to count up the outcomes that match the given event.

4-2 Group

Let's continue our study of using tables to help us with probability by looking at an example where we need to construct our own table. When two dice are rolled, how many outcomes are in the sample space?

To get a handle on this question, let's pretend that one die is red and the other is green, and let's say that we roll the red die first, then the green one. If we get 1 on the red die, there are six possible overall outcomes: 1 on red and 1 on green, 1 on red and 2 on green, 1 on red and 3 on green, and so on. If we represent each possible outcome with an ordered pair, like (1, 3), then we can systematically list the possibilities in table form.

1. Complete the sample space table below for rolling a red die followed by a green one. The first few outcomes have been done for you because that's the way we roll.

Green Die

Red Die	1	2	3	4	5	6
1	(1, 1)	(1, 2)	(1, 3)	(1, 4)	(1, 5)	(1, 6)
2	(2, 1)	(2, 2)	(2, 3)	(2, 4)	(2, 5)	(2, 6)
3	(3, 1)	(3, 2)	(3, 3)	(3, 4)	(3, 5)	(3, 6)
4	(4, 1)	(4, 2)	(4, 3)	(4, 4)	(4, 5)	(4, 6)
5	(5, 1)	(5, 2)	(5, 3)	(5, 4)	(5, 5)	(5, 6)
6	(6, 1)	(6, 2)	(6, 3)	(6, 4)	(6, 5)	(6, 6)

Now that we have our sample space, we can calculate tons of probabilities. Notice, by the way, that the sum of the two dice gives us the total for the roll, so the outcome (1, 3) would be called "rolling a 4." Find each probability for a roll of two dice:

2. Rolling an 8.

5/36

3. Rolling doubles (the same number on each die).

$6/36 = 1/6$

4. Rolling less than 5.

$6/36 = 1/6$

Did You Get It

Try this problem to see if you understand the concepts we just studied. The answer can be found at the bottom of the portfolio page.

4. Two dice are rolled. Use your sample space from Question 1 to find the probability of
   a. Rolling a 9.
   b. Rolling an even number.
   c. Rolling a number greater than 6.

After getting a cool new job, you naturally want a new apartment, and furniture to go along with it. The hip furniture boutique around the corner has the couch you want in either leather or micro suede, and each comes in your choice of four colors: tan, gray, red, and black.

© Castle Rock Entertainment/ Everett Collection

5. Create a tree diagram for this situation to determine how many different couch options you have to choose from.

Note that the bottom layer of the answer here is requested in Question 9 on the next page, not this question.

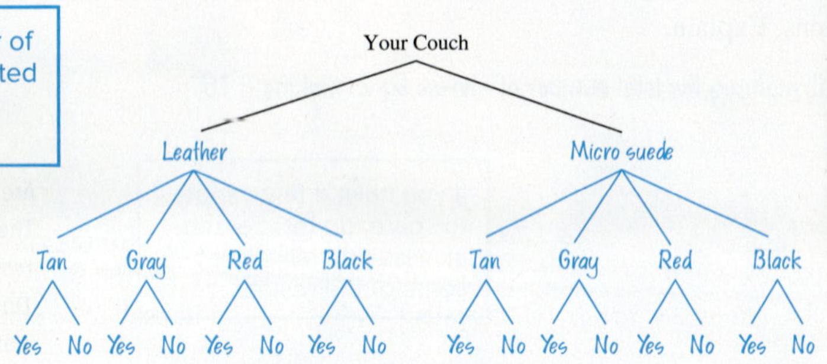

6. How many choices of couch are there? USING YOUR TREE DIAGRAM, explain why it makes perfect sense that the number of total choices is the product of the number of fabric choices and the number of color choices.

There are two branches on the tree for choice of material, then each of those branches has four branches coming off of it for color choice. So there are $2 \cdot 4$ choices total.

7. Here's a way to connect the first two topics of this lesson. The table below can be used to list all of the couch choices. Explain why this also justifies that the total number of choices is the product of the number of fabric choices and the number of color choices.

Color

Fabric	Tan (T)	Gray (G)	Red (R)	Black (B)
Leather (L)	LT	LG	LR	LB
Micro Suede (M)	MT	MG	MR	MB

The table has two rows and four columns, so it has 8 cells total.

8. More choices is better, right? Suppose the furniture salesperson calls you and points out that your couch also comes in a version with an end recliner, and one without. Based on what we've learned so far, make a conjecture as to how this will change the number of options. Explain.

Two more choices at the end will multiply the total number of choices by 2, making it 16.

If you have a few minutes to spare, do this search in class and talk about some of the results.

Math Note

Are more choices *really* better? There's some interesting research that says maybe not. Do a Google search for "choice paralysis" if you're interested.

9. Add this extra layer of choices to your tree diagram on the previous page. Does this confirm your answer to Question 8? Explain.

It does. Adding an extra two choices under each of the 8 outcomes doubled that number to 16.

We've now discovered an idea so important in counting that it's literally called "The Fundamental Counting Principle."

The Fundamental Counting Principle

In a sequence of n procedures, if the first can occur in k_1 ways, the second in k_2 ways, the third in k_3 ways, and so on, the total number of ways the sequence of procedures can occur is

$$k_1 \cdot k_2 \cdot k_3 \cdot \cdots k_n$$

The basic blood types are represented by letters: A, B, AB, and O. There is also an Rh factor represented by either positive or negative. If a local blood bank labels donations according to type, Rh factor, and biological gender of the donor, it would be useful to look at how many different ways a blood sample can be labeled, as this would have a significant impact on preparing for the labeling.

© Dynamic Graphics/ JupiterImages RF

10. How many different blood types are listed?

Four

11. How many different Rh factors are possible?

Two

12. How many different biological genders are likely to be listed?

Two

13. Based on these answers and the fundamental counting principle, how many different ways can the blood sample be labeled?

$4 \cdot 2 \cdot 2 = 16$

14. If you choose one of the labels randomly, what is the probability that it corresponds to male and AB-negative?

1/16

15. In reality, the probability of a randomly selected person in the U.S. population being a male with AB-negative blood is about 0.0035. What does that fact and your answer to Question 14 tell you about blood type?

That blood types are not all equally likely. Since AB-negative males are far fewer than you'd expect based on theoretical probability, it must be the case that AB-negative blood is less common than other types, at least among males.

IMPORTANT NOTE: The fundamental counting principle can only be used when the occurrence of the first event in no way affects the occurrence of the second event, which in turn does not affect the occurrence of the third event, and so on.

16. A company with 412 employees is designing new ID badges for its employees. Each badge will be coded with a four-letter sequence chosen from the letters A, B, C, D, and E. Will there be enough badges so that everyone gets a different code? Use the fundamental counting principle to decide. Note that letters can be used more than once.

There are five choices at each stage (stages corresponding to the letters in the ID), so using the fundamental counting principle, we get $5 \cdot 5 \cdot 5 \cdot 5 \cdot 5 = 3{,}125$. There are plenty of IDs to go around.

17. Will there be enough badges to go around if no letter can appear more than once in the code? (Hint: Think about the number of choices at each stage.)

This time, there are five choices at the first stage, four at the second (since we can't reuse the first letter), three at the third stage, two at the fourth stage, and only one at the last. Using the fundamental counting principle, we get $5 \cdot 4 \cdot 3 \cdot 2 \cdot 1 = 120$. Now there aren't nearly enough badges.

18. You could absolutely use a tree diagram to list out the possible codes. Explain why that's a tremendously bad idea.

You'd need an awful big piece of paper to list out that many different choices—and a lot of ink, and even more patience.

Did You Get It?

Try these problems to see if you understand the concepts we just studied. The answers can be found at the bottom of the portfolio page.

5. A discount paint manufacturer plans to make several different paints. The categories include

   Color: Red, blue, white, black, green, brown, yellow
   Type: Latex, oil
   Texture: Flat, semi gloss, high gloss
   Use: Outdoor, indoor
   How many different kinds of paint can be made?

6. The lock on a storage facility is controlled by a keypad containing digits 1 through 5.

   How many three-digit codes are possible if digits can be repeated?
   How many three-digit codes are possible if digits cannot be repeated?

4-2 Portfolio

Name ______________________________

Check each box when you've completed the task. Remember that your instructor will want you to turn in the portfolio pages you create.

Technology

1. ☐ There's a feature in Excel that can be used to build tree diagrams. On the Insert menu, click Smart Art, then Hierarchy, and choose one of the displayed options. It will give you a starting diagram. You'll find further instructions on the technology template for this lesson, which illustrates the tree diagram you drew on page 322. Your job is to build a new tree diagram for possible results of a four-question quiz that has two true-false questions, and two multiple choice questions with four choices each. Then use your tree diagram to answer the probability questions on the tech template.

Online Practice

1. ☐ Include any written work from the online assignment along with any notes or questions about this lesson's content.

Applications

1. ☐ Complete the applications problems.

Reflections

Type a short answer to each question.

1. ☐ What is meant by the term "counting" in this lesson?
2. ☐ Explain how counting techniques are important to the study of probability.
3. ☐ Why is the fundamental counting principle more widely applicable than tree diagrams?
4. ☐ Is the fundamental counting principle useful for empirical or theoretical probabilities? Why?
5. ☐ Take another look at your answer to Question 0 at the beginning of this lesson. Would you change your answer now that you've finished the lesson? How would you summarize the topic of this lesson now?
6. ☐ What questions do you have about this lesson?

Looking Ahead

1. ☐ Complete the Lesson 4-3 Prep Skills and read the opening paragraph in Lesson 4-3 carefully, then answer Question 0 in preparation for that lesson.

Answers to "Did You Get It?"

1. Sample space: {Coke, can, cola; Coke, can, diet cola; Coke, can, lemon-lime; Coke, bottle, cola; Coke, bottle, diet cola; Coke, bottle, lemon-lime; Pepsi, can, cola; Pepsi, can, diet cola; Pepsi, can, lemon-lime; Pepsi, bottle, cola; Pepsi, bottle, diet cola; Pepsi, bottle, lemon-lime}

2. $\frac{1}{3}$ **3.** $\frac{1}{4}$ **4. a.** $\frac{1}{9}$ **b.** $\frac{1}{2}$ **c.** $\frac{7}{12}$ **5.** 84 **6.** 125 repeated, 60 not repeated

4-2 Applications

Name ______________________________

1. Use the sample space for a standard deck of cards (refer back to page 323 if necessary) to answer each question.

   a. We know that there are 52 cards in a standard deck. Explain how you can use the fundamental counting principle to arrive at this same number.

   Four suits, 13 cards in each suit: $4 \cdot 13 = 52$.

   b. Before the cards are dealt for a game of poker, what's the probability that the 31st card in the deck is a spade?

   1/4

   c. After 5 cards have been dealt to each of six players, with eight spades among those dealt, what's the probability that the 31st card in the deck is a spade?

   22 cards remain, and 5 spades remain, so the probability is 5/22.

2. Akira wants to buy his first new car and can select one option from each category:

Model	Engine Type	Color
Ford Focus	Hybrid	Burnt Copper
Honda Civic	E-85	Cobalt blue
Toyota Corolla		Metallic green

   a. Using the fundamental counting principle, how many outcomes should there be in the sample space?

   $3 \cdot 2 \cdot 3 = 18$

4-2 Applications

Name ____________________

b. Draw a tree diagram and find the sample space for all possible choices.

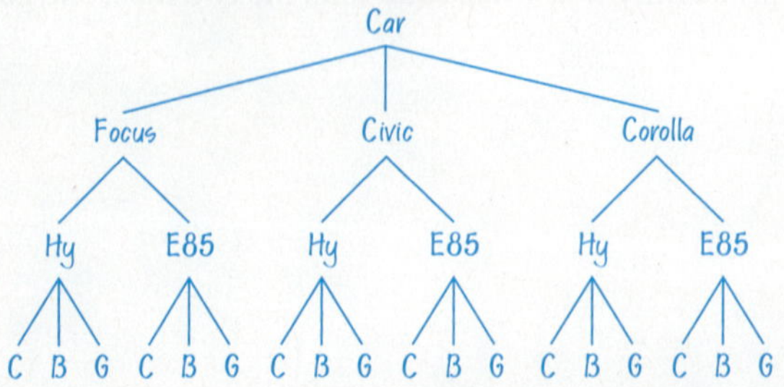

Sample space: {Focus hybrid copper, Focus hybrid blue, Focus hybrid green, Focus E-85 copper, Focus E-85 blue, Focus E-85 green, Civic hybrid copper, Civic hybrid blue, Civic hybrid green, Civic E-85 copper, Civic E-85 blue, Civic E-85 green, Corolla hybrid copper, Corolla hybrid blue, Corolla hybrid green, Corolla E-85 copper, Corolla E-85 blue, Corolla E-85 green}

In parts c–g, find the probability that the car, if chosen at random,

c. Has an E-85 engine.

1/2

d. Is a burnt copper Ford Focus.

2/18 = 1/9

e. Is a metallic green hybrid Honda Civic.

1/18

f. Is cobalt blue.

1/3

g. Is either a metallic green or cobalt blue Toyota Corolla.

4/18 = 2/9

4-2 Applications

Name ____________________

3. In between classes, Jade plays a game of online Monopoly on her laptop. Using the sample space for rolling two dice that you created in the Group portion of this lesson, find the probability that when Jade rolls the two dice, she gets a

a. Sum of 5.

1/9

b. Sum of 7 or 11.

2/9

c. Sum greater than 9.

1/6

d. Sum less than or equal to 5.

5/18

e. Three on one die or on both dice.

11/36

f. Sum that is odd.

1/2

g. Prime number on one or both dice.

3/4

h. Sum greater than 1.

1

Lesson 4-3 Prep Skills

This is a short review of skills that will come in handy in the next lesson. In each case, answer the question, then rate your confidence level by checking one of the boxes. If you feel like you're struggling with these skills, consult the online resources provided by your instructor for extra practice.

I KNOW IT | THINK SO | UNSURE | NO IDEA

SKILL 1: DIVIDE FRACTIONS AND REDUCE THE RESULT

1. Perform the division. Make sure your answer is in lowest terms.

$$\frac{\frac{1}{5}}{\frac{3}{5}}$$

$$\frac{\frac{8}{13}}{\frac{10}{13}}$$

SKILL 2: COMPUTE THEORETICAL PROBABILITIES

2. What is the probability of rolling a total of 6 with two dice? What about a total greater than 6?

3. There are 2 seniors, 1 junior, 7 sophomores, and 11 freshmen in a math course. If one student is randomly picked for extra credit, what's the probability that it's either a freshman or senior?

SKILL 3: CALCULATE A SUM OF PRODUCTS

4. Find the value of the expression below in both fractional and decimal form:

$$-10 \cdot \frac{4}{9} + 2 \cdot \frac{2}{9} + 4 \cdot \frac{2}{9} + 15 \cdot \frac{1}{9}$$

Lesson 4-3 Odds and Ends (Odds and Expected Value)

LEARNING OBJECTIVES

- ☐ 1. Distinguish between odds and probability.
- ☐ 2. Compute and interpret the odds in favor of and odds against an event.
- ☐ 3. Compute odds from probability and vice versa.
- ☐ 4. Develop a procedure for finding expected value.
- ☐ 5. Compute and interpret expected values.

© Focus on Sport/Getty Images

When something is important enough, you do it even if the odds are not in your favor.

—Elon Musk

The New England Patriots won the Super Bowl on February 1, 2015, and by the time the ink dried on newspapers reporting the victory, oddsmakers in Las Vegas had listed the odds against the Patriots winning it again in 2016 as 7 to 1. But what exactly does that mean? The term "odds" is used all the time in describing the likelihood of something happening, but a lot of people don't understand exactly what a given set of odds means. Odds are used by casinos, racetracks, and other gambling establishments to determine the payoffs when bets are made or lottery tickets are purchased. They're also used by insurance companies in determining the amount to charge for premiums. In a very real sense, odds are a way of stating what you expect will happen. So in this lesson, we'll study odds and expectations.

0. After reading the opening paragraph, what do you think the main topic of this lesson will be?

4-3 Class

The formulas for computing odds are similar to the formula we've been using for classical probability, and shortly we will see a strong connection between the two concepts.

If an event E has a favorable outcomes and b unfavorable outcomes, then

- The **odds in favor** of event E occurring are $\frac{a}{b}$ (also written as $a{:}b$).
- The **odds against** event E occurring are $\frac{b}{a}$ (also written as $b{:}a$).

Odds can be expressed as a fraction or a ratio.

1. Write the odds against New England repeating as Super Bowl champion as a fraction, and as a ratio.

7/1 and 7:1

Math Note

In common usage, the phrase "the odds of" means the same thing as "the odds against." Odds are also commonly written with a hyphen rather than a colon: 7-1, for example.

2. Write the odds in favor of New England repeating as Super Bowl champion as a fraction, and as a ratio.

1/7 and 1:7

3. By setting New England's odds at 7 to 1, the oddsmakers were predicting that if the season were played eight times, New England would win the Super Bowl __1__ times, and not win it __7__ times.

Did You Get It

Try this problem to see if you understand the concepts we just studied. The answer can be found at the bottom of the portfolio page.

1. What are the odds in favor of rolling a prime number sum with a roll of two dice? What are the odds against? Look back at your answer to Question 1 in the Group portion of Lesson 4-2 for reference.

4. Based on the Question 3, write the probability of New England winning the Super Bowl and the probability of New England not winning the Super Bowl in fraction form. Then compare your answers with the formula below.

$P(\text{winning}) =$ 1/8

$P(\text{not winning}) =$ 7/8

Formula for Probability in Terms of Odds

If the odds in favor of an event E are *a:b*, then the probability that the event will occur is

$$P(E) = \frac{a}{a+b}$$

5. Fill in the blanks to make this a true statement: P(winning) = <u>1</u> – P(<u>not winning</u>).

6. Explain why your answer to Question 5 makes perfect sense based on a fact about probability that we studied earlier in this unit.

Not winning the Super Bowl is the complement of winning the Super Bowl. The formula for probability of the complement of some event is 1 – the probability of that event occurring.

7. Compute and simplify $\frac{\text{P(winning)}}{\text{P(not winning)}}$. What does your answer represent in terms of odds?

$$\frac{\frac{1}{8}}{\frac{7}{8}} = \frac{1}{8} \cdot \frac{8}{7} = \frac{1}{7}$$

This is the same as the odds in favor of winning.

8. Compute $\frac{\text{P(not winning)}}{\text{P(winning)}}$. What does your answer represent in terms of odds?

$$\frac{\frac{7}{8}}{\frac{1}{8}} = \frac{7}{8} \cdot \frac{8}{1} = \frac{7}{1}$$

This is the same as the odds against winning.

The results of Questions 7 and 8 can be summarized in this formulas:

Formula for Odds in Terms of Probability

If the probability that an event E will occur is $P(E)$, then

$$\text{Odds in favor of } E = \frac{P(E)}{1 - P(E)} \text{ and Odds against } E = \frac{1 - P(E)}{P(E)}$$

9. According to the National Safety Council, the odds in favor of dying due to injury at some point in your life are about 10:237. Find the probability of dying from injury.

$$P(\text{Dying}) = \frac{10}{10 + 237} = \frac{10}{247}$$

10. The probability of getting exactly one pair in a five-card poker hand is 0.423. Find the odds in favor of getting exactly one pair, and the odds against.

$$\text{Odds in favor} = \frac{P(\text{getting one pair})}{1 - P(\text{getting one pair})} = \frac{0.423}{1 - 0.423} = \frac{0.423}{0.577} \text{ or } \frac{423}{577}$$

$$\text{Odds against} = \frac{577}{423}$$

This is a good time to remind students that the odds in favor of an event and the odds against the same event are reciprocals. You should probably also help with converting the odds in favor to a form where both numerator and denominator are whole numbers.

Did You Get It

Try these problems to see if you understand the concepts we just studied. The answers can be found at the bottom of the portfolio page.

2. According to the American Cancer Society, the probability of an American female developing some type of cancer at some point in her life is about 1/3. Find the odds in favor of and against an American woman developing cancer.
3. When two dice are rolled, the odds in favor of getting a sum of 9 are 1:8. Find the probability of not getting a sum of 9 when two dice are rolled.

11. On the morning of the 2015 Kentucky Derby, American Pharoah was listed as the favorite, with odds of 5:2. Write a sentence or two describing what this means.

This says that in the opinion of the oddsmakers, if the race were run seven times, American Pharoah would win twice.

12. If someone had bet $10 on American Pharoah to win (which he did), about how much money do you think they would have gotten back? Justify your answer.

This is to some extent an opinion question. The odds, however, say that you would expect to get back about 2.5 times what you bet, so $25 is a reasonable answer.

4-3 Group

Expected Value

If you play lottery games regularly (something we completely advise against, by the way), how much money do you expect to win? Obviously, this is not the same question as "How much money do you hope to win?" The concept of **expectation,** or **expected value,** is used to determine the result that would be expected over the long term in some sort of gamble. It's used not only for games of chance, but in areas like insurance, management, engineering, and others.

Here's the key thing to remember as we study expected value: *it only makes sense for events that have numerical outcomes.* For example, rolling a die has a numerical outcome (1 through 6), and expected value can be used to determine what the average long-run result is likely to be. But it doesn't make sense to ask what the long-term average of flipping a coin is. (There's nothing between heads or tails, while an average roll of 2.7 at least has some meaning.)

Now let's illustrate a procedure for computing the expected value of a game, and then we'll summarize that procedure for future reference.

Suppose you pay a dollar to roll two dice. If you roll 5 or 6, you get your dollar back plus two more just like it. If not, you get nothing and like it. The goal will be to find the amount of money you can expect to win or lose if you play this game 100 times. First, we'll find the probability of winning and losing. We know that there are 36 possible rolls of two dice. (Refer back at your answer to Question 1 in the Group portion of Lesson 4-2 if you need to.)

1. How many outcomes result in 5? List them.

Four: (1, 4), (2, 3), (3, 2), and (4, 1).

2. How many outcomes result in 6? List them.

Five: (1, 5), (2, 4), (3, 3), (4, 2), and (5, 1).

3. What's the probability of winning the game?

9/36 or 1/4

4. What's the probability of losing the game?

$1 - (1/4) = 3/4$

5. Based on Questions 3 and 4, on average, you can expect to win __25%__ of the time.

6. So if you play the game 100 times, on average you'd expect to win __25__ times and lose __75__.

7. With each win, you make __$2__, and with each loss, you lose __$1__.

8. Multiply the number of times you expect to win by the amount you'd win each time.

$25 \times \$2 = \$50$

9. Multiply the number of times you expect to lose by the amount you'd lose each time.

$75 \times \$1 = \$75$

The next three questions are the key points in this part of the lesson.

10. After playing the game 100 times, find how much you would expect to win or lose total. Show how you got your answer.

You'd expect to lose $25, because the total expected winnings are $50 and the total expected loss is $75.

11. Divide your answer to Question 10 by 100, and use your result to answer the next question.

$-\$25/100 = -\$0.25$

12. You would expect to __lose__ (win/lose) on average __25 cents__ every time you play this game. This amount is the **expected value** of the game.

Did You Get It

Try this problem to see if you understand the concepts we just studied. The answer can be found at the bottom of the portfolio page.

4. On a roulette wheel, there are 38 slots, 18 of which are colored red. If you bet $5 on red and win, you get $10 back. If red doesn't come up, you lose your $5. Find the expected value of playing the game 100 times. If you need help, follow Questions 3–12, which built the procedure for finding expected value.

Your answer to Question 12 might seem really puzzling if you think about it. How can you lose 25 cents on one trial of the game? The problem is that you can't: You either win \$2 or lose \$1. The expected value never tells you what will happen on any given outcome: It tells you what the average result will be in the long run. That's a key difference.

The procedure that we built for computing expected value was based on 100 trials of a probability experiment. If we apply that same procedure to just one trial, and extend to cases with more than two possible outcomes, we get a general formula for the expected value of a probability experiment.

Formula for Expected Value

The expected value for the outcomes of a probability experiment is

$$E = X_1 \cdot P(X_1) + X_2 \cdot P(X_2) + \cdots X_n \cdot P(X_n)$$

where the *X*'s correspond to the numerical outcomes and the *P*(*X*)'s are the corresponding probabilities of the outcomes.

Verbally, the expected value formula tells us to multiply each outcome by its probability, then add the results.

13. When a single die is rolled, make a guess as to what the expected value of the roll will be.

Answers vary.

14. To help us calculate the expected value, complete the table below, listing the possible outcomes and their probabilities:

Number on die (*X*)	1	2	3	4	5	6
P(*X*)	$\frac{1}{6}$	$\frac{1}{6}$	$\frac{1}{6}$	$\frac{1}{6}$	$\frac{1}{6}$	$\frac{1}{6}$

15. Using information from your table and the formula in the box above, calculate the expected value of one roll.

$$E = 1 \cdot \frac{1}{6} + 2 \cdot \frac{1}{6} + 3 \cdot \frac{1}{6} + 4 \cdot \frac{1}{6} + 5 \cdot \frac{1}{6} + 6 \cdot \frac{1}{6} = \frac{21}{6} = 3.5$$

Did You Get It

Try this problem to see if you understand the concepts we just studied. The answer can be found at the bottom of the portfolio page.

5. If seven cards are numbered with integers from –2 to 4, then placed into a box and picked out at random, find the expected value.

In gambling games, the expected value equation amounts to multiplying net gain for each outcome by the corresponding probabilities and then finding the sum.

The prize in a raffle is a flat-screen TV valued at $350, and 1,000 tickets are sold for a dollar each. Let's find the expected value if you buy 1 ticket.

16. For a win, the net gain is __\$349__, since you don't get the cost of the ticket ($1) back.

17. For a loss, the net gain is represented by a negative number, in this case, __−$1__.

18. Organize the outcomes and probabilities in a table and compute the expected value.

Outcome	Probability
$349	1/1,000
−$1	999/1,000

$$E = -\$1 \cdot \frac{999}{1{,}000} + \$349 \cdot \frac{1}{1{,}000} = \frac{-\$650}{1{,}000}$$
$$= -\$0.65$$

19. Can you lose 65 cents from buying one ticket?

You sure can't. Nope.

20. If that 65 cents looks familiar, your result isn't incorrect. What does that result mean?

Again, this doesn't mean that you'd lose 65 cents from buying a ticket, because you can only lose a dollar or win a $350 prize. It means that if you were to buy tickets to a similar raffle repeatedly, in the long run you'd average a 65 cent loss for each ticket bought.

A favorite game for small children at fairs and carnivals is a game some folks call Pick-a-Duck. It's a simple game, which makes it ideal for little ones. A small pool is full of rubber ducks. Children get to pick one of the ducks and their prize is determined by the color of the dot on the bottom of the duck. Of course, the probability of winning a certain prize is based on how many ducks are in the pool, and how many have each color.

Courtesy of Brian Mercer

In one such game, you pay $1 to select one of 50 ducks from the pool and prizes are determined by the following list:

Number of ducks	Outcome
20 ducks with red dots	Lose the money you paid to play*
10 ducks with orange dots	Win a sticker valued at $0.05
10 ducks with green dots	Win a small toy valued at $0.10
6 ducks with blue dots	Get to play again
4 ducks with purple dots	Win a stuffed animal valued at $4.00

* Then threaten to cry unless Mom lets you play again.

21. Fill in the table.

Color of dot	Net gain X	Probability $P(X)$
Red	−$1	2/5
Orange	−$0.95	1/5
Green	−$0.90	1/5
Blue	0	3/25
Purple	$3	2/25

22. Find the expected value of playing the game once.

$$E = -\$1 \cdot \frac{2}{5} + -\$0.95 \cdot \frac{1}{5} + -\$0.90 \cdot \frac{1}{5} + \$0 \cdot \frac{3}{25} + \$3 \cdot \frac{2}{25}$$

$$= \frac{-\$10 - \$4.75 - \$4.50 + \$6}{25} = \frac{-\$13.25}{25} = -\$0.53$$

23. Is the owner of the game making money? Explain.

Yes they are. For every dollar that mom spends, on average the owner of the game is keeping 53 cents. We know this because on average the loss per game for players is 53 cents.

Did You Get It

Try this problem to see if you understand the concepts we just studied. The answer can be found at the bottom of the portfolio page.

6. The profit made by a small ski resort, not surprisingly, depends largely on the seasonal weather. In a season with more than 75 inches of snow, it makes an average of $250,000. If snowfall is between 40 and 75 inches, the average profit is $160,000, and if snowfall is less than 40 inches, it loses $70,000. The resort gets over 75 inches of snow 40% of years, between 40 and 75 inches 45% of years, and less than 40 inches 15% of years. Find the resort's expected yearly profit.

In gambling games, if the expected value of the gain is 0, the game is said to be fair. If the expected value of the gain of a game is positive, then the game is in favor of the player. That is, the player has a better-than-even chance of winning. If the expected value of the gain is negative, then the game is said to be in favor of the house. That is, in the long run, the players will lose money.

24. The expected value of different casino games varies, but the expected values of every game in every casino have one thing in common. What is it?

They're all negative. Casinos don't run games that will lose them money in the long run.

Math Note

In American roulette, there are 22 different types of bets you can place, but it's interesting to note that all but one of them have the exact same expected value: –$0.053 on a $1 bet. (Betting on 0, 00, 1, 2, and 3, called a five-number bet, is worse, at –$0.079.)

A FINAL NOTE: The computation for expected value can be programmed into a spreadsheet. In fact, that's what you'll be doing in the technology assignment. There's also a command for computing expected value on a TI-84 graphing calculator. There's a document in the online resources for this lesson that walks you through that process.

4-3 Portfolio

Name ____

Check each box when you've completed the task. Remember that your instructor will want you to turn in the portfolio pages you create.

Technology

1. ☐ In computing expected value, we set up a table, then did some calculations using the numbers in that table. That sure sounds an awful lot like what a spreadsheet is designed to do. Your job is to set up a spreadsheet that will calculate expected value when you input numerical outcomes in one column, and the corresponding probabilities in the column next to it. Then you can use formulas to multiply each outcome by its probability, then add the results. First, do so for the experiment of rolling one die. Then repeat the process, this time for the experiment of rolling two die. The table from Question 1 in the Group portion of Lesson 4-2 will be a big help. A template to help you get started can be found in the online resources for this lesson.

Online Practice

1. ☐ Include any written work from the online assignment along with any notes or questions about this lesson's content.

Applications

1. ☐ Complete the applications problems.

Reflections

Type a short answer to each question.

1. ☐ It's illegal to bet on political races in the United States, but overseas betting on American elections is common. As of this writing, Hillary Clinton is the favorite to win the presidency in 2016, with odds listed as 8/11. Explain what that means, including relating it to probability.
2. ☐ How is it possible for the expected value of a game to be losing 65 cents if the only outcomes are win \$2 or lose \$1?
3. ☐ Take another look at your answer to Question 0 at the beginning of this lesson. Would you revise your answer now that you've completed the lesson? How would you summarize the topic of this lesson now?
4. ☐ What questions do you have about this lesson?

Looking Ahead

1. ☐ Complete the Lesson 4-4 Prep Skills and read the opening paragraph in Lesson 4-4 carefully, then answer Question 0 in preparation for that lesson.

Answers to "Did You Get It?"

1. In favor 5:7; against 7:5 **2.** In favor 1:2, against 2:1 **3.** 8/9 **4.** –\$26.32 **5.** 1 **6.** \$161,500

4-3 Applications

Name ______________________________

1. A charity is running an auction. One thousand tickets are sold at $2 each for four prizes of $100, four of $50, four of $25, and four of $10. What are the odds in favor of winning $100?

There are 4 favorable outcomes and 996 unfavorable, so the odds in favor of winning $100 are 4:996 or 1:249.

2. What are the odds against winning any prize at all?

There are 16 favorable outcomes and 984 unfavorable, so the odds against winning a prize are 984:16 or 123:2.

3. What is the expected value if you buy one ticket? Fill in the table to help organize the needed information.

$$E = -\$2 \cdot \frac{123}{125} + \$8 \cdot \frac{1}{250} + \$23 \cdot \frac{1}{250} + \$48 \cdot \frac{1}{250} + \$98 \cdot \frac{1}{250}$$
$$= \frac{-\$246 + \$8 + \$23 + \$48 + \$98}{250} = -\$0.276$$

Outcome	Gain	Probability
Lose	−$2	123/125
Win $10	$8	1/250
Win $25	$23	1/250
Win $50	$48	1/250
Win $100	$98	1/250

4. What does your answer to Question 3 mean?

If you bought tickets repeatedly, in the long run on average you'd lose almost 28 cents for each ticket purchased.

4-3 Applications

Name ______________________________

5. A stock you bought two years ago with high hopes is now selling for less than you paid, and things look grim for the company. Do you sell, or hold on and hope it will come back to the original price before you sell? A model economists use for such situations is a game no one wants to play. Suppose you have a choice: lose \$100 or take a 50-50 chance between losing nothing and losing \$300. Which do you choose? Find the expected value for each strategy.

You can lose \$100 with probability 1, or lose \$300 with probability 0.5. If you take the first choice, your expected value is easy: it's just $-\$100$. With the second choice, you get $E = \$0 \cdot 0.5 + -\$300 \cdot 0.5$ which is $-\$150$. Ultimately you'd be better off sucking it up and losing the \$100.

There's an old joke among math and stats professors that lotteries are a tax on people that are bad at math. Let's take a look at one lottery game and use what we've learned about odds and expected value to study how good of an idea it would be to play the game. The image below is the back of a \$5 scratch-off ticket from the Illinois Lottery.

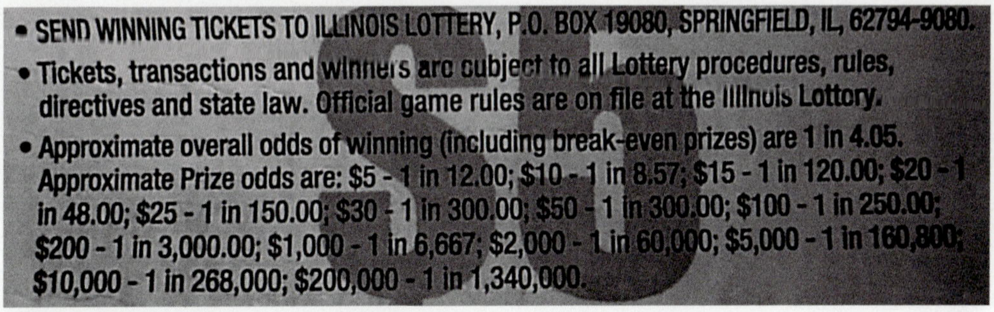

Courtesy of Brian Mercer

6. Some of the data above are described as "odds." Are they really odds? Describe exactly what these numbers represent, using terminology from this lesson.

They're actually probabilities. For the \$5 prize, for example, "One in 12" means one out of twelve times you will win that prize. In odds terms, that means the odds against winning are 11:1 and the odds in favor of winning are 1:11.

4-3 Applications

Name ______________________________

7. Fill in the table below by using the "odds" on the back of the ticket to calculate probabilities for each numerical outcome (from the perspective of a player, not the lottery commission). Remember, you have to pay $5 for the ticket. Round probabilities to three digits after any beginning zeroes. We strongly suggest using a spreadsheet to do the calculations for this question and the next one.

Outcome	Probability
-$5	0.753
$0	0.0833
$5	0.117
$10	0.00833
$15	0.0208
$20	0.00667
$25	0.00333
$45	0.00333

Outcome	Probability
$95	0.00400
$195	0.000333
$995	0.000150
$1,995	0.000 0167
$4,995	0.00 000 622
$9,995	0.00 000 373
$199,995	0.000 000 746

8. Use your table to find the expected value of buying one ticket. Again, a spreadsheet would be a great idea.

−$5	0.753	−$3.77
$0	0.0833	$0
$5	0.117	$0.59
$10	0.00833	$0.08
$15	0.0208	$0.31
$20	0.00667	$0.13
$25	0.00333	$0.08
$45	0.00333	$0.15
		−$2.42

$95	0.00400	$0.38
$195	0.000333	$0.06
$995	0.000150	$0.15
$1,995	0.0000167	$0.03
$4,995	0.00000622	$0.03
$9,995	0.00000373	$0.04
$199,995	0.000000746	$0.15
		$0.85

$E = -\$2.42 + \$0.85 = -\$1.57$

4-3 Applications

Name ______________________________

9. Carefully explain exactly what your answer to Question 8 represents. If you don't have an answer for Question 8, make an educated guess as to what you think the expected value would be and explain what it would mean.

If you play this game a lot, in the long run you'll lose \$1.57 every time you play.

10. How much profit would the state expect to make if they sold 5,000,000 of these tickets?

$5{,}000{,}000 \times \$1.57 = \$7{,}850{,}000$

11. Some people think that lotteries are a bad idea because they encourage gambling among people that can least afford to lose money. Others feel like the revenue that states make from lotteries is important to education and other social services. What are your thoughts?

Answers will vary.

Lesson 4-4 Prep Skills

This is a short review of skills that will come in handy in the next lesson. In each case, answer the question, then rate your confidence level by checking one of the boxes. If you feel like you're struggling with these skills, consult the online resources provided by your instructor for extra practice.

I KNOW IT | THINK SO | UNSURE | NO IDEA

SKILL 1: ORGANIZE INFORMATION FROM A TABLE

The cost (in thousands of dollars) of all houses sold in one town for the last week are below.

146	197	112	124	133	219	245	150	165	179
119	133	202	107	198	348	167	209	124	123

1. How many houses sold for less than $125,000?

2. What was the cost of the most expensive house that sold?

3. What was the cost of the least expensive house that sold?

4. How many houses sold last week?

5. How many houses had a sale price between $150,000 and $175,000?

Lesson 4-4 Crunching the Numbers (Gathering and Organizing Data)

© Chad Baker/Getty Images RF

LEARNING OBJECTIVES

- ☐ 1. Explain the difference between a population and a sample.
- ☐ 2. Compare and contrast different sampling methods.
- ☐ 3. Organize data with frequency distributions.
- ☐ 4. Analyze data with stem and leaf plots.

Errors using inadequate data are much less than those using no data at all.
—Charles Babbage

One of the big issues being discussed over the last 10 years or so in higher education is grade inflation—the perception that college students in general are getting much higher grades than they used to. But perception and reality are often two different things, so how could we decide if this is actually taking place? And if so, what are some possible reasons? Are students just getting smarter? Are professors lowering their standards?

In order to examine questions like this, it would be very valuable to gather some **data,** which are measurements or observations that are gathered for an event under study. The term "data age" is coming into vogue to describe this era because of the vast amount of data that is available to anyone with Internet access and a web browser. But the expense of having access to so much data is that knowing how to gather data appropriately, and how to organize it into a useful form, is more important than it's ever been. So that's what we'll study in this lesson.

0. After reading the opening paragraph, what do you think the main topic of this lesson will be?

4-4 Class

As we begin our study of statistics, it would seem reasonable to have a really good idea of what that actually means. **Statistics** is a branch of math that involves collecting, organizing, summarizing, and presenting data and drawing general conclusions from that data. If we're talking about grade inflation, then relevant data might be average grade point averages for students in several different years. The table below shows some GPA data from the website Gradeinflation.com.

School Year	Average Undergraduate GPA
1991–1992	2.93
1996–1997	2.99
2001–2002	3.07
2006–2007	3.11

Math Note

The data in this table caused so much negative publicity that many schools stopped sharing GPA data after they were published in 2009, making it harder to see if the trend continued.

In statistics, we *start* with a question we'd like to answer, like "Is grade inflation a real phenomenon?" *Then* we gather data, and try to interpret it in order to answer that question.

1. Write a verbal description of the information provided in the table. Do you think it supports the idea of grade inflation?

Average GPAs have been going up pretty steadily every five years. The increase in each five-year period was 0.06, 0.08, and 0.04. Certainly GPAs have gone up, but the numbers don't say anything about why that might be happening.

2. How do you think these numbers were calculated? What students' GPAs were considered? Discuss.

Answers vary. It's good to have students discuss this before we study sampling vs. population, and sampling methods.

© Royalty Free/Corbis

The ideal way to gather GPA data would be to add the GPAs of every college student in America during each year, then divide by the number of students. Needless to say, that's not particularly realistic. Aside from legal ramifications, in any given year there are more than 20 million students enrolled in American colleges, and by the time all of that GPA data were gathered and recorded, many of them would be contemplating retirement. Instead, GPA data is obtained from a *sample* of college students.

A **population** consists of all subjects under study. A **sample** is a representative subgroup or subset of a population.

3. In the case of the GPA data we've been studying, the population is *every college student*.

4. The sample is *the group of students whose GPAs were included in the averages*.

In this case, GPA data from a sample of 70 different colleges were used to compute the average. The thing that makes the data compelling is that the sample is representative of the population as a whole. The schools chosen vary in size, cost, and geographic location. Some are public, some are private.

5. Discuss what exactly is meant by a representative sample, and why it's so important in answering questions using statistics.

Answers vary. Get students to focus on the fact that since you're not getting data from every individual in a population, you have to do everything you can to ensure that the individuals you sample provide data that's consistent with data you'd get if you were able to sample the entire population.

This is probably the key point in understanding what sampling is all about, and is the main reason this portion of the lesson is listed as Class. Make sure that you discuss good answers to these questions.

Sampling Methods

> **Math Note**
>
> The word "data" is plural, so we say "data are," not "data is." The singular version is "datum."

We will study four basic sampling methods that can be used to obtain a representative sample: random, systematic, stratified, and cluster.

In order to obtain a **random sample,** every subject of the population has to have an equal chance of being selected. The best way to obtain a random sample is to use a list of random numbers. These can be generated from a variety of Internet sources, or from a computer or calculator. Subjects in the population are numbered, and then the ones to be studied are picked by using the corresponding random numbers.

6. If you were going to do a survey with every student at your campus as the population, describe how you might obtain a random sample to survey. More specific is better.

Answers can vary for all of Questions 6–9. Some sample answers provided.

Use student ID numbers, and have a computer choose random numbers from that list.

A **systematic sample** is taken by numbering each member of the population and then selecting every kth member, where k is some natural number. For example, the researcher from Gradeinflation.com might have numbered all of the colleges willing to release grade information and chosen every tenth one. In this case, $k = 10$. When using systematic sampling, it's important that the starting number is chosen randomly.

7. Let's get back to the campus survey mentioned in Question 6. Describe how you might develop a systematic sample.

Use student ID numbers in order, then choose every 10, or every 100, or something like that.

When a population is divided into groups where the members of each group have similar characteristics (like large public schools, large private schools, small public schools, and small private schools) and a certain number of members from each group are chosen at random, the result is called a **stratified sample.** For example, the grade inflation researcher might have decided to choose five schools from each of those groups. Medical researchers often stratify samples by dividing subjects into gender groups, age groups, racial groups, or many other categories. They would then choose a similar number of subjects from each group to compose the stratified sample.

8. Describe how you'd put together a stratified sample for that campus survey we've been talking about.

One obvious choice would be to divide students up by class standing, then choose the same number from each class.

When an existing group of subjects that represent the population is used for a sample, it is called a **cluster sample.** For example, an inspector might pick at random one carton of laptops from a large shipment and test each one to see how many are defective. The group in this carton represents a cluster. In this case, the researcher assumes that the laptops in the carton represent the population of all laptops manufactured by the company. In the case of the grade inflation survey, researchers might have looked for an existing organization that many colleges of different types already belong to, then used contact information from that organization to reach out to the schools.

9. Describe how you might develop a cluster sample for our campus survey.

Depending on what the survey was geared toward, you might be able to use students taking a particular course, in a certain major, or in a particular student organization.

Math Note

In the last lesson of this unit, we'll study some ways that samples can be poorly chosen, either intentionally or not, which can invalidate the conclusions of a study.

Think about four different sampling methods the researcher from Gradeinflation.com could have used. For Questions 10–13, write which of the sampling methods we've studied was used for each situation. Explain your choice.

10. The researcher randomly selected three states and then asked all colleges willing to release grade information from those states.

Cluster sample: the states are the clusters.

11. The researcher numbered all of the colleges willing to release grade information and chose every tenth one.

Systematic sample, because every tenth school in an ordered list was chosen.

12. The researcher numbered all of the colleges willing to release grade information and chose several of them using a random number generator.

Random sample. There was no order for the assigned numbers: just a random group chosen.

13. The researcher grouped the schools by two-year state schools, four-year state schools, two-year private schools, and four-year private schools. The researcher then decided to choose five schools from each of those groups.

Stratified sample. The key is that a certain number of schools were chosen from each category.

14. Rank the four sampling methods in Questions 10–13 from best to worst in terms of how representative you think the sample would be. Explain your rankings.

This is completely opinion.

Did You Get It

Try this problem to see if you understand the concepts we just studied. The answer can be found at the bottom of the portfolio page.

1. To study the number of credit hours taken by a typical student, Shawna asks the registrar to provide e-mail addresses for 10 freshmen, 10 sophomores, 10 juniors, and 10 seniors. From each group, she asks for one whose student ID ends in 0, one whose ends in 1, and so forth.
   a. What method of sampling did she use?
   b. Do you think the sample will be representative?

4-4 Group

Frequency Distributions

The data collected for a statistical study are called **raw data.** Raw data can be really difficult to interpret; in order to describe situations and draw conclusions, we need to organize the data in a meaningful way. Two methods that we will use are *frequency distributions* and *stem and leaf plots.* First up, we'll study **categorical frequency distributions,** which are a way to organize data that are divided into distinct categories, like gender, your class standing in school, or conferences for college football teams.

A categorical frequency distribution is a table in which you list different categories, then count up how many individuals in a sample or population fall into each category and list those frequencies. In Questions 1 and 2, you'll create a categorical frequency distribution.

1. Twenty-five volunteers for a medical research study were given a blood test to obtain their blood types. The data are shown below. What would the categories for this data be? Why?

The blood types, because that's the thing in this scenario that can be divided into categories.

2. Build a categorical frequency distribution by writing in your categories, then putting tally marks in the second column as you count the number of individuals in each category.

A	B	B	AB	O
O	O	B	AB	B
B	B	O	A	O
AB	A	O	B	A
A	O	O	O	AB

Categories	Tally Marks	Frequency
A	𝍸	5
B	𝍸 II	7
AB	IIII	4
O	𝍸 IIII	9

3. Use your distribution to write some conclusions on blood types in the population. Do you think that there's enough data for your conclusions to be reliable?

Based on these data, O is the most common blood type, occurring in more than twice as many individuals as the rarest, AB. But with only 25 individuals out of 7 billion on the planet, these results shouldn't be considered reliable.

Did You Get It

Try this problem to see if you understand the concepts we just studied. The answer can be found at the bottom of the portfolio page.

2. A health-food store recorded the type of vitamin pills 35 customers purchased during a 1-day sale. Construct a categorical frequency distribution for the data.

C C C A D E C E E A B D C E C E C C

C D A B B C C A A E E E E A B C B

Another type of frequency distribution that can be constructed uses numerical data and is called a **grouped frequency distribution.** In a grouped frequency distribution, the numerical data are divided into **classes.** For example, if you gathered data on the weights of people in your class, there's a decent chance that no two people have the exact same weight. In that case, everyone would be in a separate category, and each category would have one weight in it. This, of course would make the frequency distribution no more useful than the original list of raw data!

So it would be reasonable to group people into weight ranges, like 100–119 pounds, 120–139 pounds, and so forth. In the 100-119 pound class, we call 100 the **lower limit** and 119 the **upper limit.**

4. With these ranges, where would be put someone that weighs 119.8 pounds?

Nowhere, which is a problem.

As Question 4 illustrates, the way we choose classes is dependent on how the data values are rounded. The weight classes we chose above would be fine if all the weights are rounded to the nearest pound. But if weights were rounded to the tenth of a pound, we'd want to make the classes 100–119.9, 120–120.9, and so on.

Constructing a Grouped Frequency Distribution

Step 1: Decide how many classes you want to divide the data into. Between 5 and 15 is acceptable, depending on the number of data values in the set.

Step 2: Determine the width of each class:

$$\text{Class width} = \frac{\text{largest data value} - \text{smallest data value}}{\text{number of classes}}$$

Then round this value up to the next biggest whole number.

Step 3: Use the lowest number in the data set or a "nice" number lower than that number for the lower limit of the first class.

Step 4: Add the class width to the first lower class limit to get the second lower class limit. Do this for all classes.

Step 5: Find the upper class limit for each class by subtracting one from each lower class limit.

Step 6: Tally up the number of data values in each class, then put the frequencies next to each class in a table.

We'll illustrate by organizing the following data, which are the record high temperatures for each of the 50 states in degrees Fahrenheit.

State	Temp.	State	Temp.	State	Temp.	State	Temp
Alabama	112	Indiana	116	Nebraska	118	South Carolina	113
Alaska	100	Iowa	118	Nevada	125	South Dakota	120
Arizona	128	Kansas	121	New Hampshire	106	Tennessee	113
Arkansas	120	Kentucky	114	New Jersey	110	Texas	120
California	134	Louisiana	114	New Mexico	122	Utah	117
Colorado	118	Maine	105	New York	108	Vermont	105
Connecticut	106	Maryland	109	North Carolina	110	Virginia	110
Delaware	110	Massachusetts	107	North Dakota	121	Washington	118
Florida	109	Michigan	112	Ohio	113	West Virginia	112
Georgia	112	Minnesota	114	Oklahoma	120	Wisconsin	114
Hawaii	100	Mississippi	115	Oregon	119	Wyoming	115
Idaho	118	Missouri	118	Pennsylvania	111		
Illinois	117	Montana	117	Rhode Island	104		

Source: National Climactic Data Center

5. Why would a categorical frequency distribution be a bad idea for this data? (Hint: What would the categories be?)

Individual temperatures would be the categories. Data values range from 100 to 134, so there would be 20 or more categories.

Step 1: Decide how many classes we want.

6. How far apart are the largest and smallest values in the data set?

34 degrees

7. How wide would this make each class if we went with 15 classes? Does this sound like a good idea? Explain.

34/15 is 2.27, and rounding up to the next whole number we'd get classes of width 3. In that case, 15 classes would cover a 45 degree range, but the total range is only 34 degrees. Bad idea.

8. I think seven classes seems like a really good idea for this data. Why do you think I chose seven?

With a range of 34 degrees, seven classes would make each category cover 5 degrees, and that will make it easy to organize the data.

Step 2: Find the class width.

9. Divide the range of values you found in Question 6 by the number of classes we chose, then round up to get the class width.

See previous question: 5.

Step 3: Decide on a lower limit for the first class.

10. Why is the smallest value in the data set a very good choice for the lower limit of the first class in this case? (As opposed to some value less than it, that is.)

Our class width is 5, so starting with 100 will make all classes start with either 0 or 5. Nice and organized.

Step 4: Find the lower limit for all the other classes.

11. Starting with the lower limit we chose in the last question, keep adding the class width you found in Question 9, recording each result, to make a list of the lower limit for each of the seven classes.

100, 105, 110, 115, 120, 125, 130

Step 5: Find upper limits for each class.

12. Subtract 1 from each lower class limit you found in Question 11 to get the upper class limit for the previous class. For example, if the first two classes in a grouped frequency distribution are 50 and 58, the upper limit for the first class would be 57. (Remember, if a data set contains numbers with decimal parts, we would adjust the upper limits to reflect that. For example, if values in the data set are rounded to two decimal places, and a class limit is 2, the upper class limit below it would be 1.99.)

104, 109, 114, 119, 124, 129, 134

Step 6: Set up a table and build the grouped frequency distribution.

13. Fill in the table below, using the class limits you decided on in Questions 11 and 12.

Class	Tally Marks	Frequency
100-104	III	3
105-109	~~IIII~~ III	8
110-114	~~IIII~~ ~~IIII~~ ~~IIII~~ I	16
115-119	~~IIII~~ ~~IIII~~ III	13
120-124	~~IIII~~ II	7
125-129	II	2
130-134	I	1

14. Write down at least three observations you can make from analyzing your frequency distribution.

Answers vary.

Did You Get It

Try this problem to see if you understand the concepts we just studied. The answer can be found at the bottom of the portfolio page.

3. The online homework system I use in my calculus classes allows me to track how much time students spend working on homework. In my two Calc 2 sections last semester, here are the average number of hours spent per week for each student. Construct a grouped frequency distribution for the data.

1	2	6	7	12	13	2	6	9	5
18	7	3	15	15	4	17	1	14	5
4	16	4	5	8	6	5	18	5	2
9	11	12	1	9	2	10	11	4	10
9	18	8	8	4	14	7	3	2	6

Stem and Leaf Plots

Another way to organize data is to use a stem and leaf plot (sometimes called a stem plot). Each data value or number is separated into two parts. For a two-digit number such as 53, the tens digit, 5, is called the **stem,** and the ones digit, 3, is called its **leaf.** For the number 72, the stem is 7, and the leaf is 2. For a three-digit number, say 138, the first two digits, 13, are used as the stem, and the third digit, 8, is used as the leaf. For values rounded to the tenths place, like 8.4, you can use the value to the left of the decimal place as stem and the tenths place as leaf, as in our next example. In any case, the very last digit is used as the leaf, and what comes before is the stem. Note that we'll include a key with our plot that clarifies what the stems and leaves represent.

The data below are the July 2015 unemployment rates for each state.

State	Rate	State	Rate	State	Rate	State	Rate
Alabama	6.2	Indiana	4.6	Nebraska	2.8	South Carolina	6.0
Alaska	6.6	Iowa	3.7	Nevada	6.8	South Dakota	3.7
Arizona	6.3	Kansas	4.6	New Hampshire	3.6	Tennessee	5.7
Arkansas	5.4	Kentucky	5.2	New Jersey	5.7	Texas	4.1
California	6.1	Louisiana	6.0	New Mexico	6.7	Utah	3.7
Colorado	4.2	Maine	4.5	New York	5.2	Vermont	3.6
Connecticut	5.3	Maryland	5.1	North Carolina	5.9	Virginia	4.5
Delaware	4.9	Massachusetts	4.7	North Dakota	2.9	Washington	5.3
Florida	5.3	Michigan	5.1	Ohio	4.7	West Virginia	7.6
Georgia	5.9	Minnesota	4.0	Oklahoma	4.6	Wisconsin	4.5
Hawaii	3.5	Mississippi	6.3	Oregon	6.1	Wyoming	4.0
Idaho	4.2	Missouri	5.6	Pennsylvania	5.4		
Illinois	5.6	Montana	4.1	Rhode Island	5.6		

Source: U.S. Bureau of Labor Statistics

15. If you were to build a grouped frequency distribution, what classes would you choose?

Answers vary. I'd choose 2-2.9, 3-3.9, 4-4.9, 5-5.9, 6-6.9, and 7-7.9.

16. What will you choose to be the stems for this data?

The first digits of the unemployment rates.

17. What will you choose to be the leaves for this data?

The second digit, which is the tenths place.

18. List all whole number parts that appear in the unemployment data; that is, the digits to the left of the decimal point. (You don't need to repeat them if they appear more than once.)

2, 3, 4, 5, 6, 7

19. Write your answers from Question 18 in order under "Stems" in the table below. Write the lowest value first, then larger values as you work your way downward.

20. Go through the data values in the table, and one by one write the tenths digit in each value under "Leaves," next to the corresponding stem. For example, since the first data value is 6.2, you'd write a 2 in the leaves column next to 6 in the stems column. Separate leaves that correspond to the same stem with some space, but not with commas. Put in ALL tenths place digits, including repeats.

Stems	Leaves
2	8 9
3	5 7 6 7 7 6
4	2 9 2 6 6 5 7 0 1 7 6 1 5 5 0
5	4 3 3 9 6 2 1 1 6 7 2 9 4 6 7 3
6	2 6 3 1 0 3 8 7 1 0
7	6

Key: 6|2 means 6.2

If you want to be adventurous, you can use this as a first introduction to data that may be normally distributed.

21. Write at least three observations you can make about the unemployment data from looking at your stem and leaf plot.

Answers vary.

Deciding on stems and leaves for the unemployment data was fairly simple because there was a natural choice of stems (the whole number part of each rate) that gave us a reasonable number of them (6). That's not always the case: For the high temperature data earlier in this lesson, the natural choice of stem would seem to be the first two digits. But here's what the stem and leaf plot would look like in that case.

Stems	Leaves
10	0 6 6 9 0 5 9 7 6 8 4 5
11	2 8 0 2 8 7 6 8 4 4 2 4 5 8 7 8 0 0 9 1 3 3 7 0 8 2 4 5
12	8 0 1 5 2 1 0 0 0
13	4

Key: 13|4 means 134

22. Explain why this stem and leaf plot isn't really very helpful in organizing the data.

There aren't enough categories. Forty-nine of the fifty data values are in just three categories.

In general, it's a good idea to choose stems so that there are between 5 and 15 categories. In the case of the high temperature data, we'd get a better organization if we divided the categories in half, so that the first category covers 100 to 104, the second 105 to 109, and so on. To signify this, we can use the symbols $10^{(0)}$ to represent a stem of 10 with leaves that are digits from 0 to 4, and $10^{(5)}$ to represent a stem of 10 with leaves that are digits from 5 to 9. In that case, this is what the high temperature stem and leaf plot would look like

Stems	Leaves
$10^{(0)}$	0 0 4
$10^{(5)}$	6 6 9 5 9 7 6 8 5
$11^{(0)}$	2 0 2 4 4 2 4 0 0 1 3 3 0 2 4
$11^{(5)}$	8 8 7 6 8 5 8 7 8 9 7 8 5
$12^{(0)}$	0 1 2 1 0 0 0
$12^{(5)}$	8 5
13	4

Key: $10^{(0)}$|4 means 104; $10^{(5)}$|6 means 106

23. Describe why this stem and leaf plot makes it easier to analyze the organized data than the one above.

With more categories, we can see that the largest number of states have a high in the 110-114 range, and that the 115-119 range is a close second. Just a sample answer, of course.

Did You Get It ?

Try this problem to see if you understand the concepts we just studied. The answer can be found at the bottom of the portfolio page.

4. It's no secret that gas prices have fluctuated wildly in the last few years, but from 1980–1999, they were surprisingly stable. According to the Energy Information Administration, the following data represent the average price (in cents) per gallon of regular unleaded gas for those years. Draw a stem and leaf plot of the data.

119	131	122	113	112	86	90	90	100	115
114	113	111	111	115	123	123	123	106	117

24. Compare grouped frequency distributions and stem and leaf plots. How are they similar? Which do you think makes it easier to analyze a set of data?

Answers vary.

4-4 Portfolio

Name ____

Check each box when you've completed the task. Remember that your instructor will want you to turn in the portfolio pages you create.

Technology

1. ☐ Excel can be used to do two tasks from this lesson that are both useful and tedious to do by hand: sorting data and building a grouped frequency distribution. In the online resources for this lesson, there's a spreadsheet that has two of the data sets from this lesson entered: blood types and record high temperature by state. Your first job is to sort the data based on the instructions that appear on the spreadsheet. Note that there's a different worksheet for each data set, and you can change from one to another using the tabs along the bottom. After sorting, there will be further instructions on the template for building the frequency distribution. There are also two short videos in the class resources for this lesson that cover sorting and frequency distributions.

Online Practice

1. ☐ Include any written work from the online assignment along with any notes or questions about this lesson's content.

Applications

1. ☐ Complete the applications problems.

Reflections

Type a short answer to each question.

1. ☐ What is the point of sampling in statistics? What's the difference between a sample and a population?
2. ☐ What makes cluster sampling different from stratified sampling?
3. ☐ Why is it bad to have too few or too many categories in a grouped frequency distribution?
4. ☐ Take another look at your answer to Question 0 at the beginning of this lesson. Would you change your answer now that you've completed the lesson? How would you summarize the topic of this lesson now?
5. ☐ What questions do you have about this lesson?

Looking Ahead

1. ☐ Complete the Lesson 4-5 Prep Skills and read the opening paragraph in Lesson 4-5 carefully, then answer Question 0 in preparation for that lesson.

Answers to "Did You Get It?"

1. a. Stratified sample **b.** The last digit of a student ID number is likely to be totally random, so this should lead to a representative sample. **2–4 can be found at the end of the applications section.**

4-4 Applications

Name ____________________

1. Twenty-five fans of reality TV were asked to rate four shows, and the data below reflect which one each rated highest. (S = Survivor, D = Dancing with the Stars, B = Big Brother, A = The Amazing Race)

S	S	B	A	D	B	S	D	A	B	A	A	S
B	A	D	B	A	S	A	D	B	B	S	A	

a. Construct a frequency distribution for the data.

Class	Frequency
S	6
D	4
B	7
A	8

b. Which show did those 25 fans rate the highest?

The Amazing Race

2. The data shown are the cigarette taxes per pack in cents imposed by each state as of Oct 1, 2015.

68	84	320	129	351	170	166	103	153	30
200	365	57	60	200	64	435	131	62	303
200	160	198	86	290	180	45	160	141	55
115	134	100	200	68	178	44	375	170	252
87	37	136	200	17	270	160	57	308	60

Source: tobaccofreekids.org

a. Construct a grouped frequency distribution for this data using seven classes with the first one being 0–69.

Class	Frequency
0-69	14
70-139	10
140-209	16
210-279	2
280-349	4
350-419	3
420-489	1

b. Which class contains the most number of values? Write a sentence explaining what this means using units on any numbers used.

140-209. This means that the largest number of states impose a tax between \$1.40 and \$2.09.

4-4 Applications

Name ______________________________

3. As an experiment in a botany class, plants are placed in a greenhouse, and their growth in centimeters after 25 days is recorded, with the results shown below in a stem and leaf plot.

Stems	Leaves
1	2
2	0 3
3	2 5 8 8 9
4	1 3 3
5	0 1 2 3 3 5 8 9 9

Key: 1|2 means 12

a. How many plants grew 203 cm?

None

b. How many plants grew 38 cm?

Two

c. How many plants had their growth measured and recorded?

Twenty

d. What was the largest growth recorded for any plant?

59 cm

4. The National Insurance Crime Bureau reported that these data represent the number of registered vehicles per car stolen for 35 selected cities in the United States. For example, in Miami, one automobile is stolen for every 38 registered vehicles in the city. (The data have been rounded to the nearest whole number.)

38	53	53	56	69	89	94
41	58	68	66	69	89	52
50	70	83	81	80	90	74
50	70	83	59	75	78	73
92	84	87	84	85	84	89

a. Build a stem and leaf plot for the data.

Stems	Leaves
3	8
4	1
5	0 0 3 8 3 6 9 2
6	8 6 9 9
7	0 0 5 8 4 3
8	4 3 3 7 1 4 0 5 9 9 4 9
9	2 0 4

4-4 Applications

Name ______________________________

b. Build a grouped frequency distribution for the data, where the classes correspond to the stems that you chose.

Class	Frequency
30-39	1
40-49	1
50-59	8
60-69	4
70-79	6
80-89	12
90-99	3

c. A **histogram** is very much like a bar graph that represents the information in a frequency distribution. There are two main differences between a typical bar graph and a histogram. First, the width of the bars has meaning, and is representative of the class width. Second, the bars are drawn right next to each other with no space in between. Create a histogram from the grouped frequency distribution above.

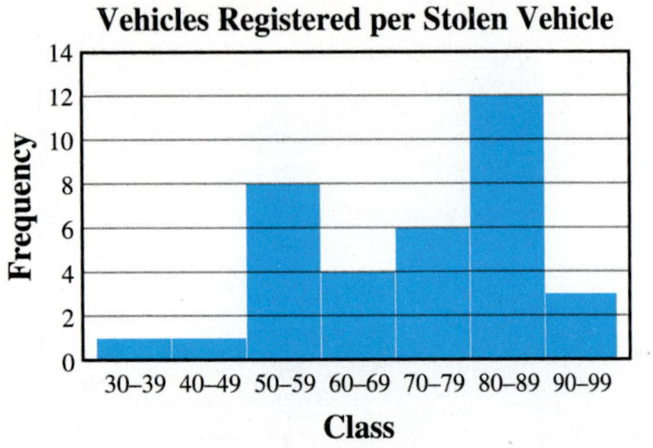

d. Which class contains the most values? Explain what this tells us.

The 80-89 class has the most values. This means that the most common rates of stolen cars among these cities are between 1 in 80 and 1 in 89.

e. Which class would you want to have your city listed in? Why?

The 90-99 class, because I like my car. Higher is better, as 1 in 99 means that one out of every 99 cars will get stolen, while 30 is 1 in every 30 cars. That's bad.

4-4 Applications

Name ______________________________

f. Of the three methods you used to represent the data, which was the easiest to interpret in your opinion? Explain.

This is completely an opinion, but very important for students to think about in terms of analyzing the different methods we've used.

Additional Answers to "Did You Get It?"

2.

Type	Tally Marks	Frequency
A	~~IIII~~ I	6
B	~~IIII~~	5
C	~~IIII~~ ~~IIII~~ II	12
D	III	3
E	~~IIII~~ IIII	9

3.

Class	Tally Marks	Frequency
1–3	~~IIII~~ ~~IIII~~	10
4–6	~~IIII~~ ~~IIII~~ IIII	14
7–9	~~IIII~~ ~~IIII~~	10
10–12	~~IIII~~ I	6
13–15	~~IIII~~	5
16–18	~~IIII~~	5

4.

Stems	Leaves
8	6
9	0 0
10	0 6
11	9 3 2 5 4 3 1 1 5 7
12	2 3 3 3
13	1

Key: 13 | 1 means 131

Lesson 4-5 Prep Skills

This is a short review of skills that will come in handy in the next lesson. In each case, answer the question, then rate your confidence level by checking one of the boxes. If you feel like you're struggling with these skills, consult the online resources provided by your instructor for extra practice.

I KNOW IT | THINK SO | UNSURE | NO IDEA

SKILL 1: ORDERING A LIST OF NUMBERS

1. Arrange the list of numbers below in order from least to greatest.

   2.4 3.7 1.7 1.7 2.3 3.1 1.2 2.4 1.3 1.7 2.5 2.4 2.8 2.1 1.3 2.5

SKILL 2: COMPUTING LENGTHY SUMS

2. Find the sum of the numbers in the list above.

SKILL 3: ORDER OF OPERATIONS

Complete each calculation below.

3. $\dfrac{8 + 11 + 13 + 5 + 6}{5}$

4. $\dfrac{2 \cdot 98 + 4 \cdot 112 + 12 \cdot 13 + 10 \cdot 59 + 6 \cdot 80}{34}$

SKILL 4: INTERPRET SUBSCRIPT NOTATION

5. If we use the symbols X_1, X_2, X_3, and so on to represent the numbers in the list in Question 1, what number corresponds to X_7? (In the original list, not the ordered list.)

6. Refer to Question 5. If the entire list is written as $X_1, X_2, X_3, \ldots, X_n$, what number is represented by n?

Lesson 4-5 An Average Joe (Measures of Average)

LEARNING OBJECTIVES

- ☐ 1. Compute measures of average for given data.
- ☐ 2. Interpret the story told by measures of average.
- ☐ 3. Compute and interpret the mean for grouped data.
- ☐ 4. Compute weighted grades.
- ☐ 5. Use technology to compute measures of average.

© Ryan McVay/Getty Images RF

The average dog is a nicer person than the average person.
—Andy Rooney

Are you an average college student? What does that even mean? According to a variety of sources found on the Web, the average college student is 20 years old, lives on campus and works off campus during the school year. She is female, talks to her parents every day, and comes from a family with an annual income between $50,000 and $100,000. She sleeps 6 hours a night, has $2,700 in credit card debt, and will leave college with $21,000 in total debt.

If this doesn't sound exactly like anyone you know, don't feel bad. As we will learn in this lesson, the word average is ambiguous—there are a variety of ways to describe an average. So what one source considers average might not match someone else's thoughts, and with so many different facets to a person, it's possible that nobody exactly meets all of the average criteria.

0. After reading the opening paragraph, what do you think the main topic of this lesson will be?

4-5 Group

Our goal in this lesson will be to understand the different measures of average. In casual terms, average means the most typical case, or the center of where the values in a data set are distributed. Measures of average include the mean, median, mode, and midrange.

Let's suppose that an individual, let's just call him George, is about to finish college and step out into the work force. George finds a small company, Vandelay Industries, which he's considering going to work for. While researching this company, George finds a flier on which that company advertises that its average employee makes almost $150,000 per year. This sounds pretty darn good to George, so he accepts an entry-level position at a low salary, betting on quick advancement to that alleged average.

Shortly after starting his career, while emptying the recycling bin in the copy room (just one of George's glamorous duties), he stumbles across a list of salaries for all employees at his new company. Let's investigate just how true the company's average salary claim is.

Here's the salary list for Vandelay Industries, which we'll use to study measures of average, which are more technically called **measures of central tendency.**

Employee	Salary
Jerry	$58,000
Kramer	$65,000
Newman	$944,000
George	$20,000
Elaine	$52,000
Susan	$51,000
Tim	$53,000
Estelle	$55,000
Frank	$50,000

The "balance point" analogy for the mean, in the paragraph below, is an ideal opportunity for a class demonstration. Coins, a ruler or yardstick, and half of a toilet paper tube work nicely.

The Mean

There's a pretty good chance that when you think of "average" in a math class, what you're thinking of is actually called the **mean.** The mean, also known as the arithmetic average, is found by adding the values of the data and dividing by the total number of values. Here's a good way to think of the mean: if all of the numbers were equal weights put on a number line, the mean is where the number line could be balanced without tipping in one direction or the other.

The Greek letter Σ (sigma) is used to represent the sum of a list of numbers. If we use the letter *X* to represent data values, then Σ*X* means to find the sum of all values in a data set. Using this notation, we can write a formula for finding the mean.

The **mean** is the sum of values in a data set divided by the number of values. If $X_1, X_2, X_3, \ldots, X_n$, are the data values (which means there are *n* of them), we use the symbol $\overline{X}$ to represent the mean, and

$$\overline{X} = \frac{X_1 + X_2 + X_3 + \cdots + X_n}{n} = \frac{\Sigma X}{n}$$

The symbol $\overline{X}$ is read "*X* bar."

1. Find the mean salary for the company.

$$\frac{58 + 65 + 944 + 20 + 52 + 51 + 53 + 55 + 50}{9} = 149.8$$ Mean salary is $149,800.

2. Newman owns the company, and obviously pays himself a handsome salary. If we ignore Newman's salary, what is the new mean? Based on this, do you feel like the mean is a good measure of what the typical employee makes?

$$\frac{58 + 65 + 20 + 52 + 51 + 53 + 55 + 50}{8} = 50.5$$, so the mean salary is $50,500.

The second mean is a good measure, but not the first. The one super-large salary inflates the mean way higher than what we'd think of as the typical salary.

Did You Get It

Try this problem to see if you understand the concepts we just studied. The answer can be found at the bottom of the portfolio page.

1. The total areas, in thousands of square miles, of the 15 largest states in the U.S. are provided in the list below. Find the mean area for these 15 states.

665	269	164	147	122	114	111	104
98	98	97	87	85	84	82	

The Median

According to the National Organization of Realtors, the mean selling price for existing homes nationally in August of 2015 was \$271,600. The **median** selling price, on the other hand, was \$228,700. Since both of these are measures of central tendency, how can they be so far apart? What exactly is the median? In this case, the median tells us that half of all homes sold during that month were priced above \$228,700, and half were priced below that amount. In short, the median of a data set is the value in the middle if all values are arranged in order. The median will either be a specific data value in the set, or will fall in between two values. Here's how we find a median:

3. Arrange all of the salaries at Vandelay Industries in order from smallest to largest.

\$20,000, \$50,000, \$51,000, \$52,000, \$53,000, \$55,000, \$58,000, \$65,000, \$944,000

4. Find the salary that's right in the middle of your ordered list. That's the median salary. How does it compare to the mean?

\$20,000, \$50,000, \$51,000, \$52,000, \$53,000, \$55,000, \$58,000, \$65,000, \$944,000

1 2 3 4 5 6 7 8 9

The value in the middle is the fifth one, which is \$53,000. This is WAY less than the mean. Barely a third as large, in fact.

5. If there is an even number of values in a data set, there will be two values in the middle, not one like in Question 4. How do you think you'd find the median in that case?

We split the difference of the two values in the middle, choosing the median to be the number halfway between them. (Which is the mean of those two numbers.)

Finding the Median of a Data Set

Step 1: Arrange the data in order, from smallest to largest. (Actually, largest to smallest is fine too. Whatever makes you happy.)

Step 2: If the number of data values is odd, there will be one value in the middle of the ordered list. That's the median. If the number of data values is even, there will be two numbers in the middle of the list: the median is the mean of these two values.

6. Find the median of the salaries at Vandelay Industries (without Newman's). How does it compare to the median of the salaries with Newman's included?

$20,000, $50,000, $51,000, $52,000, $53,000, $55,000, $58,000, $65,000

1 2 3 4 5 6 7 8

The values in the middle are $52,000 and $53,000, so the median is $52,500. It's almost the same with or without Newman's salary.

7. What does this tell you about median when there are **outliers** (sample data values that are very far away from the majority of values)?

It's barely affected at all, unlike the mean, in which the one huge salary almost tripled the mean.

8. Find the median of the salaries at Vandelay Industries (without Newman's) if Kramer quits to start his own company (The Peterman Reality Tour). What can you observe?

$20,000, $50,000, $51,000, $52,000, $53,000, $55,000, $58,000

1 2 3 4 5 6 7

The value in the middle $52,000. Even throwing away the TWO biggest salaries, the median isn't affected all that much.

© Columbia TriStar Television/ Everett Collection

Did You Get It

Try these problems to see if you understand the concepts we just studied. The answers can be found at the bottom of the portfolio page.

2. Find the median of the land areas provided in Did You Get It 1. Compare to the mean, and discuss the difference. Try to come up with an explanation of the discrepancy.
3. Find the mean and median if you throw out Alaska, which has the largest area by far. What does this tell you?

The Midrange

The **midrange** is another measure of average. The advantage of the midrange is that it's very quick and easy to calculate. The disadvantage is that it totally ignores most of the data values, so it's not a particularly reliable measure. But in a pinch, it can give you a rough idea of average.

9. Find the number halfway between the lowest and highest salaries at Vandelay Industries (including Newman's). That's the midrange.

Lowest \$20,000, highest \$944,000. Halfway between: $\frac{\$20{,}000 + \$944{,}000}{2} = \$482{,}000$

Finding the Midrange of a Data Set

Step 1: Find the smallest and largest values in the data set.

Step 2: Find the mean of those two values.

10. Find the midrange of the salaries at Vandelay Industries (excluding Newman's). How does it compare to the midrange of all salaries?

Lowest \$20,000, highest \$68,000. Halfway between: $\frac{\$20{,}000 + \$68{,}000}{2} = \$44{,}000$. It's less than one tenth as much!

11. Explain why the midrange isn't very reliable as a measure of central tendency.

It's HUGELY affected by the one very large salary. \$482,000 is a ridiculous measure of average here, as there's not a single salary within \$400,000 of that number.

Did You Get It

Try this problem to see if you understand the concepts we just studied. The answer can be found at the bottom of the portfolio page.

4. Find the midrange area for the top fifteen largest states from Did You Get It 1.

The Mode

The third measure of average is called the mode. The **mode** is sometimes said to be the most typical case.

NASA

12. These data represent the duration (in days) of the final 20 U.S. space shuttle voyages. Which number of days appears most often on the list? That's the mode.

11 12 13 12 15 12 15 13 15 12 12 15 13 10 13 15 11 12 15 12

12 appears 7 times, so it's the mode.

13. How does the mode compare to the mean? Which do you think describes the typical case best?

The mean is 12.9. Second part is an opinion.

The **mode** is the value that occurs most often in a data set.

A data set can have more than one mode (if two or more values are tied for most appearances), or no mode at all (if every value appears only once). While modes aren't extremely useful for small data sets with numerical data, we'll begin there to help get an understanding of what the mode is about.

14. The number of wins in a 16-game season for the Cincinnati Bengals from 1996–2014 is listed below. Find the mode.

8 7 3 4 4 6 2 8 8 11 8 7 4 10 4 9 10 11 10

Four and eight wins, which both appear on the list four times.

15. How does the mode compare to the mean? Discuss again. Include a description of which measure of central tendency you think is more reliable in general, and why.

The mean is about 7.1 wins. Second question is again an opinion.

16. Do the employee salaries at Vandelay Industries have a mode? Why or why not?

They do not. Every employee has a different salary.

The mode is probably most useful when data are classified by groups or categories. In fact, it's the only measure of central tendency that even makes sense when the data values aren't numbers.

17. A survey of the junior class at Fiesta State University shows the following number of students majoring in each field. Find the mode.

Major	Number
Computer Science	632
Liberal Arts	878
Business	1,425
General Studies	95
Education	471

The mode is majoring in business.

This is a GREAT example to discuss when the mode is useful. Students will focus on the numbers and want to say that there is no mode. Point out that you can't find an "average" of something qualitative, like major, but the mode makes perfect sense in that setting.

18. Now let's evaluate the "average salary" claim made by Vandelay Industries at the beginning of this lesson. How truthful do you think that claim was? Use the measures of average you found in this lesson to discuss whether or not you think they were being deceptive in any way.

Answers vary. Of course, the point is that if you take "average" to mean "mean," then the claim is technically truthful, but it's still deceiving in terms of what a reasonable person would think of as the average salary.

Did You Get It ?

Try this problem to see if you understand the concepts we just studied. The answer can be found at the bottom of the portfolio page.

5. The table below lists the average high temperature in degrees Fahrenheit for each month of the year on the island of Antigua (which is delightful, by the way). Find the mode and the median.

Month	Jan	Feb	Mar	Apr	May	Jun	Jul	Aug	Sep	Oct	Nov	Dec
High	81	82	82	83	85	86	87	87	87	86	84	82

Did You Get It

Try this problem to see if you understand the concepts we just studied. The answer can be found at the bottom of the portfolio page.

6. Five hundred college graduates were asked how much they donate to their alma mater on an annual basis. Find the mode of their responses, summarized in the table.

Amount	Number
$500 or more	45
Between $0 and $500	150
Nothing	275
Declined to answer	30

4-5 Class

We now return to the mean, this time in a situation that's a bit more complicated: when we have grouped data in a frequency distribution. The procedure for finding the mean for grouped data uses the midpoints and the frequencies of the classes. This procedure will give only an approximate value for the mean, so it's typically used in two situations. One is when the data set is very large and calculating the exact mean using all the data is impractical. The other is when the original raw data are unavailable but have been grouped by someone else.

Finding the Mean for Grouped Data

Step 1: Find the midpoint of each class in the grouped data.
Step 2: Multiply the frequency for each class by the midpoint of that class.
Step 3: Add up all of the products from step 2.
Step 4: Divide by the sum of all frequencies (which is the total number of data values).

If you prefer formulas to procedures:

$$\overline{X} = \frac{\Sigma(f \cdot X_m)}{n}$$

where f is the frequency for each class, X_m is the midpoint of each class, and n is the sum of all frequencies.

1. Here's another look at the grouped frequency distribution for the record high temperature in each state from earlier in this unit. Find the mean of these high temperatures. Begin by filling in the midpoints for each class.

Class	Midpoint	Frequency
100–104	102	2
105–109	107	8
110–114	112	18
115–119	117	13
120–124	122	7
125–129	127	1
130–134	132	1

$$\frac{2 \cdot 102 + 8 \cdot 107 + 18 \cdot 112 + 13 \cdot 117 + 7 \cdot 122 + 127 + 132}{2 + 8 + 18 + 13 + 7 + 1 + 1} = 114.2°$$

2. Explain why this formula only gives an approximation to the mean, not the actual mean.

Because we're not actually using all of the data values. By using the midpoint of each class, we're basically changing every value in that class to the midpoint. This of course has an effect on the mean.

This is a tough question. I give extra credit for groups that get it in my classes.

> **Math Note**
>
> Thinking of the mean as a "balance point" is really helpful in understanding why this way of approximating the mean makes sense. Think of stacking a number of coins corresponding to the frequency at the center of each class on a seesaw.

Finding the mean for grouped data is similar to finding a **weighted mean.** We can illustrate the concept by finding a student's average score in a course where different components contribute different amounts toward the final grade. For example, let's look at the grading structure for one course:

Homework average: 50 points
Four tests: 30 points each
Final exam: 50 points
Group project: 20 points
Attendance/Participation: 10 points

Tong had an 87% homework average, got 91%, 82%, 86%, and 94% on the four tests, 88% on the final exam, 92% on the group project, and 100% of attendance/participation points. We can treat this as a weighted mean problem to find her overall average percentage. The number of points assigned to each component of the course acts as the frequencies for each class, and the percentage earned on each components plays the role of the midpoints for each class. Then we can use our procedure for approximating the mean for grouped data.

3. With that in mind, what role does the sum of the points for each component play in the weighted mean calculation? Why?

The sum of the points corresponds to the total number of data values, or n, in the formula. Since each number of points corresponds to a frequency, the sum of points is the sum of all frequencies.

4. Use the mean for grouped data formula to find Tong's average for the course.

$$\frac{50 \cdot 0.87 + 30 \cdot 0.91 + 30 \cdot 0.82 + 30 \cdot 0.86 + 30 \cdot 0.94 + 50 \cdot 0.88 + 20 \cdot 0.92 + 10 \cdot 1}{50 + 4 \cdot 30 + 50 + 20 + 10} = 0.887$$

Her average is 88.7%. Note that you can also use the scores in percentage form rather than decimal form.

Did You Get It ?

Try this problem to see if you understand the concepts we just studied. The answer can be found at the bottom of the portfolio page.

7. The data below is the number of losses by the team that won the NCAA men's basketball championship for the years from 1939–2012. Find the mean number of losses.

Losses	Frequency
0–1	13
2–3	28
4–5	18

Losses	Frequency
6–7	9
8–9	3
10–11	3

5. What are the pros and cons of each of the measures of average we studied?

Answers vary. Our answers are summarized at the end of the lesson in the colored box.

6. (This is my favorite question in the whole lesson.) Look back at the description of the "average" college student in the opening paragraph. For each characteristic listed, write which measure of average you think was used.

Answers can vary. Our guesses: 20 years old-mean. Lives on campus-mode. Works off campus-mode. Female-mode. Talks to parents every day-mode. Family income-median. Sleeps 6 hours a night-mean. Has $2,700 in credit card debt-median. Leave with $21,000 in total debt-median.

Using Technology to Compute Measures of Central Tendency

Technology is certainly a useful tool in computing measure of average, so let's have a look at using spreadsheets and graphing calculators to do the heavy lifting for us. (For what it's worth, when a list of data has just a few numbers on it, it's probably quicker to find the measures of average by hand. It's the large data sets that make technology incredibly useful.)

To illustrate these features, we'll use the following data set. According to deathpenaltyinfo.org, there were 12 death row inmates that were exonerated and freed in 2003. The table below shows the number freed each year since then.

Year	'03	'04	'05	'06	'07	'08	'09	'10	'11	'12	'13	'14	'15
Number	12	6	2	1	3	4	9	1	1	3	1	7	6

Excel

There are built-in commands for finding the mean and median in Excel. There's no midrange command, but we can build one, and ordering the list allows us to find the mode.

First, the data should be entered into a single column, as shown in Figure 1. To calculate the mean, choose an empty cell and enter "=AVERAGE(". Then use the cursor to select all of the cells containing the data, enter a close parentheses, and hit enter. This displays the mean, shown in cell B15. To find the median, use the same routine, but enter "MEDIAN" rather than "AVERAGE". This is in cell B16.

There's no direct command for midrange, but there are commands to find the highest and lowest numbers on the list; we can embed them in a simple calculation that finds the midrange. The string we need to enter looks like

= (MAX(range) +MIN(range))/2

where "range" represents the range of cells that the data lives in. Again, the simplest way to enter that range is to just use the cursor to select it after typing "MAX(" and "MIN(". The resulting midrange in shown in cell B17.

To find the mode, we'll first sort the list. Select the cells containing the data (B1 through B13 in this case), then choose "Sort . . ." from the "Data" menu. You can choose to sort in either ascending or descending order. The result, shown in Figure 2, makes it a simple matter to find the value or values that occur most often. Notice that (as we'd hope) sorting didn't affect mean, median, or midrange. Figure 3 shows the commands entered for finding the mean, median, and midrange.

	A	B
1		12
2		6
3		2
4		1
5		3
6		4
7		9
8		1
9		1
10		3
11		1
12		7
13		6
14		
15	Mean	4.30769231
16	Median	3
17	Midrange	6.5

Figure 1

	A	B
1		1
2		1
3		1
4		1
5		2
6		3
7		3
8		4
9		6
10		6
11		7
12		9
13		12
14		
15	Mean	4.30769231
16	Median	3
17	Midrange	6.5

Figure 2

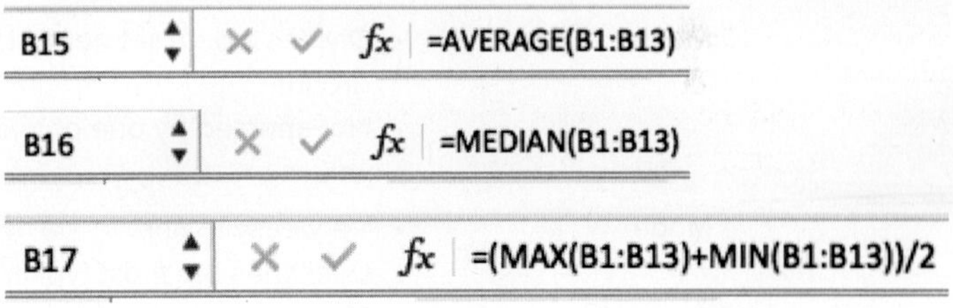

Figure 3

TI-84 Graphing Calculator

In order to find measures of average for a group of numbers, the first step is entering those numbers in a list. To do this, we need to use the list editor: hit the STAT key, then choose 1. Enter the numbers one at a time, hitting ENTER after each. The above data is shown entered in Figure 4 below (most of it, anyhow). Then hit STAT again, and use the right arrow to select CALC at the top of the screen, and pick choice 1: 1-Var Stats (Fig. 5). After hitting ENTER twice, you get the 1-Variable Stats menu, shown in Figures 6 and 7.

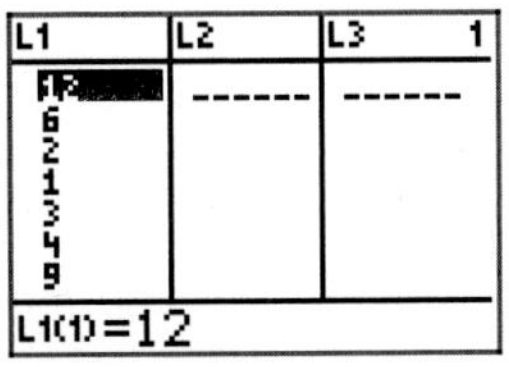

Figure 4

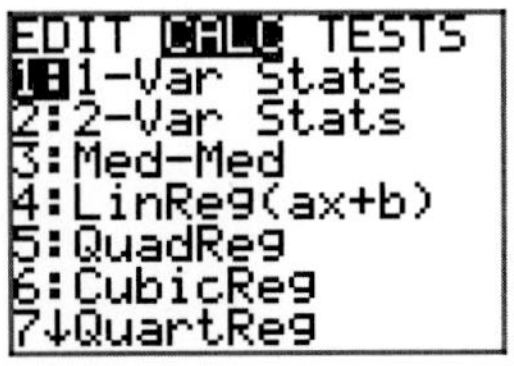

Figure 5

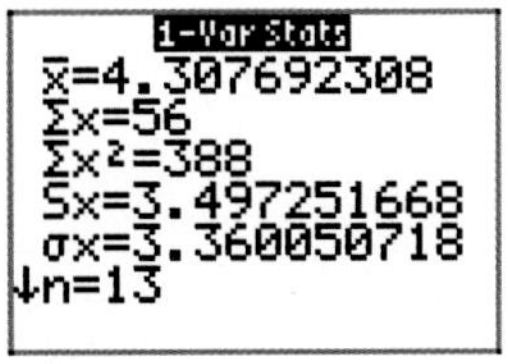

Figure 6

Figure 7

There's actually quite a bit more information on the screen in Fig. 7 than we're looking for, and some of it will be discussed in later lessons. For now, the mean is the very first thing on the list: $\bar{x} = 4.307692308$. Using the down arrow to scroll down, we find the median: Med = 3 (Fig. 7). The midrange isn't displayed, but minX = 1 and maxX = 12 provide the lowest and highest values. When you have those, it's simple to add them and divide by 2 to get the midrange of 6.5.

As with Excel, there's no specific command for finding the mode, but the calculator will order the data, which makes it a lot easier to find the mode. Hit the STAT key again and choose 2, which is SortA (for "sort ascending"). Then you need to tell it which list to sort. Since we entered the data in list L1, enter that by hitting 2nd 1 (Fig. 8). After you press ENTER, the calculator displays Done, and the list is ordered. When you go back to the list editor, you can easily find which value or values appear most often, giving you the mode (Fig. 9).

Figure 8

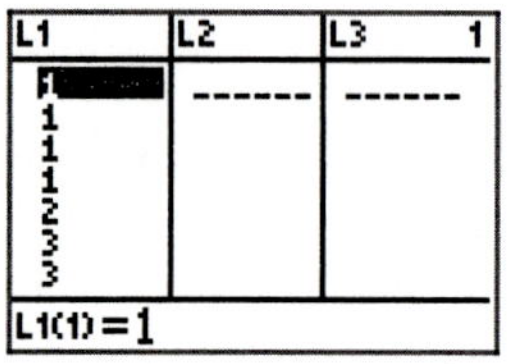

Figure 9

To approximate the mean for grouped data, enter the midpoint of each class in list L1, and the associated frequency in list L2. The choose 1-Var Stats, but before hitting ENTER, key in L1,L2 (which is 2nd 1, 2nd 2). Then hit ENTER. See the Lesson 4-5-1 and 4-5-2 videos in class resources for further instruction.

Earlier, we asked for your thoughts on the pros and cons of each measure of central tendency. We'll summarize the lesson with our thoughts.

A Comparison of Measures of Average

Measure	Strengths	Weaknesses
Mean	• Unique—there's exactly one mean for any data set • Factors in all values in set • Easy to understand	• Can be adversely affected by a few unusually high or low values • Can be time-consuming to calculate for large data sets
Median	• Divides a data set neatly into two groups • Not affected by one or two extreme values	• Can ignore the effects of large or small values even if they are important to consider
Mode	• Very easy to find • Describes the most typical case • Can be used with categorical data like candidate preference, choice of major, etc.	• May not exist for a data set • May not be unique • Can be very different from the mean and median if the most typical case happens to be near the low or high end of the range
Midrange	• Very quick and easy to compute • Provides a simple look at average	• Dramatically affected by extremely high or low values in the data set • Ignores all but two values in the set

4-5 Portfolio

Name ____

Check each box when you've completed the task. Remember that your instructor will want you to turn in the portfolio pages you create.

Technology

1. ☐ In the last problem of the Applications homework, you'll approximate the mean for a grouped data set using the procedure within this lesson. The tech template for this lesson has the full data set, and your job is to use Excel to calculate the true mean, the median, the mode, and the midrange using commands from the tech box in the lesson. Then answer these questions: (a) How accurate was the approximation using grouped data? (b) Why was it so useful to use technology to get the true mean? (c) Describe what you can learn from looking at all four measures of average for this data set.

Online Practice

1. ☐ Include any written work from the online assignment along with any notes or questions about this lesson's content.

Applications

1. ☐ Complete the applications problems.

Encourage students to use technology to find the measures of average in this assignment!

Reflections

Type a short answer to each question.

1. ☐ What did this lesson do to change your view of what the word "average" means?
2. ☐ Which measure of average do you think is the most consistent and the best to use? Why?
3. ☐ Which measures of average are more reliable when you have extreme outliers in the data set?
4. ☐ Take another look at your answer to Question 0 at the beginning of this lesson. Would you change your answer now that you've finished the lesson? How would you summarize the topic of this lesson now?
5. ☐ What questions do you have about this lesson?

Looking Ahead

1. ☐ Complete the Lesson 4-6 Prep Skills and read the opening paragraph in Lesson 4-6 carefully, then answer Question 0 in preparation for that lesson.

Answers to "Did You Get It?"

1. 155,133 sq. mi. **2.** 104,000 sq. mi.; the one extremely large state raises the mean considerably **3.** Mean: 118,714 sq. mi.; median: 101,000 sq. mi.; that one large value did in fact skew the mean upward **4.** 373,500 sq. mi. **5.** Modes: 82° and 87°; median: 84.5° **6.** Nothing

4-5 Applications

Name ______________________________

For Questions 1–4, describe which measure of average—mean, median, or mode—was most likely to have been used in each situation. Provide a brief justification.

1. Half of the factory workers make more than $11.37 per hour and half make less than $11.37 per hour.

   Median: describes the value in the middle of the spread

2. The average number of children per family in the Plaza Heights is 1.8.

   Almost certainly mean. It can't be mode because it's not a whole number, and unlikely to be median because it's not halfway between two whole numbers.

3. Most people prefer red convertibles over any other color.

   Has to be mode. There's no numerical information to find mean or median.

4. The average age of college professors is 42.3 years.

   Could be either mean or median. If ages are listed as whole numbers, it's most likely mean. But ages could be broken down into years and months, in which case it could be either.

For Questions 5 and 6, find the mean, median, mode, and midrange. Then decide which you think is the most meaningful measure of average for that data set.

5. The table displays the number of murders in the United States for the years from 2000 to 2014.

Year	Murders
2000	15,586
2001	16,037
2002	16,229
2003	16,258
2004	16,148
2005	16,740
2006	17,309
2007	17,128
2008	16,465
2009	15,399
2010	14,722
2011	14,661
2012	14,886
2013	14,319
2014	14,249

Source: FBI Uniform Crime Reports

   Mean: 15,742 murders. Median: 16,037 murders. Mode: None. Midrange: 15,779 murders
   Most meaningful is an opinion.

4-5 Applications

Name ______________________________

6. The 12 states with the highest unemployment rates in December 2015 and their rates of unemployment are given below:

State	Rate
New Mexico	6.8
West Virginia	6.6
Nevada	6.5
Alaska	6.4
Louisiana	6.3
Mississippi	6.0

State	Rate
Arizona	6.0
Alabama	6.0
Oregon	5.7
North Carolina	5.7
Illinois	5.7
California	5.7

Source: U.S. Bureau of Labor Statistics

Mean: 6.12%. Median: 6.0%. Mode: 5.7%. Midrange: 6.25%.
Most meaningful is an opinion.

Forty new automobiles were tested for fuel efficiency by the Environmental Protection Agency (in miles per gallon). The individual MPG values are displayed below. Use this information for Questions 7–12.

37	27	20	30	24	17	32	26	17	32	25	28	20	18	27	32	36	25	9	26
34	30	25	23	25	24	20	13	31	20	20	25	20	22	37	27	27	25	37	30

7. Find the mean miles per gallon.

25.6 MPG

8. Complete the frequency distribution.

Class	Frequency
8–12	1
13–17	3
18–22	8
23–27	15
28–32	8
33–37	5

4-5 Applications

Name ____________________

9. Use the frequency distribution to approximate the mean. How does your approximation compare to the actual mean?

Approximation = 25.1 mpg. This is a pretty good approximation of the actual mean.

10. Find the median, mode, and midrange of this data set.

Median = 25 mpg, Mode = 20 mpg, Midrange = 23 mpg

11. Which measure of central tendency would a government official choose to show if she wanted to make the case that gas mileage was getting better for newer vehicles? Why?

She would want to pick the largest measure of average in order to give the appearance that gas mileage was improving. She would probably go with the actual mean of 25.6 mpg.

12. Which measure of central tendency would an environmental lobbyist choose to show if he wanted to make the case that gas mileage was still not very good for newer vehicles? Why?

He would want to pick the smallest measure of average in order to give the appearance that gas mileage still had room to improve. He would probably go with the mode of 20 mpg.

4-5 Applications

Name ______________________________

13. Eighty randomly selected compact fluorescent light-bulbs were tested to determine their lifetimes (in hours). The grouped frequency distribution was obtained as shown. The full data set for this distribution is in the technology template for this lesson. You'll further analyze this data in the technology assignment. Use the frequency distribution to approximate the mean. Write a sentence explaining what your answer means. Be sure to include units.

Class	Frequency
651–800	6
801–950	12
951–1,100	25
1,101–1,250	18
1,251–1,400	14
1,401–1,550	5

The approximate mean is 1,094.9 hours. This means that the average light bulb should last for a touch less than 1,095 hours before burning out.

Lesson 4-6 Prep Skills

This is a short review of skills that will come in handy in the next lesson. In each case, answer the question, then rate your confidence level by checking one of the boxes. If you feel like you're struggling with these skills, consult the online resources provided by your instructor for extra practice.

I KNOW IT | THINK SO | UNSURE | NO IDEA

SKILL 1: FIND THE MEAN

1. Find the mean for the list of numbers below.

26 32 18 65 12 51 162 30

SKILL 2: COMPUTING SUMS OF SQUARES

2. Compute the sum:

$(8 - 13)^2 + (22 - 13)^2 + (13 - 13)^2 + (20 - 13)^2 + (7 - 13)^2$

SKILL 3: COMPUTE SQUARE ROOTS

3. Find the square root of each number. Round to two decimal places.

112 14,873

SKILL 4: USE SUMMATION NOTATION

4. If $x_1 = 18$, $x_2 = 23$, $x_3 = 50$, and $x_4 = 19$, find Σx_i.

Lesson 4-6 Your Results May Vary (Measures of Variation)

LEARNING OBJECTIVES

- ☐ 1. Compute measures of variation for a given data set.
- ☐ 2. Interpret standard deviation for a data set.
- ☐ 3. Make meaningful comparisons of standard deviation for two data sets.
- ☐ 4. Analyze the procedure for computing standard deviation.

© Tim Davis/Corbis

Research shows that there is only half as much variation in student achievement between schools as there is among classrooms in the same school. If you want your child to get the best education possible, it is actually more important to get him assigned to a great teacher than to a great school.

—Bill Gates

© Chris Collins/Corbis

Now we know quite a lot about measures of average. But just how much information are we getting when we study measures of average? Take a look at the two pictures of dogs on this page. If we look only at measures of average, particularly the mean, we might be fooled into thinking that the two groups are very similar, when clearly they are not. The difference, of course, is that all of the dogs in the first picture are of similar size, while those in the second picture have many different weights.

Because there are some small dogs and one very large one, the mean weight in both groups is probably similar. This leads us to the conclusion that there's a lot more to the story told by a data set than simply measures of average. There can be a lot of variation within a group.

0. After reading the opening paragraph, what do you think the main topic of this lesson will be?

4-6 Class

The simplest measure of variation that we'll study is the **range.**

The **range** of a data set is the difference between the largest and smallest values in the set.

Range = Largest value − Smallest value

1. The first list below is the weights of the dogs in the first picture, and the second is the weights of the dogs in the second picture. Find the mean and range for each list.

First picture: 70 73 58 60
Second picture: 30 85 40 125 42 75 60 55

First: Mean 65.3 lbs, range 15 lbs
Second: Mean 64 lbs, range 95 lbs

2. Describe what you notice about the range and mean for each group of dogs.

The means are very similar, but the range shows that the dogs in the second picture have weights that vary much more widely.

3. Why is the range somewhat limited as a measure of variation for a data set? (Not these particular data sets: ANY data set.)

It has to be limited based on the fact that it only uses the smallest and largest values in the data set, and completely ignores all others.

Did You Get It

Try this problem to see if you understand the concepts we just studied. The answer can be found at the bottom of the portfolio page.

1. The monthly high temperatures from January to December in Aruba and St. Louis are shown below. Find the median and range for each, then write any observations you make based on this information.

   Aruba: 85° 85° 86° 87° 88° 89° 88° 89° 89° 89° 87° 85°

   St. Louis: 38° 45° 55° 66° 77° 86° 91° 88° 81° 69° 54° 42°

Standard Deviation

The range is a limited measure of variation because it ignores all the data except the highest and lowest values. If most of the values are similar, but there's just one unusually high value, the range will make it look like there's a lot more variation than there actually is. For this reason, we will next study **standard deviation,** which is a much more reliable (and commonly used) measure of variation.

Standard deviation is a little tricky to compute. Eventually, we'll learn how to use technology to handle the computation, which will allow us to focus on interpretation. But first, we'll develop a procedure for computing standard deviation by hand. This turns out to be useful even if you never actually DO compute it by hand, as it really helps you to understand what exactly standard deviation measures. This is a bit of an oversimplification, but it's still very useful in understanding what standard deviation is designed to measure: *standard deviation essentially describes the average distance that each data value is away from the mean.*

4. Read the description of standard deviation in italics above carefully again. Why does standard deviation provide a very reasonable way of describing how spread out a data set is?

Since the mean is kind of the "center" of the data values, finding how far away most values are from the mean is exactly what you'd typically think of when considering how spread out the values are.

Interpretation of Standard Deviation

A standard deviation that's large relative to the mean tells us that the data values vary pretty widely from the mean. A relatively small standard deviation means that most data values are clustered fairly close to the mean.

Standard deviation comes in two delightful flavors: population and sample. As we saw in Lesson 4-4, in many cases, data are based not on every single member of a population, but rather a representative sample. When the standard deviation is computed for a data set derived from an entire population under study, it's called the **population standard deviation.** When it's based on data derived from a representative sample of a larger population, it's called the **sample standard deviation.**

As we begin our efforts to develop a formula that accomplishes what we want standard deviation to do, try to focus not so much on just doing the arithmetic, but the interpretation of what each step is accomplishing. The plan is to find the standard deviation for the weights of the eight dogs pictured in the lesson opener.

Step 1. Find the mean weight.

5. Find the mean of the dog weights: (Or, better still, copy it from Question 1.)

30 85 40 125 42 75 60 55

The mean is 64 lbs.

Interpretation: Standard deviation is a measure of how far the data values vary from the mean, so it makes sense to begin by computing the mean.

Step 2: Subtract the mean from each data value.

6. Fill in the table below by subtracting the mean weight from each value.

X	$X - \overline{X}$
30	−34
85	21
40	−24
125	61

X	$X - \overline{X}$
42	−22
75	11
60	−4
55	−9

Interpretation: We are literally measuring exactly how far each data value is from the mean. We'd like to average them, which means we'd like to first add them. But if we try, something bad happens . . .

7. Add up all of the differences from the mean. Why is the result not helpful in finding an average deviation from the mean?

The sum is zero. If you think about it, that will always happen. Obviously, getting zero all the time isn't going to tell us anything useful about the spread of data. It's a good critical thinking exercise to ask students to figure out if the sum will always be zero, and why.

Step 3: Square the differences.

8. Fill in the table below by squaring all of your results from the table in Question 6.

X	$(X - \overline{X})^2$
30	1,156
85	441
40	576
125	3,721

X	$(X - \overline{X})^2$
42	484
75	121
60	16
55	81

Interpretation: Squaring gets rid of the negatives so that the sum is no longer zero. It also makes the differences too big, and changes the units (our values are no longer in pounds), but we'll deal with that later.

Step 4: Add up the squares of the differences.

9. Add your answers from Question 8.

1,156 + 441 + 576 + 3,721 + 484 + 121 + 16 + 81 = 6,596

Interpretation: In this and the next step, we're calculating an approximate average of the squares of the differences, so we want to first add them.

Step 5: Divide by $n - 1$, which is the number of data values in the sample minus 1.

10. Divide your result from Question 9 by one less than the number of dogs.

6,596/7 = 942.29

Interpretation: It seems like dividing by the number of values (n) here is a good idea, but it turns out that when we're using a sample from a larger population to compute mean and standard deviation, dividing by $n - 1$ makes the sample standard deviation more likely to be a true reflection of the population standard deviation. In any case, at this point we have an approximate average of the squares of the individual variations from the mean.

Step 6: Calculate the square root of the approximate average.

11. Find the square root of your answer from Question 10. This is the sample standard deviation.

$\sqrt{942.29} \approx 30.70$

Interpretation: This is the step that "undoes" the squaring we did in step 3. When we squared the results earlier, it changed the units from pounds to pounds squared: the square root returns the units to pounds, giving us a good measure of how far the typical data value varies from the mean.

Did You Get It

Try this problem to see if you understand the concepts we just studied. The answer can be found at the bottom of the portfolio page.

2. Find the sample standard deviation for the four dogs in the first picture at the beginning of the lesson. How does it compare to the standard deviation for the dogs in the second picture? Explain why this makes sense.

   70 73 58 60

12. Carefully describe exactly what your answer to Question 11 tells you about the dogs in that picture.

On average, the dogs' weights are a little less than 31 pounds away from the mean weight of 64 pounds. That tells us that the weights vary quite a lot.

13. Briefly describe how the step-by-step procedure helps to answer Question 12.

Answers vary. Basically, this should be a summary of the "interpretation" comments in the development of standard deviation.

The formulas below summarize how to find sample standard deviation, and also show how to find population standard deviation. For population, the only difference in the procedure is dividing by the total number of data values, rather than one less as we did in finding sample standard deviation.

Formulas for Sample and Population Standard Deviation

Sample: $s = \sqrt{\dfrac{\Sigma(X - \overline{X})^2}{n - 1}}$ **Population:** $\sigma = \sqrt{\dfrac{\Sigma(X - \mu)^2}{N}}$

X represents each data value.

$\overline{X}$ is the mean of the sample.

μ is the mean of the population.

n is the number of data values in the sample.

N is the number of data values in the population.

Math Note

The square of the standard deviation is called the *variance* for a data set.

4-6 Group

Now that we have a handle on not only WHAT standard deviation tells us, but WHY it tells us that, we'll look at using technology to do the calculation for us. This allows us to expend more mental energy on interpreting the result. The procedures are summarized in the box below.

Using Technology: Finding Standard Deviation

TI-84 Calculator

The 1-Var Stats screen, which we used to compute the mean and median in Lesson 4-5, also calculates the standard deviation. Below is another look at a screen shot from Lesson 4-5, where the data on death row exonerations were entered in the list editor, and then the 1-Variable Stats command was chosen. The quantities we're interested in are Sx and σx: these are the sample standard deviation (Sx) and the population standard deviation (σx).

```
1-Var Stats
 x̄=5.538461538
 Σx=72
 Σx²=1022
 Sx=7.206656467
 σx=6.923931571
↓n=13
```

Excel

To use Excel to calculate standard deviation, the data are entered in a single column, just like we did for computing measures of average in Lesson 4-5. Here are sample calculations for the second group of dog weights from Question 1 in this lesson, with the formulas used to calculate each statistic illustrated to the right.

	A	B
1		30
2		85
3		40
4		125
5		42
6		75
7		60
8		55
9		
10	Range	95
11	Sample st. dev.	30.696673
12	Population st. dev.	28.714108

Range

B10 | fx =MAX(B1:B8)-MIN(B1:B8)

Sample Standard Deviation

B11 | fx =STDEV(B1:B8)

Population Standard Deviation

B12 | fx =STDEVP(B1:B8)

See the Lesson 4-6-1 and 4-6-2 video in class resources for further information.

A word on notation: We use a lower-case *s* to represent the sample standard deviation for a data set; the symbol Sx is simply what's used on the calculator screen. Similarly, we use a lower-case Greek sigma (σ) to represent population standard deviation, not σx.

1. The table below shows the number of arrests in the United States for possession of marijuana, in thousands, for the years from 2005 to 2014. All data was gathered from the annual Uniform Crime Report published by the Federal Bureau of Investigation. Find the mean and sample standard deviation for the number of arrests. Include units.

Year	2005	2006	2007	2008	2009	2010	2011	2012	2013	2014
Arrests (Thousands)	569	547	640	620	620	614	541	543	503	516

We get 571.3 and 48.7, so the mean is about 571,300 arrests, and the standard deviation is about 48,700 arrests.

2. Why was it appropriate to find sample standard deviation, rather than population?

Because the data came from a sampling of years, not every year ever.

Standard deviation is easiest to interpret when compared to the size of the mean. For example, a standard deviation of 10 in a data set with mean 20 indicates a much wider spread than a standard deviation of 10 in a data set with mean 200. To help make this easier, we can find the **relative standard deviation** (also known as the **coefficient of variation).** This number, usually denoted c_v, is the standard deviation divided by the mean: it's literally what percent of the mean the standard deviation is. In symbols:

The Coefficient of Variation

$c_v = \frac{s}{\overline{X}}$ for sample data, and $c_v = \frac{\sigma}{\mu}$ for population data.

3. Find the coefficient of variation for the preceding arrest data. What percent of the mean is the standard deviation?

$c_v = 48.7/571.3 \approx 0.085$. The standard deviation is 8.5% of the mean.

4. Do you think the standard deviation and coefficient of variation for this data set would be more likely to be used in an argument by someone that is in favor of legalizing marijuana, or against? Explain.

This is an opinion. You could kind of use it to argue on both sides. The consistency in these numbers could be interpreted to mean that the war on drugs is a complete failure, or it could be interpreted to mean that marijuana use remains a big societal problem.

5. Find the coefficient of variation for the 8 dogs from Questions 5–11 in the Class portion. Compare it to your answer to Question 3 on the previous page. What can you conclude?

$30.7/64 \approx 0.48$. The standard deviation for the dogs is 48% of the mean, while the standard deviation for the arrests is only 8.5% of the mean. This shows that the dog data is much more spread out than the arrest data.

Key point of emphasis here: comparing 30.7 pounds and 48.7 arrests is meaningless. But we can make an intelligent comparison of the two spreads by comparing the coefficient of variation.

Did You Get It ?

Try this problem to see if you understand the concepts we just studied. The answer can be found at the bottom of the portfolio page.

3. The numbers below are the percentages of all drug possession arrests that were for marijuana for the years from 2005 to 2014. Find the sample mean, standard deviation, and coefficient of variation.

   37.7 39.1 42.1 44.3 45.6 45.8 43.3 42.4 40.6 39.7

For Questions 6–8, identify the data set that is likely to have a larger standard deviation. Make sure you EXPLAIN your reasoning. Like with sentences and stuff.

6. Data set 1: 12 15 13 10 16 13 12 13
Data set 2: 5 26 31 2 10 25 6 33

Data set 2. The values in set 1 are all within 6 units of each other, while the ones in set 2 are very spread out.

7. Data set 1: 40-yard dash times for starting running backs in the National Football League
Data set 2: 40-yard dash times for every student in your math class

Data set 2. Everyone that becomes a professional running back has to be at least reasonably fast, so you'd expect the times to be pretty consistent. But a group of students in a math class would undoubtedly have some pretty fast runners, and some dreadfully slow runners.

8. Data set 1: Average monthly high temperature in Chicago
Data set 2: Average monthly high temperature in Los Angeles

Data set 1. Los Angeles has a very consistent climate, and is fairly warm even in winter. Chicago, on the other hand, gets very warm in the summer and very cold in the winter, so the average high will vary a lot.

Did You Get It

Try this problem to see if you understand the concepts we just studied. The answer can be found at the bottom of the portfolio page.

4. Which data set would be likely to have the larger standard deviation? EXPLAIN.

   Data set 1: Weights of all dogs at a golden retriever rescue shelter
   Data set 2: Weights of all dogs for sale in a pet store

9. A professor has two sections of Math 115 this semester. The 8:30 A.M. class has a mean score of 74% with a standard deviation of 3.6%. The 2 P.M. class also has a mean score of 74%, but a standard deviation of 19.2%. What can we conclude about the students' averages in these two sections?

The first class has a very, very low coefficient of variation, meaning the scores are really consistent. You're going to see a boatload of Cs, with relatively few As or Fs. The second class has much more variation, so you'd see more As and Fs, with grades spread across the spectrum.

10. In the Technology portion of the Portfolio, you'll be analyzing the heights of every student in your class, so before class is over, make sure that you collect that data. Record heights in inches, and make sure you indicate whether each height is a male or female.

4-6 Portfolio

Name ______________________________

Check each box when you've completed the task. Remember that your instructor will want you to turn in the portfolio pages you create.

Technology

1. ☐ Enter the heights in inches of all students in your class in three separate columns of a spreadsheet: in column A, include everyone. In column B, list only the males, and in column C list only the females. Then use formulas to calculate the mean and sample standard deviation for each data set. Explain any differences you notice in the standard deviations. A template to help you get started can be found in the online resources for this lesson. It has a text box set up where you can enter your discussion of the results.

Online Practice

1. ☐ Include any written work from the online assignment along with any notes or questions about this lesson's content.

Applications

1. ☐ Complete the applications problems.

Reflections

Type a short answer to each question.

1. ☐ If someone asked you informally to describe what standard deviation is, what would you say?
2. ☐ In many ways, standard deviation is most useful when you're comparing the standard deviations of two data sets. Explain why this is reasonable.
3. ☐ What do the mean and standard deviation together tell you about a data set? What do they not tell you?
4. ☐ Take another look at your answer to Question 0 at the beginning of this lesson. Would you change your answer now that you've finished the lesson? How would you summarize the topic of this lesson now?
5. ☐ What questions do you have about this lesson?

Looking Ahead

1. ☐ Complete the Lesson 4-7 Prep Skills and read the opening paragraph in Lesson 4-7 carefully, then answer Question 0 in preparation for that lesson.

Answers to "Did You Get It?"

1. Aruba, median 87.5°, range 4°; St. Louis, median 67.5°, range 53°; Aruba has a much more consistent climate. **2.** 7.37 lb; much less; the dogs are similar in size **3.** Mean 42.06%, standard deviation 2.76%, coefficient of variation 0.066 **4.** Data set 2: All of the dogs in the first set are the same breed, so you'd expect them to be reasonably similar in size.

4-6 Applications

Name ______________________________

The table shows the number of class periods missed over the last 10 courses by two students. Use this information for Questions 1–5.

Helena	3	4	3	2	2	4	3	2	2	3
Juanita	0	9	2	1	0	6	0	4	0	8

1. Find the mean number of classes missed for each student.

   Helena: 2.8 classes
   Juanita: 3.0 classes

2. Find the range of the number of classes missed for each student.

   Helena: 2 classes
   Juanita: 9 classes

3. Without doing the calculations, how do you think the standard deviations compare for the two students? Explain.

   Clearly the numbers for Juanita are much more spread out, so we'd expect a considerably larger standard deviation

4. Now calculate the standard deviation for these two students. Do these values agree with your thoughts from Question 3?

   Helena: 0.79 classes
   Juanita: 3.53 classes
   This is just what we expected. We're pretty smart.

5. Which is a better student in your opinion? What does that say about the importance of measures of average compared to measures of variation in some situations?

   Opinion question. But a good answer should certainly point out that you need measures of variation to decide since the means are essentially the same.

4-6 Applications

Name ____

When playing golf, in most cases the player will want to get the ball as close as possible to the hole when hitting a shot approaching the green. Suppose that two players hit 10 approach shots to a green, and their results are as shown in the two figures. The black dot is the hole, and the white dots are where each shot ended up. Use the diagram for Questions 6–8.

Pat's shots   Ron's shots

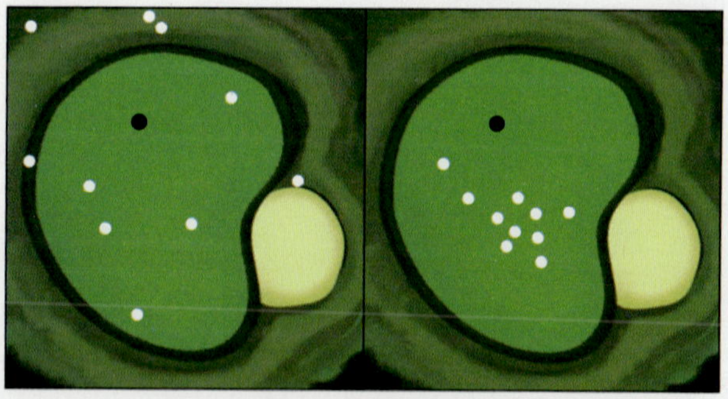

6. If you were to "average" the locations of all of the results for each player, where would you estimate the average shot would be for each?

Pat's average shot is pretty close to the hole, while Ron's is right in the middle of the green where most shots are clustered.

7. How would you describe the measures of variation for each of the players?

Pat's variation would be much higher than Ron's. His shot locations are very consistent, while Pat's are spread out all over the place, with many missing the green entirely.

8. Which is a better player in your opinion? What does that say about the importance of measures of average compared to measures of variation in some situations?

Opinion question, but the point is that Pat's average is practically in the hole, so based strictly on that you'd say that she's the better player. But Ron is much more consistent.

4-6 Applications

Name ______________________________

The heights in inches of nine army recruits (first list) and nine randomly selected people walking through a mall (second list) are shown below. Use these data for Questions 9–13.

List 1:	78	72	68	73	75	69	74	73	72
List 2:	32	75	54	48	70	63	67	46	66

9. Compute the mean, range, and standard deviation for the heights of the recruits.

Mean: 72.7 inches
Range: 10 inches
Standard deviation: 3 inches

10. Compute the mean, range, and standard deviation for the mall shoppers.

Mean: 57.9 inches
Range: 43 inches
Standard deviation: 13.9 inches

11. How do the means compare? What does this mean?

The mean for recruits is much greater. For one, this is because all of the army recruits are adults, while it looks like some of the mall shoppers were children. There are also size standards for military recruits.

12. How do the ranges compare? What does this mean?

The recruits' range is way smaller. This really confirms the idea that some of the mall shoppers are most likely children, while the recruits obviously are all adults.

13. How do the standard deviations compare? What does this mean?

The recruits' standard deviation is very small in general, and especially compared to the mall shoppers. Similar ages and military physical standards make it totally reasonable that there's not a lot of variance in the recruits' heights. But if you pick nine people totally randomly in a mall, you would expect all shapes and sizes.

4-6 Applications

Name ______________________________

14. The standard deviation for Juanita's number of classes missed (Question 1) was 3.53 classes. The standard deviation for the heights of the randomly selected shoppers in a mall was 13.9 inches. Explain why comparing these two numbers is meaningless. (Hint: Did you include units?)

The units tell the story here: These standard deviations measure two completely different things, classes missed and inches. Also, the standard deviation is only meaningful when we know what the mean is, and the means for the two data sets are very different.

15. Find the coefficient of variation for Juanita's number of classes missed and the mall shoppers' heights.

Missed classes by Juanita: $3.53/3.0 \approx 1.18$, or 118%

Mall shoppers' heights: $13.9/57.9 \approx 0.24$ or 24%

16. The numbers from Question 15 CAN be compared in a meaningful way. Explain what comparing them tells us.

Even though the actual standard deviation for the shoppers' heights is more than three times as large, the coefficient of variation shows that the variation in their heights is nowhere near as significant as the variation in the number of classes missed by Juanita.

Lesson 4-7 Prep Skills

This is a short review of skills that will come in handy in the next lesson. In each case, answer the question, then rate your confidence level by checking one of the boxes. If you feel like you're struggling with these skills, consult the online resources provided by your instructor for extra practice.

I KNOW IT | THINK SO | UNSURE | NO IDEA

SKILL 1: INTERPRET A HISTOGRAM

Use the histogram to answer Questions 1 and 2. It represents a distribution of the toll required (in cents) at various toll booths on American roads.

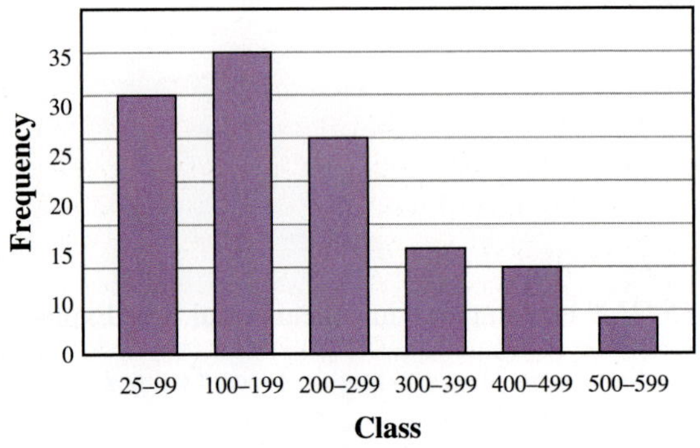

1. What range of tolls is most common?

2. How many toll booths have a toll between $4 and $6?

SKILL 2: PROBABILITY AND PERCENT CHANCE

3. If there's a 12% chance of a hurricane making landfall in Florida this year, what's the probability of that occurring?

SKILL 3: ESTIMATE AN ENCLOSED AREA

4. If the area enclosed by all boxes in the histogram above is 15 square centimeters, estimate the area enclosed by the first two boxes.

SKILL 4: FIND PERCENT OF A WHOLE

5. There are 712 people watching a play in a theater with 850 seats. What percentage of the seats are filled?

Lesson 4-7 Just a Normal Day (Normal Distributions and Z Scores)

LEARNING OBJECTIVES

- ☐ 1. Recognize characteristics of data that are normally distributed.
- ☐ 2. Understand the connection between area under a normal curve, percentage, and probability.
- ☐ 3. Make an educated guess about the empirical rule, then use the rule to calculate percentages and probabilities.
- ☐ 4. Compare data values from different sets using Z scores.

Courtesy of David Sobecki

To study the abnormal is the best way of understanding the normal.
—William James

If you grew up anywhere near maple trees, you're probably familiar with the seed pods they drop by the thousands in the spring. Like almost all living things, the individual pods vary in size—on any given tree, there's a typical size, with some pods bigger and some smaller.

On a beautiful spring day, I gathered 100 pods from the maple in my back yard, measured each, then grouped them according to their length. The smallest was 42 mm, the largest 59 mm, and in the photo they're grouped in classes 42–43, 44–45, 46–47, etc. An interesting thing happened—we see that the largest number of pods have lengths somewhere in the middle of the range, and the classes farther away from the center have less pods. A wide variety of quantities in our world tend to exhibit this same phenomenon. In fact, it's so common that frequency distributions of this type came to be known as *normal distributions*. That means that if we get really good at studying normal distributions, we'll be able to analyze tons of interesting things in our world.

0. After reading the opening paragraphs, what do you think the main topic of this lesson will be?

4-7 Class

1. List some things that you think might be normally distributed. Try not to focus strictly on physical sizes.

Answers vary. You'll probably want to pitch in on this one.

Suppose a researcher selects a random sample of 100 adult women, measures their heights, and constructs a histogram. The researcher would probably get a graph similar to the one shown below.

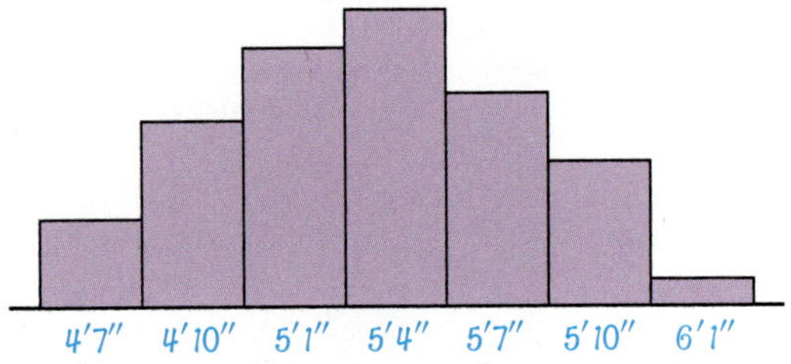

Histograms were introduced in the Applications for Lesson 4-4. If you haven't discussed those questions, now would be a good time to go over them and talk about histograms.

2. Notice that we omitted a scale on the horizontal axis. Put in a scale that you think seems reasonable based on the description of this histogram. What do you think the value right in the middle of the horizontal axis represents?

Sample answers shown. Inches are the better way to label, but most students will probably think in terms of feet, which is fine. The value in the middle should represent the mean height.

Next, the researcher decides to go big or go home, and she chooses "big." She dramatically increases the number of women in the sample, and since she has more data to go on, makes the classes narrower as well. Happy with the results, she adds even more women to her sample, and makes the classes even narrower. The results would most likely look like the next two graphs.

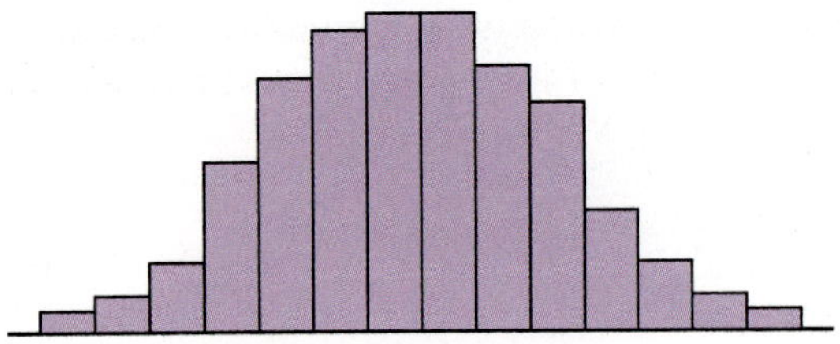

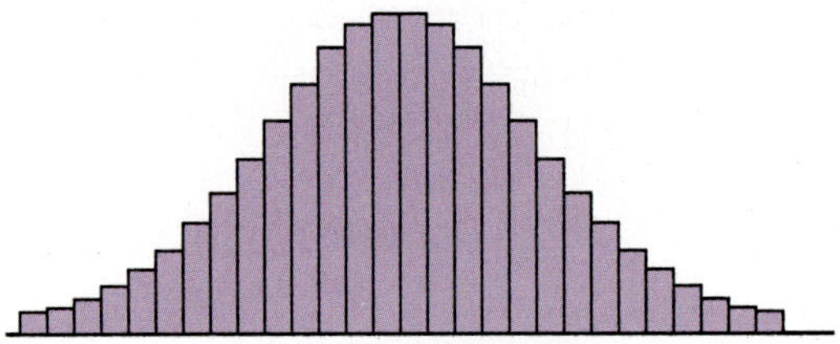

3. Now let's use some imagination. Suppose that our researcher has magical powers, and can continue adding data and improving her histogram until there are infinitely many data points. Wow, that's a lot. Instead of drawing boxes, draw a smooth curve that you think the histogram would become in the long run.

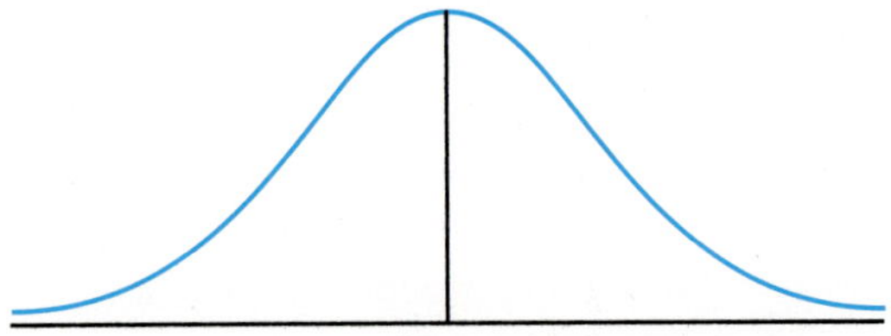

4. If all went well, you've just drawn your first normal probability distribution. (We'll see plenty of examples in the remainder of the lesson if you're unsure how well you did.) How does the drawing relate to the maple seed pod example from the opening paragraphs of this lesson?

The shape is very similar. It shows that the greatest number of women will have heights somewhere close to the mean, and as you go outward to larger and smaller heights, there will be less women at those heights.

To show that normal distributions are not even close to a purely mathematical topic, we're going to borrow our definition of normal distribution from an investment site called "Investopedia":

> A probability distribution that plots all of its values in a symmetrical fashion and most of the results are situated around the probability's mean is called a **normal distribution.** Values are equally likely to plot either above or below the mean. Grouping takes place at values that are close to the mean and then tails off symmetrically away from the mean.

Next, we'll highlight some of the important features of a normal distribution (which, in turn, give us the key features of a data set that is normally distributed).

- The value in the middle of the distribution, which appears most often in the sample, is the mean.
- The distribution is symmetric about the mean. This means that the graph has two halves that are mirror images on either side of the mean value.
- (This is the key fact): The area under any portion of the curve is the percentage (in decimal form) of data values that fall between the values that begin and end that region.

Let's explore that last key point, because it's what makes normal distributions so useful in analyzing data. The following graph shows a normal distribution for heights of women in the United States. The numbers on the horizontal axis are heights in inches, and some areas are labeled for reference.

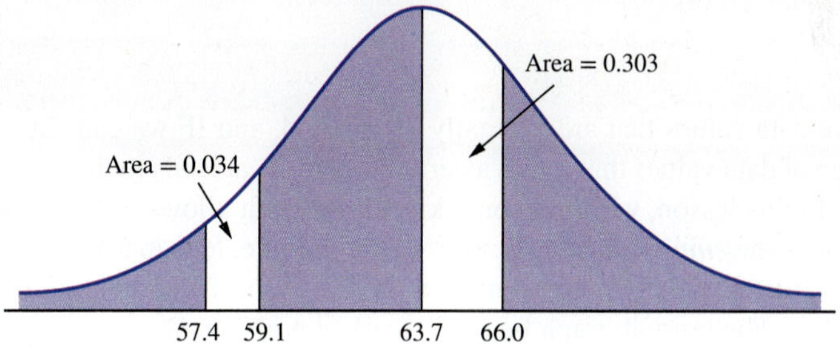

5. What is the mean height for American women?

63.7 inches

6. What percentage of American women are between 57.4 inches and 59.1 inches tall?

3.4%

7. If there are 31,806 women in the stands for a college football game, how many of them would you expect to be between 63.7 and 66.0 inches tall?

30.3% of 31,806 is 9,637.

Good teaching moment: Discuss when it is and is not appropriate to include decimal parts of percent calculations. I'll bet you five American dollars that many of the students will answer 9,637.218. Feel free to make fun of them for including leftover body parts.

Did You Get It?

Try this problem to see if you understand the concepts we just studied. The answer can be found at the bottom of the portfolio page.

1. What percentage of American women are taller than 66 inches? (Hint: Half of all women are taller than 63.7 inches.)

So now we know that if we have data values that are normally distributed, and IF we can find the area under portions of the graph, we can find the percentage of data values that fall in a certain range. That leads to the big question: HOW in the world do you find areas under the graph? In this lesson, we'll rely on a known rule that allows us to quickly find a range in which most of the data values fall. It's known as the *empirical rule*. Before seeing the rule, let's make some educated guesses.

According to recent research, the mean height for women in the U.S. is 63.7 inches (which you already knew, right?), and the standard deviation is 2.7 inches. The normal graph below is marked with the mean and heights that are one and two standard deviations above and below the mean.

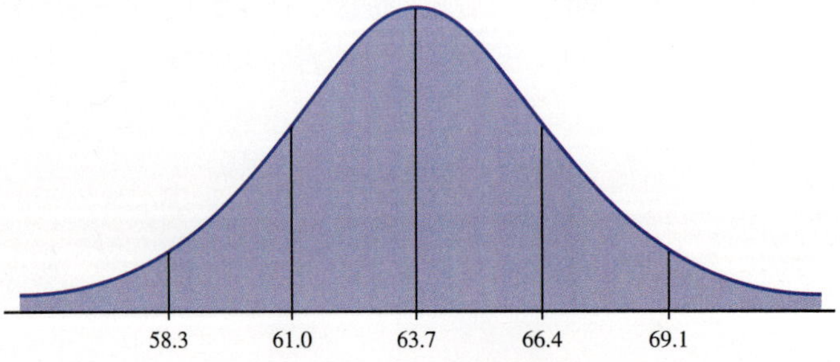

8. The total shaded area under the entire graph is 1, and this is true for ANY normal distribution. Explain why. Think carefully about what we know regarding area under a normal curve.

The area under any portion of the curve gives us the percentage of data values that fall within the vertical boundaries. If you look at the area under the entire curve, that should give you the percentage of values that fall somewhere between $-\infty$ and $+\infty$. Clearly, this is all values, or 100%. In decimal form, that's 1.

9. Make an educated guess as to what percentage of the total shaded area is between 61.0 and 66.4. What does this tell us about women's heights?

Answers vary. In any case, what it tells us is the percentage of women in the population between 61.0 and 66.4 inches in height.

10. Based on your answer to Question 9, estimate the percentage of area between 63.7 and 66.4 inches.

Whatever the student's answer to Question 9 is, this one should be half as much.

11. Now make an educated guess on the percentage of total shaded area between 58.3 and 69.1 inches.

Answers vary. The actual percentage, of course, is about 95%.

Ask students to write their estimates for percentages on the board, then see how they did when the empirical rule is introduced.

Now it's time to see how you did.

The Empirical Rule

When data are normally distributed, approximately 68% of the values are within 1 standard deviation of the mean, approximately 95% are within 2 standard deviations of the mean, and approximately 99.7% are within 3 standard deviations of the mean.

The graph below shows an illustration of the empirical rule. Recall that $\overline{X}$ is the mean, and s is the standard deviation.

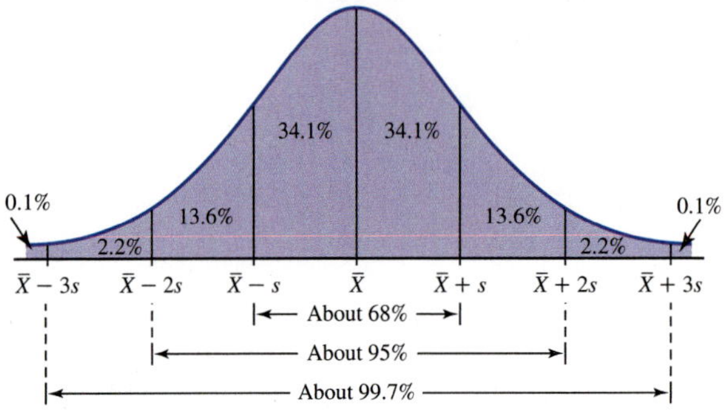

Man, I love M&M's. Can't get enough of them. When I buy a bag, the LAST thing I want is to get shorted. So how many M&M's is enough? According to a fascinating post on joshmadison.com, the mean number of M&M's in a standard vending machine bag is 54.6, with a standard deviation of 1.3.

12. Assuming that the number of M&M's per package is normally distributed (which, by the way, it more than likely is), fill in the blanks in the diagram below, which represents the distribution of M&M's in a standard package. Think number of M&M's.

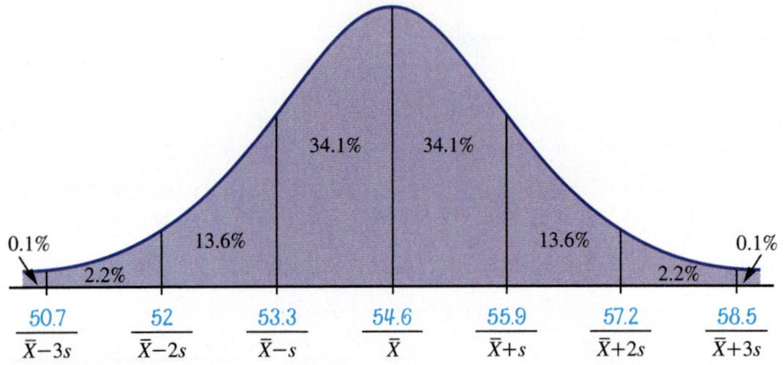

Math Note

In case you're wondering, to get the M&M data, some dude hand-counted over 2,600 M&M's from 48 packages. And I say "Good for him!"

13. If you buy 1,000 packages of M&M's (a number I assure you I've surpassed in my lifetime), how many bags would you expect to have between 53 and 56 M&M's in it?

This should be roughly 68.2% of the bags, which would be $0.682 \cdot 1{,}000 = 682$.

14. It's likely that over 99% of all bags have between ___50___ and ___59___ M&M's in them.

15. What percentage of bags would you expect to have at least 55 M&M's in them?

Pretty close to half of them. You'd expect half to have more than 54.6, and since you can't have 0.6 M&M's, that would mean half should have 55 or more.

16. What percentage of bags have less than 52 M&M's? How bad would that stink if you got one?

About 2.3%, and it would be devastating.

17. How unusual would it be to get a bag with more than 58 M&M's, or less than 52? Explain.

It would be pretty unusual. These amounts cover the regions below 2 standard deviations under the mean and above 3 standard deviations above the mean. Collectively, this area is about 2.4% of the total area, meaning that less than 2.5% of all bags should have less than 52 or more than 58 M&M's.

Did You Get It?

Try this problem to see if you understand the concepts we just studied. The answer can be found at the bottom of the portfolio page.

2. Normal distributions are very commonly used for quality control in manufacturing. For example, a manufacturer of light bulbs would surely be interested in studying how long their bulbs last. In one case, the lifetime of one model of light bulb is normally distributed and has a mean $\mu = 1{,}600$ hours and a standard deviation of $\sigma = 300$ hours. What percentage of bulbs would you expect to last longer than 1,900 hours?

4-7 Group

A quantity that can vary randomly from one individual trial to another, like the number of M&M's in a randomly selected bag, is called a **random variable.** Why? Because it's random, and it can vary. See, math DOES make sense!

If you're really sharp, you may have thought about the connection between the the percentages we've been calculating in this lesson and probability. If you haven't we're going to ask you to do so now. (There's a good reason that the normal is called a probability distribution, not a percentage distribution.)

1. Using what we know about area under a normal curve and percentage of data values, what does the area under the curve between two data values tell us about probability for a random variable?

It gives us EXACTLY the probability that a randomly selected data value will fall between those two values. We know we get the decimal form of the percent chance, which is the same thing as the probability.

This is a key idea in this lesson, so make sure that you circulate to all groups and check that theyre clear on how to use area to find probabilities.

2. Based on the work you did earlier, what's the probability that the next bag of M&M's you get from a vending machine will have either 56 or 57 M&M's in it? Briefly describe how you got your answer.

The probability is about 0.136. Those numbers cover all the possibilities between 1 and 2 standard deviations above the mean, and the empirical rules tells us that about 13.6% of the data should live in that range. This corresponds to a probability of 0.136.

Did You Get It?

Try this problem to see if you understand the concepts we just studied. The answer can be found at the bottom of the portfolio page.

3. Refer back to the light bulb manufacturer in Did You Get It 2. On the packaging for this particular bulb, the manufacturer lists the expected life of the bulb as 1,000 hours. If you buy one of those bulbs, find the probability that the manufacturer's claim will be true.

Z Scores

As you probably know, there are two main companies that offer standardized college entrance exams, ACT and SAT. Since each has a completely different scoring scale, it's really difficult to compare the scores of students that took different exams.

3. Do you think that scores for these exams are normally distributed? Discuss.

They are, but answers can vary.

One way to compare statistics drawn from different populations is to compare how far above or below the mean they are. The Z score is a statistical measure that was designed to do exactly that.

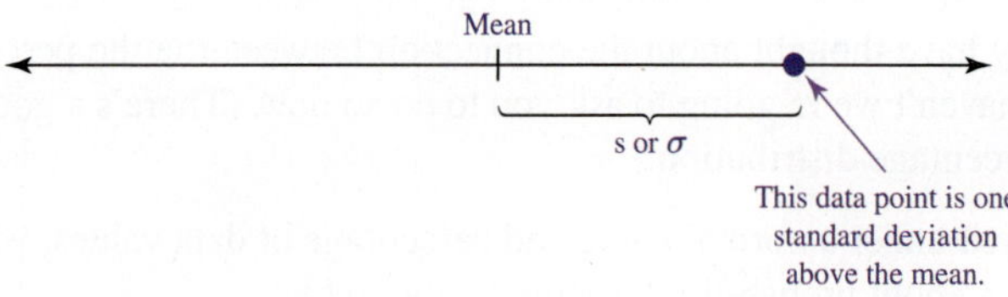

A Z score tells you how many standard deviations a data value is from the mean. The data value on the number line above would have a Z score of 1 because it is 1 standard deviation above the mean. Data values that are below the mean would have negative Z scores.

4. Fill in the blanks in the diagram with Z scores, using the description at the bottom of the previous page.

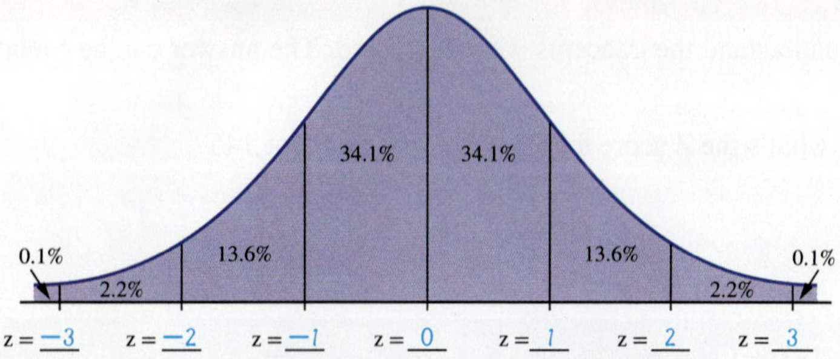

Make sure that students have this diagram labeled correctly before moving on.

A standardized test of intelligence is scaled so that the mean IQ is 100, and the standard deviation is 15. We'll be using these facts for the next few questions.

5. Find the Z score for a person with an IQ of 85.

−1

6. Find the Z score for a person with an IQ of 115.

1

7. What's the probability that a randomly selected person has an IQ between 85 and 115?

About 0.682

8. If a data value has a Z score of __0__, then half of the data are below that value, and half are above.

9. We know that the center of a normal distribution is the mean for a random variable. What does Question 8 tell you is also true about the center of the distribution?

It's also the median! The fact that half of the data lives above and half lives below is the definition of median.

10. A job applicant is told that their IQ score, measured in the screening process, is higher than 84% of the general population. What was their Z score, and their IQ?

About 16% of the area is to the right of $Z = 1$, so their Z score is right around 1, making their IQ about 115.

Did You Get It

Try this problem to see if you understand the concepts we just studied. The answer can be found at the bottom of the portfolio page.

4. On a standardized IQ test, what's the Z score for a person with an IQ of 145?

Computing Z Scores

For a data value from a sample with mean $\overline{X}$ and standard deviation s, or from a population with mean μ and standard deviation σ, the Z score is

$$Z = \frac{X - \overline{X}}{s} = \frac{X - \mu}{\sigma}$$

Some notes:

- A data point is greater than the mean if $Z > 0$ and less than the mean if $Z < 0$.
- Z scores have no units.
- Z scores are typically rounded to two decimal places

One year the ACT had a mean score of 21.2 and a standard deviation of 5.1. That same year, the SAT had a mean score of 1498 and a standard deviation of 347. Suppose that a scholarship committee is considering two students, one who scored 26 on the ACT and another who scored 1,800 on the SAT. Both are pretty good scores, but which one is better? That's not so easy to decide because the scales are totally different.

11. First instinct, which score sounds better to you?

Opinion question, so answers vary.

12. Find the Z score for each test score.

ACT of 26: $Z = \frac{26 - 21.2}{5.1} = 0.94$

SAT of 1,800: $Z = \frac{1800 - 1498}{347} = 0.87$

Math Note

Z scores are also used to approximate the area under any portion of a normal curve. This would allow you to compute probabilities more accurately than estimating based on the empirical rule.

13. Describe exactly what each Z score tells us.

A Z score of 0.94 means that the ACT score of 26 was 0.94 standard deviations above the mean, while the Z score of 0.87 shows 1800 on the SAT is 0.87 standard deviations above the mean.

14. Based on the Z scores, which would you say was a better score? Why?

The ACT score of 26 is better relative to other ACT scores, because it's farther above the mean than the SAT of 1800.

Did You Get It?

Try this problem to see if you understand the concepts we just studied. The answer can be found at the bottom of the portfolio page.

5. I'm a 6′4″ male, and my wife is 5′10″. We're both pretty tall, but clearly I am taller. I can send you pictures if you don't believe me. Anyhow, based on Z scores, which of us is taller compared to the rest of our gender? (The mean and standard deviation for male heights in the U.S. are 69.1 in. and 2.9 in.) The values for females were given earlier in the lesson.

Big sports fans tend to love arguing, and one of their favorite types of argument is "Who was better?" Michael or LeBron? Montana or Manning? Kershaw or Koufax? Blogs and airwaves are filled with arguments of this type. And what makes it so difficult to decide is that players and teams from different eras are hard to compare statistically because sports evolve, and what might have been an amazing statistic 50 years ago could be completely pedestrian today. Fortunately, this is EXACTLY the kind of thing that studying statistics analytically is good for.

The 1942, St. Louis Cardinals were an awesome baseball team. They won 106 games in an era when teams only played 154 games in a regular season. During the 1942 season, the mean number of games won by a National League team was 75.8 with a standard deviation of 20.5.

Another great Cardinals team was the 2004 squad, which won 105 out of 162 games. In 2004, the mean number of games won by a National League team was 80.9 with a standard deviation of 13.6.

© Associated Press

15. Based strictly on winning percentage (the percent of games won, that is), which was the better team?

$106/154 = 0.688$, so the 1942 team won 68.8% of their games.
$105/162 = 0.648$, so the 2004 team won 64.8% of their games. That makes the 1942 team better in terms of winning percentage.

16. Use information besides the winning percentages in the preceding paragraphs to make a good statistical argument about which team was better. Obviously, showing your work is a must. Write your final answer in the form of a few sentences, showing you know what you are talking about.

This question is intentionally left open-ended. Of course, any answer that doesn't compute and compare Z scores is no good. The Z scores work out to be:

1942 team: $Z = \dfrac{106 - 75.8}{20.5} = 01.47$

2004 team: $Z = \dfrac{105 - 80.9}{13.6} = 1.77$

17. Look back at your answer to Question 1 on the first page of this lesson. How did you do? Would you be able to come up with more examples now?

Answers vary.

4-7 Portfolio

Name ____

Check each box when you've completed the task. Remember that your instructor will want you to turn in the portfolio pages you create.

Technology

1. ☐ Spreadsheets can be used to efficiently do many comparisons between data values from different sets, because you can set up a formula to quickly calculate Z scores. The plan here is to build a Z score calculator that allows you to input mean and standard deviation for two different data sets, then use formulas to calculate the Z scores for different data values. In the template you can find in online resources for this lesson, there are a variety of ACT and SAT scores listed. Fill in the appropriate means and standard deviations from within the lesson, then use a formula to find Z scores and copy it down to find Z scores for all of the test scores for comparison. There are further instructions on the template.

Online Practice

1. ☐ Include any written work from the online assignment along with any notes or questions about this lesson's content.

Applications

1. ☐ Complete the applications problems.

Reflections

Type a short answer to each question.

1. ☐ Describe a random variable that is normally distributed.

2. ☐ Why are measures of average and variation so important in studying normal distributions?

3. ☐ What's the connection between a normal distribution and probability?

4. ☐ In what setting are Z scores useful? Why?

5. ☐ Take another look at your answer to Question 0 at the beginning of this lesson. Would you change your answer now that you've finished the lesson? How would you summarize the topic of this lesson now?

6. ☐ What questions do you have about this lesson?

Looking Ahead

1. ☐ Complete the Lesson 4-8 Prep Skills and read the opening paragraph in Lesson 4-8 carefully, then answer Question 0 in preparation for that lesson.

2. ☐ You'll need access to the Internet for Lesson 4-8, so make sure that someone in your group brings a laptop or a web-capable smart phone to class.

Answers to "Did You Get It?"

1. 19.7% **2.** About 16% **3.** About 0.9765 **4.** 3 **5.** I am, by just a bit.

4-7 Applications

Name ______________________________

The College Board reported that SAT scores in the early 2000s were normally distributed with mean 1026 and standard deviation 210. Use this information for Questions 1–3.

1. Fill in the blanks in the diagram.

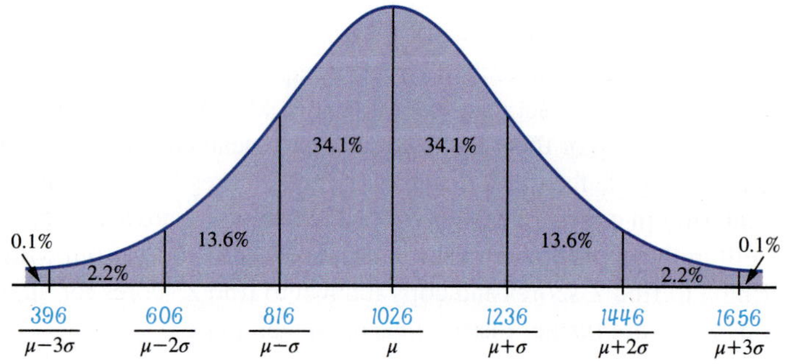

Math Note

Notice that in this case we're using μ and σ, which indicates that the mean and standard deviation are population, not sample, statistics.

2. What percentage of students scored over 606?

About 97.7%

3. Out of every 500 students, how many would we expect to score between 816 and 1236?

About 68.2% of them, which is 341.

4. If a student from that era scored 1400 on the SAT and her IQ was tested at 144, use Z scores to decide which test she did better on. Refer to Questions 5–10 in the Group portion of the lesson for information on IQ scores.

1400 SAT: $Z = \frac{1400 - 1026}{210} \approx 1.78$

144 IQ: $Z = \frac{144 - 100}{15} \approx 2.93$

She did MUCH better on the IQ test.

4-7 Applications

Name ____________________

According to 9monthsafter.com, the lengths of human pregnancies are normally distributed with mean 268 days and standard deviation 15 days. Use this for Questions 5–7.

5. Fill in the blanks in the diagram.

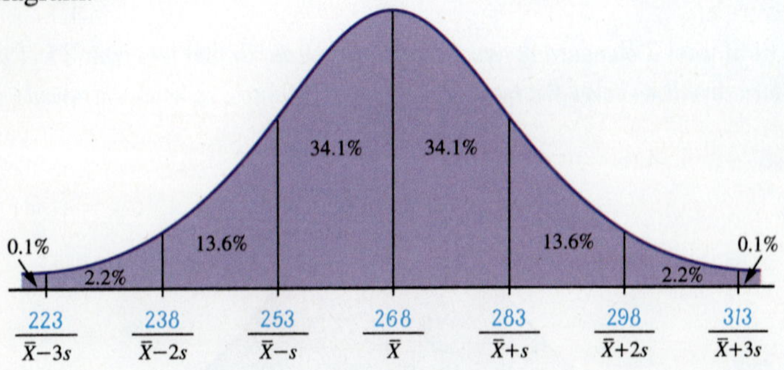

6. What is the probability that you were born more than 298 days after you were conceived?

About 0.023.

7. If a child is conceived on April 19, what is the probability that it will be born on or before January 27?

11 days in April; 31 in May, July, August, October, and December; 30 in June, September, and November.
$11 + 31 \cdot 5 + 30 \cdot 3 + 27 = 283$. Probability is 0.841.

According to the label, a bag of Kettle brand Tias tortilla chips contains 8 ounces of chips. Suppose that the bagging process is normally distributed with a standard deviation of 0.14 ounce.

8. If the machine that does the bagging is set to put 8 ounces of chips in each bag, this would be the mean of the distribution, with standard deviation as mentioned above. What percentage of all bags would have less than the amount stated on the package in this case?

50%

9. Explain why it would be a bad idea for the manufacturer to set the bagging machine at 8 ounces.

Half of every bag containing below the listed weight is a class-action lawsuit waiting to happen. At the very least, they risk a lot of unhappy customers.

4-7 Applications

Name ______________________________

10. If the manager wants there to be greater than a 97% chance that any given bag has at least 8 ounces, and the weight they set their filling machine for provides the mean weight, how should they set the machine? Complete the diagram as part of showing your work.

They would want 8 ounces to be at least 2 standard deviations below the mean, so that less than 3% of all bags will be likely to have less than 8 ounces. Two standard deviations below the mean would be 0.28 ounces, so he or she should set the machine to put 8.28 ounces in each bag.

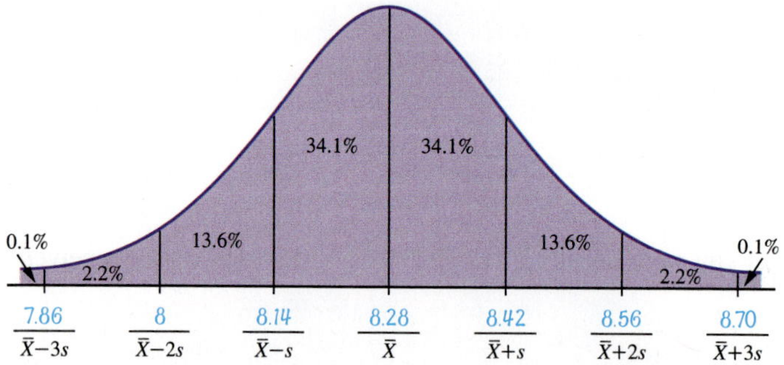

For Questions 11–14, we'll assume that the finishing times for a group of runners in a marathon race are likely to be normally distributed. The mean finish time for male marathon (26.2 miles) runners is 278.42 min with a standard deviation of 39.4 min. The mean finish time for female marathon runners is 294.45 min with a standard deviation of 39.9 min. At the 2014 Boston Marathon, Meb Keflezighi won the men's division with a time of 128.62 minutes and Rita Jeptoo won the women's division at 138.95 minutes.

11. Who finished faster, relative to their gender? Explain why. (Remember that in a race you want to finish in a short time.)

Meb: $Z = \dfrac{128.62 - 278.42}{39.4} \approx -3.80$

Rita: $Z = \dfrac{138.95 - 294.45}{39.9} \approx -3.90$

Rita's time is farther below the mean than Meb's, so she had a better time relative to her gender than Meb did.

4-7 Applications

Name ______________________________

12. Out of 10,000 runners, where would you expect a woman who finishes in a time of 294.45 min to finish? Complete the diagram to help show your work.

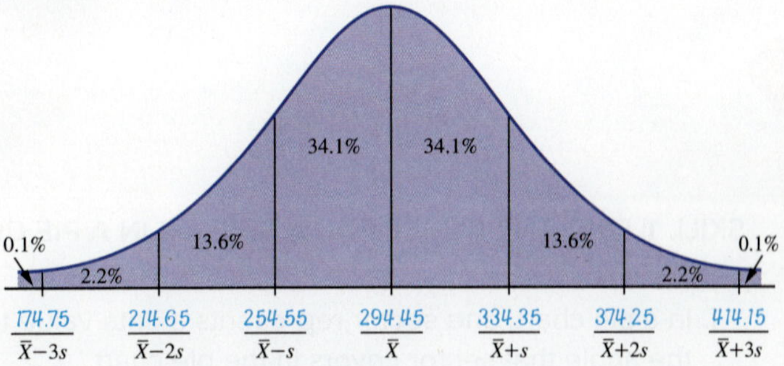

5,000th. Since 294.95 minutes is the mean, it's also the median, meaning half of the data values will be higher.

13. A man who finished in a time of three and a half hours would finish in the fastest __5__ percent of male racers. Complete the diagram to help you show your work. You'll have to make an estimate of area since that time doesn't correspond to one of the labeled times. (Interesting note after you have the percentage: This guy would finish well over an HOUR after the winner. Wow.) Hint: A time of 160.22 minutes means you're in the fastest 0.1%, so you finished before 99.9% of the runners.

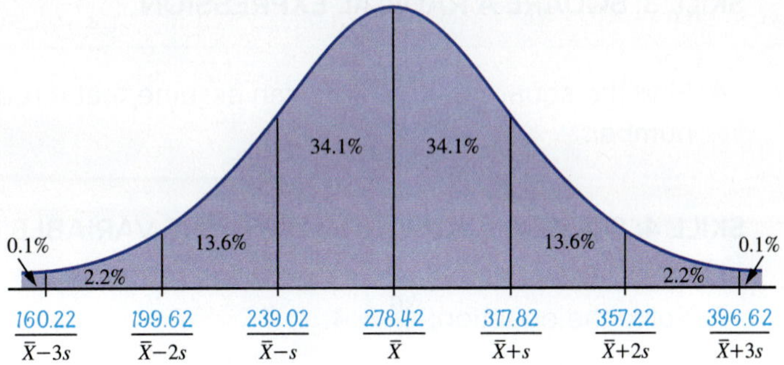

Three and a half hours, which is 210 minutes, is a fourth of the way between 199.62 and 239.02. Estimates will vary but any good answer at least has to be between 2.3% and 15.9%. I think something around 5% is a pretty good guess.

Lesson 4-8 Prep Skills

This is a short review of skills that will come in handy in the next lesson. In each case, answer the question, then rate your confidence level by checking one of the boxes. If you feel like you're struggling with these skills, consult the online resources provided by your instructor for extra practice.

I KNOW IT	THINK SO	UNSURE	NO IDEA	
				SKILL 1: FIND THE ANGLE FOR A SECTOR IN A PIE CHART
				1. In a pie chart, one sector represents a data value that's 47% of a whole. Find the angle that sector covers in the pie chart.
				SKILL 2: APPROXIMATING QUANTITIES INVOLVING SQUARE ROOTS
				2. Use a calculator to find an approximate value for $\sqrt{18.4}$. Round to two decimal places.
				3. Use a calculator to find an approximate value for $\sqrt{\frac{0.4(1-0.4)}{800}}$. Round to two decimal places.
				SKILL 3: SQUARE A RADICAL EXPRESSION
				4. Find the square of $\sqrt{\frac{12}{n}}$. You can assume that n represents a positive number.
				SKILL 4: SOLVE AN EQUATION WITH THE VARIABLE IN A DENOMINATOR
				5. Solve the equation: $\frac{36}{n} = 4$.
				6. Solve the equation: $\frac{0.027}{n} = 3.5$.

Lesson 4-8 The Error of Your Ways (Polling and Margin of Error)

LEARNING OBJECTIVES

- ☐ 1. Identify margin of error in a poll.
- ☐ 2. Calculate and interpret the meaning of margin of error.
- ☐ 3. Calculate the number of respondents needed for a given margin of error.

Better to trust the man who is frequently in error than the one who is never in doubt.

—Eric Sevareid

Courtesy of David Sobecki

Do you know what percentage of Americans think that Bigfoot exists? Me neither. But I bet you can find that information online somewhere. In our information society, it seems like there's a poll for everything, from approval of the president to favorite type of underwear. And since so much information is provided to citizens and consumers in the form of polls, an understanding of how polls operate has become a pretty important survival skill. In this lesson, we'll study a key aspect of interpreting the results of a poll, the margin of error. And we'll find out that understanding error in polls can prevent errors in judgment.

0. After reading the opening paragraph, what do you think the main topic of this lesson will be?

4-8 Group

Polling has become big business, and two of the biggest names in that business are Gallup and Pollster. One of the more widely quoted polls they run is used to determine the president's current approval rating among citizens. But of course, they can't ask everyone in the country every week whether they approve or disapprove of the president's performance, so instead they ask a sample of citizens and use sampling and statistical techniques to try to decide if the results are truly indicative of the general population.

1. Go to the website www.pollster.com, then find the link for the presidential approval survey and click on it. What is the president's most current approval rating? Write a couple of sentences explaining what that means.

Poll numbers of course vary. The rating describes the percentage of Americans that approve of the job currently being done by the president.

> Pollster is a very well-established site, but the Web is a fluid place. Check the site before class to make sure that the needed information is readily available, and if it's not, find a site that has approval rating polls for the president and Congress.

2. How do you think the approval rating was calculated? Be specific. Make sure your answer contains both of these key words: "population" and "sampler."

There's no way they could survey the entire population: instead, they survey a sample of the population, then use the results to draw general conclusions about the opinions of the overall population.

3. Draw a pie chart describing the most current approval rating. Your chart should have three sectors.

Charts will vary depending on the current poll numbers. The three sectors should be approve, disapprove, and undecided.

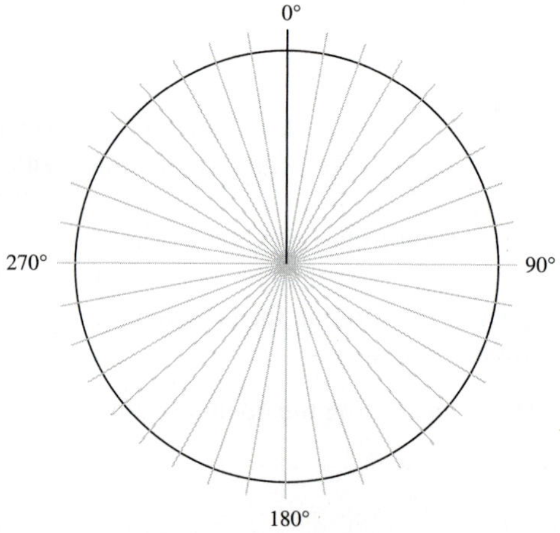

4. How would you describe the trend in approval rating over the last six months?

Answers vary depending on latest polling results.

5. Scroll down a bit and find the first link to the Gallup survey and click on it. You'll see an approval rating and a brief description of how the poll was conducted, including a line that says "Margin of error," followed by something like "±3 percentage points." What do you think the term "margin of error" means in this context?

Since we have not yet defined margin of error, these answers should vary a lot. In fact, we hope they do.

Use this question to discuss what the students think margin of error means. This is a great opportunity to call on each group and have a student read their group's answer, then compare.

6. Can you think of another situation in life, school, or work where you've heard either "margin of error" or "margin for error"? Describe.

Answers vary.

7. What do you think is better to have, a small margin of error or a big one? Why? Does it depend on the situation?

Hopefully, most groups will answer smaller. Whether or not they know specifically what margin of error means, the term 'error' should certainly make everyone think that less is better.

Did You Get It

Try this problem to see if you understand the concepts we just studied. The answer can be found at the bottom of the portfolio page.

1. Back at the Pollster website, click on the link for "Congress Job Approval."
   a. What is the approval rating for Congress?
   b. What's the margin of error?
   c. Compare the approval rating to the one for the president and discuss what you think that says.

4-8 Class

Because it's not possible to poll every individual all the time to ask their opinions, sampling plays a really important role in polling. As you know, sampling is the process of choosing a portion of a population in such a way that the individuals chosen are in some way representative of an entire population. The downside of sampling is that it introduces uncertainty in the results of a poll, since to some extent the pollster is making an educated guess.

The **margin of error** for a poll is a statistical measure that provides a range of values that the true outcome of the poll is likely to be inside. A margin of error of 5% means that the true value of the statistic reported by the poll is likely to be within 5% of the reported value.

Let's look at an example. If someone reports that 84% of Americans like pizza (which sounds awful low to me, by the way), and the margin of error is ±3%, this means they're confident that the actual percentage of Americans that like pizza is somewhere between 81% and 87% (since those are the percentages that are within 3% of 84.)

Here's the catch: how confident is confident? For margins of error, confidence is measured as a percentage. One way of computing margin of error is with a 95% confidence level. In our example above, a 95% confidence level would mean that the polling institution is 95% sure that the accurate percentage of Americans that like pizza is between 81% and 87%.

1. Take another look at the Gallup presidential approval rating, and find the margin of error. Then describe exactly what the poll result means, referring to both the approval rating and the margin of error.

   Answers vary depending on latest poll numbers and margin of error. Look for answers that not only add and subtract the margin of error from the reported approval rating, but that also say something about 95% confidence.

2. An Associated Press poll indicated that 63% of teachers below the level of college feel like the amount of homework students get is not excessive. The poll had a margin of error of ±3.5% with a 95% confidence level. Write a sentence or two explaining what that result means.

   The researchers are 95% sure that if all teachers below college level had been surveyed, between 59.5% and 66.5% of them would say that students do not get an excessive amount of homework.

A formula for computing margin of error is provided in the colored box below. Sadly, it means absolutely nothing until you know what all of the symbols stand for, which is why you should read everything in the colored box.

Computing Margin of Error

When n people are polled, and $\hat{p}$ give a particular response, the margin of error with a 95% confidence level is given by the formula

$$\text{Margin of Error} = 2\sqrt{\frac{\hat{p}(1-\hat{p})}{n}}$$

Note that both the margin of error and $\hat{p}$ (which is often called the **sample proportion**) are percentages written in decimal form. The symbol $\hat{p}$ is usually read "p hat."

The margin of error means that if this poll were conducted many times, we would expect that at least 95% of the time the actual number of people in the population that would give a particular response is somewhere between $\hat{p}$ − the margin of error and $\hat{p}$ + the margin of error.

© Marka/Alamy Stock Photo

3. On December 18, 2015, an organization called Public Policy Polling released the results of a poll with 532 responses. Of those respondents, 181 supported one particular candidate for the presidency in 2016. Among those folks, 41% were in favor of bombing Agrabah, which as it turns out is the fictional country from the Disney movie *Aladdin.* Use the formula in the colored box to find the margin of error for this result.

$$2\sqrt{\frac{0.41(1-0.41)}{181}} \approx 0.073 \text{ or } 7.3\%$$

4. Write a sentence describing exactly what your answer to Question 3 tells us about all American voters who supported that candidate. More detail is always better.

We are 95% confident that of all people in the country who supported that particular candidate, somewhere between 33.7% and 48.3% were in favor of bombing a fictional country. Use this as a rebuttal to anyone who says that education isn't important.

5. Among all respondents to the survey, 30% supported bombing a cartoon country, and Public Policy Polling reported a margin of error of 4.3%. Does that number match what you get from the formula above?

$$2\sqrt{\frac{0.30(1-0.30)}{532}} \approx 0.040 \text{ or } 4.0\%.$$ The reported margin of error is a bit high.

Did You Get It

Try this problem to see if you understand the concepts we just studied. The answer can be found at the bottom of the portfolio page.

2. A *Wall Street Journal* poll conducted in December of 2014 asked 1,000 adult Americans if they or a family member had ever been notified by a credit card company of a possible data breach involving their personal information. Forty-five percent reported that they had been notified. Use the formula in the colored box to find the margin of error for this result, then describe what the poll tells us.

6. For this question, don't look at the margin of error formula. Seriously, no peeking or I'll know. Based on what margin of error measures, do you think the margin of error for a poll should go up or down as the number of people surveyed gets larger? Justify your answer.

Since we're using a sample to draw conclusions about an overall population, it stands to reason that sampling a higher percentage of the population would make it more likely that our conclusions are accurate. So the margin of error should go down.

7. Now look at the margin of error formula. Based strictly on that formula, what should happen to the size of the margin of error if the number of people surveyed gets larger? Does this match your answer to Question 6? Explain.

The only place the number of people surveyed appears is in the denominator. If the denominator of a fraction gets bigger, the overall value of the fraction gets smaller. So surveying more people should result in a smaller margin of error, which is exactly what we reasoned in Question 6. Is it just me, or are we getting smarter?

8. What if the folks from Public Policy Polling who conducted the Agrabah poll got a little bit lazy, and decided to only survey 100 people, then go bowling. Recalculate the margin of error for the overall poll. Did it get bigger or smaller?

$2\sqrt{\frac{0.30(1-0.30)}{100}} \approx 0.092$ or 9.2%. The margin of error gets a LOT bigger.

While it's absolutely fine and dandy to calculate the margin of error after conducting a poll, in many cases it's more useful to think about what the margin of error might be like BEFORE you're done sampling. This would allow you to decide how many people need to be surveyed in order to make your result reliable.

Suppose that you begin surveying students on your campus about their attitudes on race relations in the United States. After 70 surveys are returned, you find that 32% of respondents think that race relations are a serious problem facing our society.

9. What is the margin of error at that point in the survey?

$$2\sqrt{\frac{0.32(1-0.32)}{70}} \approx 0.112 \text{ or } 11.2\%$$

10. If you decide that you'll only consider your survey a success if the margin of error is at most 4%, it would be perfectly reasonable to want to know how many more people you'd have to survey. Assuming that the result holds at 32%, set up the margin of error formula with a 4% margin of error, a 32% response proportion, and variable n representing the number of people surveyed. (Remember, percentages need to be written in decimal form.)

$$2\sqrt{\frac{0.32(1-0.32)}{n}} = 0.04$$

In order to find the sample size needed to get that 4% margin of error, we'd need to be able to solve the equation you wrote in Question 10 for n. There are two issues to address: First, the quantity we want to solve for is inside a square root, and second, it's in the denominator of a fraction.

Let's begin by eliminating the radical. To accomplish this, we can square both sides of the equation, because we know that squaring a square root eliminates the radical. But first, it's helpful to rearrange so that the square root is the only thing on one side of the equation.

11. Starting with the equation $0.04 = 2\sqrt{\frac{(0.32)(0.68)}{n}}$ (Aw man, I just gave away the answer to the last question . . .), divide both sides by 2 to isolate the radical.

$$0.02 = \sqrt{\frac{0.32(0.68)}{n}}$$

12. Now square both sides of the equation.

$$(0.02)^2 = \sqrt{\frac{0.32(0.68)}{n}}^{\,2} \Rightarrow 0.0004 = \frac{0.32(0.68)}{n}$$

13. Keeping in mind that the goal is to solve for *n*, what can we do to BOTH SIDES of the equation in order to get *n* out of the denominator? Explain what we can do, then, you know, do it.

Multiply both sides of the equation by n.

$$0.0004n = \frac{0.32(0.68)}{n} \cdot n \Rightarrow 0.0004n = 0.32(0.68)$$

14. Now you should be able to solve for *n*, so write how many MORE PEOPLE you'd need to survey to get that 4% margin of error we were shooting for.

$$n = \frac{0.32(0.68)}{0.0004} = 544$$

Since 70 have already been surveyed, we'd need to survey at least 474 more.

15. Do you find that number surprising? Discuss.

Opinion question, so answers vary.

Did You Get It ?

Try this problem to see if you understand the concepts we just studied. The answer can be found at the bottom of the portfolio page.

3. Preliminary results of a survey show that 11% of Americans lack health insurance. How many people need to be surveyed in order to have a margin of error of 2%?

16. Now let's go back to the presidential poll we started this lesson with. Use the approval rating you discovered, along with the margin of error, to find the number of people that were polled. Now look again at the description of the poll. How does your result compare to the number of people that were actually surveyed? What do you think accounts for any discrepancy?

Answers vary depending on current polling numbers. Sample answer based on today's rating shown. Approval rating 45%, margin of error 3%.

$$2\sqrt{\frac{0.45(1-0.45)}{n}} = 0.03 \Rightarrow n = 1{,}100$$

The polling page shows that 1,500 adults were surveyed. The difference could be attributable to rounding error in the approval rating (rounded to whole number), or the margin of error. It could also be possible that they used a higher confidence percentage than 95%.

17. Suppose that the result of a public opinion poll shows that just 9% of people would be willing to give up their cell phone for 1 week in exchange for $100. How many people would need to be surveyed in order to have a margin of error of ±2%?

$$2\sqrt{\frac{0.09(1-0.09)}{n}} = 0.02 \Rightarrow n = 819$$

© Hero/Corbis/Glow Images RF

18. Repeat Question 17, but this time for a poll where 50% of respondents say they'd be willing to give up their phone for a week. How does this affect the number of responses needed to get the 2% margin of error?

$$2\sqrt{\frac{0.5(1-0.5)}{n}} = 0.02 \Rightarrow n = 2{,}500$$

Math Note

Remember, the formula we're using provides 95% confidence in our margin of error. For other confidence levels, the coefficient in front of the root changes.

19. Bonus Question: Look carefully at the margin of error formula. What about it, mathematically, makes your conclusion from Question 18 make perfect sense? Detailed explanation if you want bonus points!

The numerator of the formula has the product of the sample proportion and one minus that number. If one of those two numbers is close to zero, the product is going to be smaller than if both numbers are close to a half. That in turn makes the margin of error smaller. In order to keep the margin of error the same for both, you'd need a bigger denominator when the numerator is smaller.

20. Now design your own problem. Find a survey online that interests you, and using the result and the reported margin of error, calculate the number of respondents.

Answers vary.

4-8 Portfolio

Name ____

Check each box when you've completed the task. Remember that your instructor will want you to turn in the portfolio pages you create.

Technology

1. ☐ Find an article on either Gallup.com or Pollster.com that has the term "margin of error" in it, and briefly explain what that means in the context of that particular poll.
2. ☐ As we saw in this lesson, being able to find the number of people that are needed to get an acceptable margin of error is really useful. Build a spreadsheet like the one below that uses a formula to compute the value of n needed to achieve a certain margin of error given a value of $\hat{p}$. (Hint: You'll need to solve the equation that computes margin of error for n. Questions 9–14 in the Class portion will be a big help.)

	A	B	C
1	p	Margin of error	n
2	0.32	0.01	8704
3	0.32	0.02	2176
4	0.32	0.03	967
5	0.32	0.04	544

Online Practice

1. ☐ Include any written work from the online assignment along with any notes or questions about this lesson's content.

Applications

1. ☐ Complete the applications problems.

Reflections

Type a short answer to each question.

1. ☐ What is the margin of error for a poll? When is it used, and why?
2. ☐ Describe how being familiar with margin of error might help you in the future.
3. ☐ Take another look at your answer to Question 0 at the beginning of this lesson. Would you change your answer now that you've completed the lesson? How would you summarize the topic of this lesson now?
4. ☐ What questions do you have about this lesson?

Looking Ahead

1. ☐ Complete the Lesson 4-9 Prep Skills and read the opening paragraph in Lesson 4-9 carefully, then answer Question 0 in preparation for that lesson.

Answers to "Did You Get It?"

1. Answers vary depending on recent poll numbers. **2.** 3.1%; Descriptions vary.
3. 979 people

4-8 Applications

Name ______________________________

A stats class was interested in determining the percentage of students on their campus who are opposed to a new policy eliminating all paper books from the campus bookstore in favor of eBooks. After surveying 90 students, the class finds that 33% are opposed to the policy.

1. Calculate the margin of error for this survey. Round the percentage to one decimal place.

© Fancy/Veer/Corbis/Glow Images RF

$$2\sqrt{\frac{0.33(1-0.33)}{90}} \approx 0.099 \text{ or } 9.9\%$$

2. Write a sentence or two describing exactly what your answer to Question 1 tells you about the percentage of the overall student body that is opposed to the policy.

We can be 95% confident that somewhere between 23.1% and 42.9% of all students on campus are opposed to the policy.

3. The class conducting the poll was unhappy with the margin of error (as well they should), so they decide to survey another 90 students. With the additional data, the percentage of students opposed to the policy goes up to 35%. What's the margin of error now?

$$2\sqrt{\frac{0.35(1-0.35)}{180}} \approx 0.071 \text{ or } 7.1\%$$

4. If surveying more students keeps the percentage at 35%, how many more would need to be surveyed in order to be 95% confident that the correct percentage for the entire student body is between 32% and 38%? Don't forget to consider the number of students that had already been surveyed. (Hint: What would the margin of error be in this case?)

$$2\sqrt{\frac{0.35(1-0.35)}{n}} = 0.03 \Rightarrow n \approx 1{,}011$$ *They'd need to survey 831 more students.*

4-8 Applications

Name ____________________

You may have noticed that the margin of error formula only factors in the *sample* size, not the *population* size. If you think about it, this sounds kind of fishy: If you survey 30 members of a club out of a population of 40, you'd expect your results to be much more reliable than if you survey 30 people out of the entire U.S. population of 325 million. But the margin of error formula would treat both polls the same.

It turns out that the margin of error formula we've been using works consistently well as long as the sample surveyed is less than 5% of the overall population. In cases where the sample is larger than 5% of the population, we would want to adjust the margin of error to reflect the fact that we're actually polling a large sample relative to the population. For that purpose, statisticians developed a statistic called **finite population correction** (FPC) to account for the smaller margin of error that should result.

5. Suppose that the campus in Questions 1–4 is a small satellite campus, with only 400 students. After surveying 180 students (Question 3), at that point almost half of everyone on campus would have been surveyed. The formula for finite population correction is below. Find the value, and in the next question we'll learn what to do with that number.

$\text{FPC} = \sqrt{\frac{N-n}{N-1}}$ where N is the population size, and n is the sample size.

$$\text{FPC} = \sqrt{\frac{400-180}{400-1}} \approx 0.74$$

6. The FPC acts as a multiplier: To get the adjusted margin of error, multiply the value of the FPC by the margin of error that came from the usual formula. Find the new margin of error for the situation where 180 students have been polled out of a campus population of 400. (Use the margin of error you found in Question 3.)

$0.071 \cdot 0.74 \approx 0.053$

The adjusted margin of error is 5.3%.

7. Without referring to any formulas, describe what should happen to the margin of error as the size of the sample gets really close to the size of the population. Then use the FPC formula to argue that your conclusion matches what would happen mathematically.

The margin of error should basically shrink toward zero, because when you get to the point where you've polled an entire population, there IS no error. Since the numerator of the FPC is $N - n$, that's exactly what will happen. As the sample size gets close to the population size, that numerator heads toward zero, and multiplying that result by the margin of error will make the margin of error head toward zero as well.

Lesson 4-9 Prep Skills

This is a short review of skills that will come in handy in the next lesson. In each case, answer the question, then rate your confidence level by checking one of the boxes. If you feel like you're struggling with these skills, consult the online resources provided by your instructor for extra practice.

I KNOW IT | THINK SO | UNSURE | NO IDEA

SKILL 1: COMPUTE MEASURES OF AVERAGE

1. Find all four measures of average for the data set below:

   12 18 21 18 9 33 97 20 19 25 13

SKILL 2: COMPUTE RELATIVE DIFFERENCE

2. After a promising earnings report, the price of a tech stock jumps from $7.45 to $11.30 in one day. Find the change in price, and the relative change in price.

SKILL 3: DRAW A BAR GRAPH

3. Draw a bar graph that illustrates the data in the table, which are the maximum depth in feet of each of the Great Lakes. The scale on the vertical axis should go from zero to 1,500 feet.

Lake	Superior	Erie	Michigan	Ontario	Huron
Max Depth	1,333	210	923	802	750

SKILL 4: GRAPH PAIRED DATA

4. The data in the table are the percentage of total power in the United States generated by wind for selected years from 2002 to 2014. Graph the data on a rectangular coordinate system with the year on the *x* axis.

Year	2002	2005	2008	2011	2014
Percent	0.26%	0.46%	1.25%	3.20%	3.95%

Lesson 4-9 Trust No One (Misuses of Statistics)

© Ed Araquel/TM and Copyright 20th Century Fox Film Corp/Everett Collection

LEARNING OBJECTIVES

- ☐ 1. Identify misuses of sampling and evaluate their effect on statistical results.
- ☐ 2. Recognize and describe common misuses of compiling and reporting statistics that make them meaningless or deceiving.
- ☐ 3. Study ways that graphs can be manipulated to tell a desired story.

There are two ways of lying. One, not telling the truth and the other, making up statistics.

—Josefina Vazquez Mota

"Trust No One" was the unofficial motto of the classic TV show *The X Files*, in which intrepid agent Fox Mulder and his faithful assistant Dr. Scully work to find the truth behind incredible events amid vast conspiracies. In the world of that show, things were rarely what they seemed, and the best course of action for the investigators (and viewers) was to assume that anything that anyone said may have been at best a stretching of the truth. In our modern world, there's a lot of wisdom in that approach. We are constantly bombarded with information and statistics, and it would be quite a leap to expect that most of the people disseminating that information have your best interests in mind. An educated adult knows that it's a good idea to view information suspiciously, and also knows how to recognize when statistics are being manipulated to tell the story that someone wants you to hear, not the truth. But, as Mulder was fond of pointing out, the truth is out there. You just have to want to find it.

0. After reading the opening paragraph, what do you think the main topic of this lesson will be?

4-9 Group

Ideally, statistical research is used to provide knowledge and information to help us make intelligent decisions about our health and welfare, or maybe just to give us information about things we find interesting. But the world is a less-than-ideal place and there are people who will misuse statistics to sell us products that don't work; to attempt to prove something that isn't true; or to get our attention by using fear, shock, or outrage supported by bad statistics.

Just because we read or hear the results of a research study or an opinion poll in the media, this doesn't mean that these results are reliable or that they can be applied to any and all situations. For example, reporters sometimes leave out critical details like the size of the sample used or how the research subjects were selected. Without this information, you can't properly evaluate research and properly interpret the conclusions of a study or survey.

The purpose of this lesson is to show you some ways that statistics can be misused. The point isn't to make you reject any statistical study or result; it's to help you to recognize when information or conclusions you're being given might be presented in a biased or dishonest way. Throughout this unit, we've sprinkled in some information on ways that statistics can be misused: in this lesson, we'll compile and expand on those misuses.

Suspect Samples

Since there are over 300 million people in the United States (and over 7 billion in the world), the vast majority of statistical studies you run across involve sampling: This is where any study is going to begin. That's why we started our study of statistics with sampling, and it's also why that's where we'll begin our look at statistical mischief.

Unreported Sample Sizes

In the first part of the 21st century, there was a chewing gum brand that ran a large campaign claiming that "Four out of five dentists surveyed" would recommend their product. That sounds like a ringing endorsement. But what does it really mean?

1. By making this statement, what is the advertiser trying to imply? Calculate and use a percentage in your answer.

   That 80% of all dentists in the country would recommend this particular product.

2. How many dentists do you think there are in the United States? Make a guess.

   Answers vary, but the actual number is about 98,000.

3. How many dentists do you feel like they would have to survey in order to make this statistic meaningful?

   This is an opinion. But it better be more than five.

The bottom line is that the advertiser (of course) provided no information on how many dentists they actually surveyed. If they only surveyed five, their claim could still be technically true, but they may have stumbled across four that by coincidence would recommend their product, even if the vast majority of dentists would rather see their patients chewing razor blades. Worse still, they could easily have chosen to survey five that would be predisposed to recommend their product for one reason or another (like say a check with multiple zeroes on it).

If samples in a study are too small, the results could be completely meaningless. And if the size of a sample is conveniently left out, there's a pretty good chance that the person reporting the statistics would just as soon have you not know what that sample size was.

4. Why would an advertiser be likely to leave out the size of a sample on which a statistic is based?

   If they're making a claim that is unlikely to hold true with a meaningful sample size, they'd be likely to use a smaller sample, and one improperly chosen. And in that case, obviously, they'd rather keep that quiet.

Did You Get It

Try this problem to see if you understand the concepts we just studied. The answer can be found at the bottom of the portfolio page.

1. Evaluate this claim by an advertiser: 60% of people surveyed prefer our brand of Cola over the leading national brand.

Samples Relying on Volunteer Participation

5. Suppose that a national poll is conducted by telephone to gauge the favorability rating for a particular candidate for the presidency. What types of people will not be represented in this survey? List as many as you can think of.

Some potential answers: people that don't have a phone; people that don't answer calls from numbers they don't recognize; people that answered the phone but declined to answer; people on the do not call list; people that work odd hours and aren't available to answer calls during normal hours.

People who are annoyed by unsolicited phone calls won't appear in a sample of people contacted by phone.

Studies using volunteers sometimes have a built-in bias because volunteers generally don't represent the population at large. Sometimes they are recruited from a particular socioeconomic background. For example, unemployed people are far more likely to volunteer for research studies in order to get paid, and because they're likely to have more free time available to participate.

6. What groups of people would be available for studies that require the subjects to spend several days or weeks in an environment other than their home or workplace?

Almost exclusively people that don't have a job with regular hours. People that don't have school-aged children would probably be overrepresented as well.

7. In the past, many studies have used only men but have tried to generalize the results to both men and women. Do you think this is a good idea? Why? Does it depend on what is being studied?

Opinion question.

8. To conduct market research on their products, some companies will offer coupons, samples, or even cash to participate in a study. Comment on how you think this is likely to affect the outcome.

Answers vary. First, it's a classic volunteer sample situation, which automatically cuts down on randomness. Second, the inducements are likely to bias many people in favor of the product. Inducements also give the recipients a general feeling of good will toward the company, and respondents are more likely to say what they think the company wants to hear, rather than what they really think.

Samples Relying on Convenience

It can be difficult and time-consuming to conduct a study involving a truly random sample. It's not uncommon for people to cut corners, and work with a sample that's easy to get to for one reason or another.

9. A sports reporter in Chicago is asked to write an article about the popularity of certain baseball teams in the Midwest. He decides that conducting a poll to gather some data is a fine idea, and while he's at it, he may as well enjoy a beverage and an appetizer. So he hops on the L (public transportation) and goes to a very popular sports bar on the north side of the city, then asks everyone that will pay attention to him which is their favorite baseball team. Over the course of four hours, he manages to gather information from over 300 people. Do you think this was a reasonable approach to gathering information? Explain.

© Cheryl Keegan

It may have been convenient, but it's a ridiculous sample. The north side of Chicago is where the Chicago Cubs stadium is, and the vast majority of people he runs into there would likely choose the Cubs as their favorite team.

A **convenience sample** is one where members of a preexisting group are selected. This can lead to bias because there may be something that connects all the members of that group.

10. An educational researcher wants some data on the number of hours per week that average college students study outside of class. She sends around an email asking colleagues to administer a short survey in their classes, and ends up surveying two 300-level chemistry courses and students in a senior-level physics course. List several different reasons why this is likely to be a nonrepresentative sample.

First, the fact that both are upper-level courses is a bad idea. Next, both courses are in the hard sciences, so it's likely that the students will have a lot in common other than their class standing. Other answers are possible, of course.

If you're trying to interpret the results of a study done using a small sample, a volunteer sample, or a convenience sample, you have to give a lot of thought to whether the characteristics of that sample make it representative of an entire population.

11. There are many sites on the Internet where college students can rate their professors in terms of teaching effectiveness. Suppose that one instructor has seven reviews at a site, with six of them describing the teacher as being a wonderful educator. Discuss how reliable you think that would be for a student deciding whether or not to take that instructor. Back up your conclusions with evidence based on the ideas in this lesson.

Answers vary. There are many interpretations here. Some feel that only students that feel strongly one way or another will make the effort to fill out an evaluation, so it doesn't end up being a representative sample.

Did You Get It

Try this problem to see if you understand the concepts we just studied. The answer can be found at the bottom of the portfolio page.

2. In a study to determine how long it takes to fall asleep, would college students be a good sample of the overall population? Why?

Ambiguous Averages

There was a time when we all thought that "average" meant one thing: the result of adding all data values and dividing by how many there are. But we know better now, don't we? When someone uses the term "average" in making some sort of claim instead of letting us know which measure of average (mean, median, mode, midrange) that they're using, it's time to trust no one and be suspicious.

We know that the values of these different measures of average sometimes differ quite a lot. People who know that there are several types of average can, without lying, select the one for the data that most lends evidence to support their position. For example, suppose a storeowner employs four salespeople. The number of years each has been employed by the store is 22, 10, 2, and 2.

12. What is the mean number of years of service for the four sales people?

Nine years

13. What is the median?

Six years

14. What is the mode?

Two years

15. What about the midrange?

Twelve years

16. If the storeowner wanted to advertise the fact that his employees have many years of experience, which average do you think he would use? Why?

In this case, he'd pick the midrange, since it's the highest of the four measures.

Faulty Survey Questions

Surveys and opinion polls obtain information by using questionnaires. There are two types of studies: interviews and self-administered questionnaires. The interview survey requires a person to ask the questions either in person or by telephone. Self-administered questionnaire surveys require the participant to answer the questions by mail, computer, or in a group setting, such as a classroom. When analyzing the results of a survey using questionnaires, you should be sure that the questions are properly written, since the way questions are phrased can often influence the way people answer them.

© Photographer's Choice/ SuperStock RF

17. A group of people was asked two questions relating to gun control:

- Do you favor a waiting period before guns are sold?
- Do you favor a national gun-registration program costing about 20% of all dollars spent on crime control?

Both questions ask people if they're in favor of stricter gun restrictions. One question was answered "Yes" by 91% of respondents, and 6% said "No." The other was answered "Yes" by just 37% and "No" by 61%. Which is which? Explain why.

The first was answered yes by 91%, the second only 37%. The second is a very biased question: it focuses on the negative aspects of the program, rather than simply asking if the respondents are in favor of the concept.

When reading and interpreting the results obtained from questionnaire surveys, watch out for some of these common mistakes made in the writing of the survey questions.

- *Asking Biased Questions.* By asking a question in a certain way, the researcher can lead the respondents to answer the question the way the researcher wants them to.
- *Asking Double-Barreled Questions.* Sometimes two ideas are contained in one question, and the respondent may answer one or the other in his or her response.
- *Using Double Negatives.* Survey questions containing double negatives often confuse the respondent.
- *Using Confusing Words.* Using words in a survey question that are not well-defined or understood can invalidate the responses. Most people wouldn't admit that they don't understand a question, so they just answer it in some random way. Clearly, this is not optimal.

In Questions 18–21, consider each survey question. Why do you think each question is a poor one? Think about the common mistakes described above.

18. "Are you going to vote for Candidate Jones, even though the latest survey shows he will lose the election?"

The question makes it less likely people will answer yes, because (1) people generally don't like being associated with the loser, and (2) the question will lead people to conclude that it's not worth their time to even vote.

19. "Do you think people would live longer if they were on a diet?"

What sort of diet? That word is not at all well-defined.

20. "Are you in favor of a national health program and do you think it should be subsidized by a special tax as opposed to other ways to finance it, such as a national lottery?"

There are two questions here. The first is a perfectly legitimate question, but it's tied to the second, which leads people to answer no, since it brings up the prospect of a national tax. The wording is also confusing.

21. "Do you feel that it is not appropriate to have areas where people cannot smoke?"

I wrote this question and I have to read it three times to figure out what it's asking. It's clearly designed to confuse the respondents.

Other factors that could bias a survey would include anonymity of the participant, the time and place of the survey, and whether the questions were open-ended or closed-ended.

22. Participants will, in some cases, respond differently to questions based on whether or not their identity is known. Researchers try to ensure confidentiality rather than anonymity; however, many people will be suspicious in either case. List some types of issues that people might be surveyed on where lack of anonymity might affect their responses.

Answers vary, but anything involving sexuality, morality, or self-reported criminal behavior is guaranteed to lead to biased responses if anonymity is not ensured.

23. The time and place where a survey is taken can influence the results. For example, if a survey on airline safety is taken right after a major airline crash, the results may differ from those obtained in a year with no major airline disasters. Think of at least two other similar examples where time and/or place could bias results.

Answers vary.

Did You Get It ?

Try this problem to see if you understand the concepts we just studied. The answer can be found at the bottom of the portfolio page.

3. Two "researchers" (by which we mean "spin doctors") conduct surveys on people's attitudes toward health care. One survey contains the question "Do you feel that the high cost of Obamacare is justified by what it accomplishes?" The other survey has the question "Are you in favor of expanding the Affordable Health Care Act so that even more families can get quality medical care?" Describe how each "researcher" wants those surveyed to respond.

Convenient Use of Percentages

© Jim Bourg/Reuters

Another type of statistical distortion can occur when different values are used to represent the same data. For example, one political candidate who is running for reelection might say, "During my administration, expenditures increased a mere 3%." His opponent, who is trying to unseat him, might say, "During my opponent's administration, expenditures have increased a whopping $6,000,000." Both figures could be correct; but expressing $6,000,000 as "a mere 3%" makes it seem like a very small increase, while expressing the same increase as "a whopping $6,000,000" makes it sound like a very large increase. If you want to get to the truth, ask yourself, "Which measure best represents the data?"

24. In North Carolina, the individual income tax rate was 7.75% in 2013. In 2015, that rate dropped to 5.75%. What is the difference in the two tax rates? Answer as a percent.

The difference is 2%.

25. What is the relative difference in the two tax rates?

$$\frac{5.75\% - 7.75\%}{7.75\%} \approx -0.26$$

The relative difference is negative 26%.

26. If someone earned $40,000 in income, what is the difference this tax cut had on that person's income?

7.75% of $40,000 is $3,100. 5.75% of $40,000 is $2,300. So the tax cut saved that person $800.

27. Which of the three differences would be used by someone wanting to argue that this tax cut won't hurt the state's budget much?

Open to interpretation, but probably the 2%, which probably sounds the least significant.

28. Which of the three differences would be used by someone wanting to argue that this tax cut will have a devastating effect on the state's budget?

Open to interpretation, but probably −26%, which shows that the state will lose over a quarter of income tax revenue.

Did You Get It ?

Try this problem to see if you understand the concepts we just studied. The answer can be found at the bottom of the portfolio page.

4. An upcoming school levy in my town would raise the property taxes on a $150,000 home from $248 per month to $280 per month. Calculate the annual increase and the percent increase, then pick which one of these numbers would most likely be publicized by an advocacy group that opposes the tax increase.

Implied Connections

This one is a classic—If you look for it, you'll find it in advertising pretty much every day. Many claims imply that there's a connection between two variables without having any compelling evidence (or maybe any evidence at all) to back up that connection. To cover themselves legally, advertisers will use words like "may," "suggest," "some," etc. so that if the connection actually doesn't exist, they're not lying.

29. Explain why each of the following claims, found in online advertisements, are essentially meaningless.

a. "Eating fish may help to reduce your cholesterol."

Or it MAY increase it. Or it MAY have no effect whatsoever.

b. "Studies suggest that using our exercise machine will reduce your weight."

Studies 'suggest' isn't exactly a ringing endorsement. If a study PROVED it, that would mean something.

c. "Taking calcium will lower blood pressure in some people."

And in others? It could raise blood pressure, or have no effect.

Misleading Graphs

Graphs are great. We love them. In general, they're a fantastic way to illustrate information, making it easier to interpret. But there's a dark side to graphs. They can often be found hanging out with shady characters and committing crimes. Well, that's an exaggeration, but an inappropriately drawn graph can (either intentionally or not) misrepresent data and lead to false or misleading conclusions.

30. Draw a bar graph illustrating the change in North Carolina's tax rate on each of the axes below. (See Question 24.) Note that you'll be using the same information on each graph.

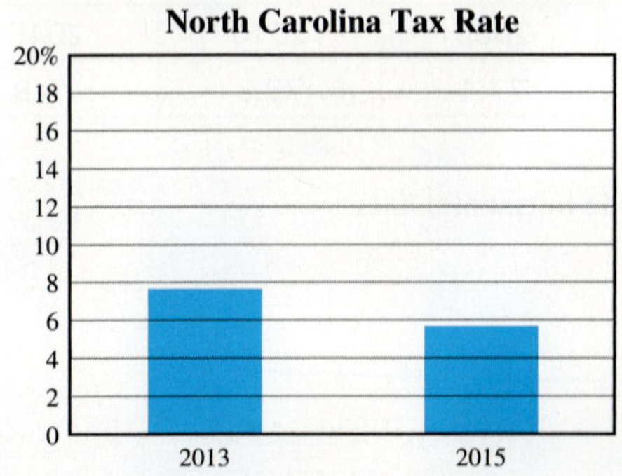

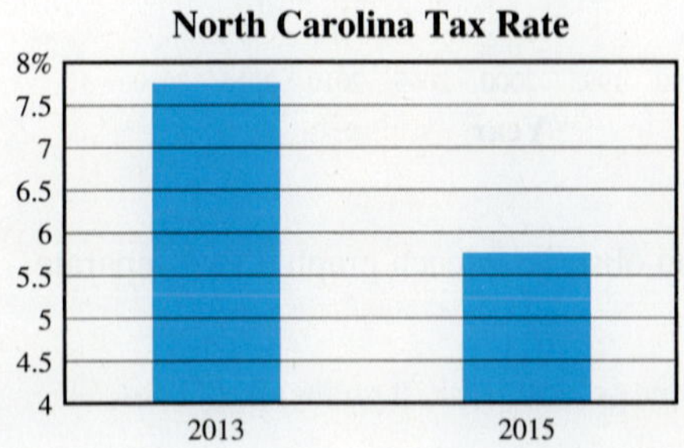

Math Note

It isn't necessarily *wrong* to truncate an axis (like the vertical axis here) when drawing a graph. Sometimes it may even be necessary. But the reader should be aware of this fact and interpret the graph accordingly. Don't be misled if an inappropriate explanation is given Trust no one!

31. Which of the two bar graphs would be used by someone wanting to argue that this tax cut won't hurt the state's budget much? Why?

The first graph makes it look like the two rates are pretty similar. You really have to read the scale carefully to see the significance in the tax cut.

32. Which of the two bar graphs would be used by someone wanting to argue that this tax cut will have a devastating effect on the state's budget? Why?

The second. If makes it look like the new tax rate is about half of the old rate, which is inaccurate.

The average lifespan for Americans has been increasing for decades (although, to be fair, we still rank 34th in the world). The data below displays the average life expectancy for Americans according to the National Center for Health Statistics.

Year	1985	1990	1995	2000	2005	2010	2015
Life Expectancy	74.7	75.4	75.8	76.8	77.4	78.3	78.8

33. Draw a graph for the pairs of data on each graph below, with the year on the horizontal axis.

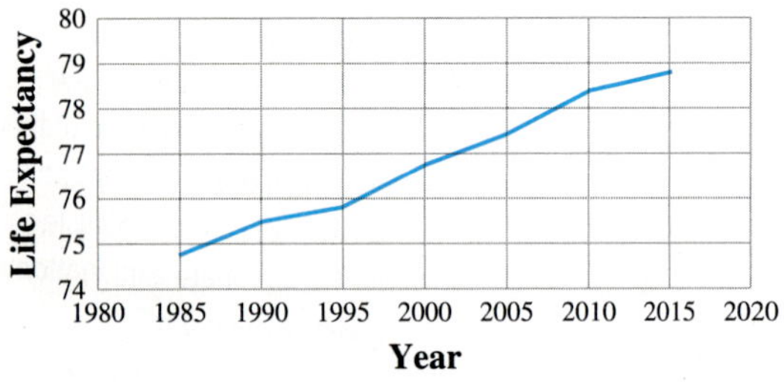

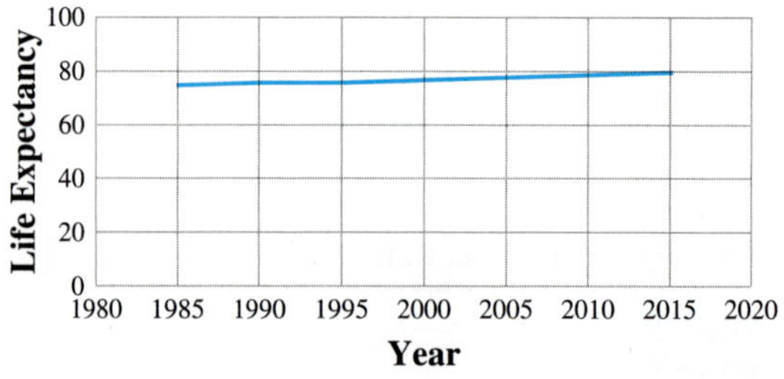

34. Describe how the life expectancy appears to be changing based on what you observe in each graph. (Two separate descriptions, please.)

The first graph makes it look like the increase in lifespan has been very dramatic. In the second graph, it pretty much looks negligble.

35. Which graph would you show if you wanted to convince someone that the advances of modern medicine will soon have us living to be 150?

The first. The increase looks so dramatic, it naturally tricks you into thinking that average lifespan is headed toward the hundreds and beyond very quickly.

36. Which graph would you show if you were arguing that we're really not living much longer than we were way, way back in 1985?

The second. With that scale, you have to look very closely to see any improvement at all.

4-9 Portfolio

Name ____

Check each box when you've completed the task. Remember that your instructor will want you to turn in the portfolio pages you create.

Technology

1. ☐ One of my favorite misuses of statistics is missing comparisons. For example, you might hear a claim like "Our brand of crackers has 1/3 fewer calories." That's completely meaningless if no comparison is given. One-third fewer calories than what? Someone else's crackers? An entire cow? Another common example is time comparisons: "Our aspirin works four times faster." Faster than what? The lady at the DMV? Your job is to search the Internet for an example of a missing comparison. Provide a link to the example, or upload a screen shot, then write an explanation of why their claim is completely ridiculous.

Online Practice

1. ☐ Include any written work from the online assignment along with any notes or questions about this lesson's content.

Applications

1. ☐ Complete the applications problems.

Reflections

Type a short answer to each question.

1. ☐ Why is a truly representative sample so important in gathering data?
2. ☐ Write about an example you've seen or heard recently where someone made a fishy claim backed either by dubious evidence, misleading evidence, or maybe no evidence at all.
3. ☐ How can you use the ideas in this lesson throughout your lifetime?
4. ☐ Take another look at your answer to Question 0 at the beginning of this lesson. Would you change your answer now that you've finished the lesson? How would you summarize the topic of this lesson now?
5. ☐ What questions do you have about this lesson?

Looking Ahead

1. ☐ Start preparing for your final exam, working on your final project, or telling your professor how much you're going to miss this course.

Answers to "Did You Get It?"

1–3. Answers can vary. Talk to your group mates and instructor to make sure you understand the key ideas! **4.** Increase: \$32; percent increase: 13%. Which the group might choose is an opinion. Justify yours!

4-9 Applications

Name ______________________________

1. According to a pilot study of 20 people conducted at the University of Minnesota, daily doses of a compound called arabinogalactan over a period of 6 months resulted in a significant increase in the beneficial lactobacillus species of bacteria. Why can't it be concluded that the compound is beneficial for the majority of people?

   First, 20 people is a very small sample. Second, we have no idea how those 20 were chosen. Are they representative of the general population, or did they all come from a group like college students?

2. Comment on this statement taken from a magazine advertisement: "In a recent clinical study, Cell-Tech was proved to be 1,950 percent better than creatine!"

   Better in what way? At building muscle? At generating big profits for the manufacturer?

3. In an ad for women, the following statement was made: "For every hundred women, 91 have taken the road less traveled." Comment on this statistic.

   "The road less traveled" is at best vague. It could mean wildly different things to different people.

4. In many ads for weight-loss products, under the product claims and in small print, the following statement is made: "These results are not typical." What does this say about the product being advertised?

   It says that for most people, it doesn't work nearly as well as for the small handful of people highlighted in the ad.

5. An article in a leading magazine stated that, "When 18 people with chronic, daily, whiplash-related headaches received steroid injections in a specific neck joint, 11% had no more headaches." Think of a possible reason why the figure 11% was used.

   The writer was probably hoping that the reader would mistake 11% of patients with 11 out of 18 patients. That's a huge difference, since 11% of 18 patients is 2 patients.

4-9 Applications

Name ____________________

6. "Vitamin E is a proven antioxidant and may help in fighting cancer and heart disease." Is there anything ambiguous about this claim?

If it "may help," then it may also not help, or it may make these conditions worse.

7. The results of a survey reported in *USA Weekend* stated that "9% would drive through a toll booth without paying," and "13% would steal cable television or inflate their resumes." Explain why these figures may not be representative of the population in general.

Each of those behaviors is illegal, and people are a lot less likely to admit to doing something illegal or clearly wrong.

8. In a recent year, there were 6,067 male fatalities in the workplace and 521 female deaths. A government official made this statement: "Over 90 percent of the fatal injuries the past year were men, although men accounted for only 54 percent of the nation's employment." Can we conclude that women are more careful on the job? Explain.

No. It's entirely possible that men are more likely to work in high-risk jobs. Also, a fatality on the job may or may not have anything to do with being careful.

9. At one point in the voting process for the 2015 baseball All-Star team, the Kansas City Royals had the leading vote getter for 7 of the 9 starting positions. Fans were allowed to vote online as many times as they want. Do you think it was likely that the vote was representative of the population as a whole? Why?

Almost certainly not. First, people have to go out of their way to vote online. Second, people would be far more likely to vote for players from a team that they are excited about rooting for, which means teams that are winning a lot of games that year.

10. A stats instructor has recorded the scores of a recent quiz worth 20 possible points below.

20 18 18 15 15 18 15 15 15 20 19 20 15 20 19 0 0 18 19 15

a. If he wanted to be a jerk and tell his students the average on the quiz was terrible, which measure of average is he likely to use? Give that value in your explanation.

Because of the two zeroes, the midrange is 10, which is 50%. Ugh. That makes it seem like the scores were awful.

4-9 Applications

Name ______________________________

b. If he wanted to be a great guy and tell them the average on the quiz was fantastic, which measure of average is he likely to use? Give that value in your explanation.

The median was 18–great job class! Most of the scores were high, but the two zeroes lowered all other measures of average.

For Questions 11–14, explain why each survey question might lead to an erroneous conclusion. Also suggest a better question.

11. "How often do you run red lights?"

Asking flat-out about something against the law would tend to lead people to underreport how often they do it. Something like "how many times per year does a light turn red before you clear an intersection" has two advantages: it specifies what's meant by "how often," and makes the activity sound much more reasonable and less illegal.

12. "Do you think gun manufacturers should put safety locks on all guns sold even though it would increase the cost of the gun by 20%?"

The focus of the question appears to be the cost, not what it's really asking about. It would be best to leave the cost part out, or at worst make it two separate questions: "Do you think gun manufacturers should put safety locks on all guns?" and "Would you still support that if it added 20% to the cost of a gun?"

13. "Do you think that it is not important to give extra tutoring to students who are not failing?"

The double-negative makes this question confusing. It should be worded "Do you think it is important to give extra tutoring to students who are passing?"

14. "Which restaurant in town has the best food and the greatest service?"

There are two questions here that may be unrelated. A lot of restaurants have fantastic food and shaky service, and many have great service and mediocre food. This should be two completely separate questions.

4-9 Applications

Name ______________________________

15. The two graphs below were used by President Obama's reelection campaign commission to illustrate certain aspects of domestic energy production during his presidency. Which graph is misleading and why?

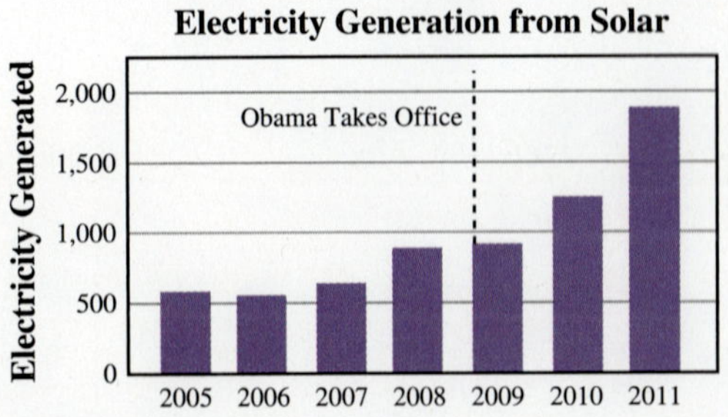

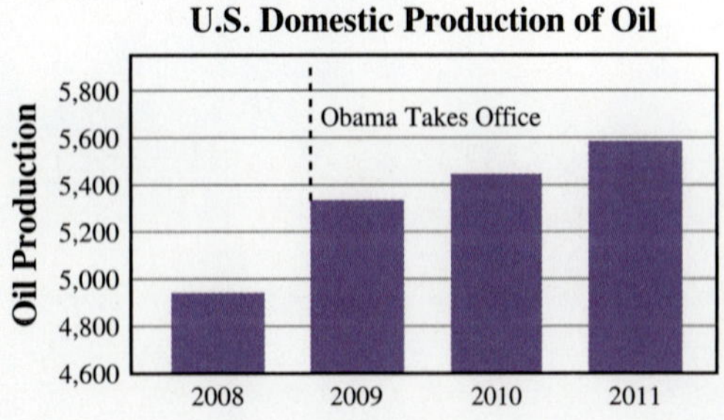

The first is fine. The second has a scale on the vertical axis that doesn't start at zero, so it makes the growth in oil production look much more significant than it actually was. It only increased by about 14%, but the last bar is more than twice as tall as the first.

Lesson 1-1

1. $\frac{3}{5}, \frac{4}{5}, \frac{17}{7}, \frac{11}{24}$

2. $\frac{3}{5}, \frac{6}{25}, \frac{49}{50}, \frac{33}{100}$

3. 0.12, 0.3, 1.375, $3.\overline{6}$ or 3.666 . . .

4. $x = 7$

5. $y = 700$

6. $x = 383.92$

Lesson 1-2

1. 37%, 12.5%, 9.1%

2. 23%, 45%, 212.5%, $566.\overline{6}\%$

3. 97.2

4. 112

5. 33.912

6. 78%, 25%

Lesson 1-3

1. 171.875

2. $53.27

3. 19%

4. 15%, Work

5. 4

6. 56

Lesson 1-4

1. >, >, <, >

2. 150, 12,360

3. 12, 146

4. 0.2, 146.4

5. 0.05, 1,893.74

Lesson 1-5

1. 12, 5,280, 3, 2

2. $\frac{21}{4}$, −440

3. $\frac{7}{10}$, −5

Lesson 1-6

1. 23%, 79.4%, 3.5%

2. 286 sq ft

3. 1,600 sq ft

4. 3,024 cu in

Lesson 1-7

1. 100,000

2. $\frac{1}{1{,}000}$

3. 8^7

4. 10^{-5}

5. 10^{17}, 8^{-1}

6. 5^4, 10^{-7}

7. $l = \frac{A}{w}$

Lesson 1-8

1. $\frac{2}{15}$

2. $\frac{ca}{b}$

3. 3, 9, 57

4. 35

5. 6.5, 22.5

Lesson 2-1

1. 36, 68

2. 22,918.33

3. $1,514.98

4. $4,501.86

5. 41, 47, 53

6. 96, 192, 384

Lesson 2-2

1. $\frac{1}{4}, \frac{9}{25}$

2. $\frac{16}{12}, \frac{160}{8}$

3. $0.7495

4. 552 hours

5. 592 ounces

6. About 54.27 feet per second

Lesson 2-3

1. $A = (-8, 12)$, $B = (4, 4)$, $C = (8, -16)$, $D = (-5, 0)$

2.

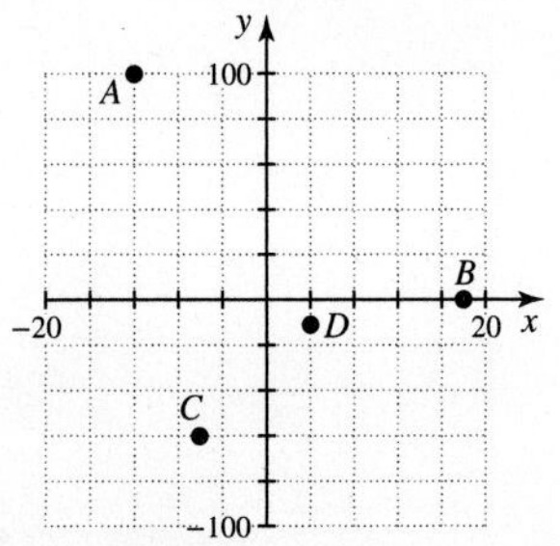

Lesson 2-4

1. The second one

2. $-\frac{5}{2}$

3. 34.1, −11.9, −46.4

4. a. 37
   b. 59, 92; 59 students came to the 5th meeting, 92 came to the 11th meeting.

Lesson 2-5

1. a. 0.04
   b. 0.81
   c. 0.114

2. a. 37%
   b. 91.4%
   c. 1.9%

3. Linear

4. Exponential

5. About 11.9%

6. About 13,128.46

Lesson 2-6

1. Close to linear

2. Neither

3. $y = 29.031x - 56.8$

4. 26, 20%

5. About −17.8%

Lesson 2-7

1. a. Linear, exponential
   b. Exponential, linear

2. About 10.7%

3. The exponential equation

Lesson 3-1

1. 0.65, 0.0425

2. $292.5

3. $1,300

4. $44,626.50

5. 1,800

6. 11,875

7. $t \approx 20.28$

8. $r \approx 0.238$

Lesson 3-2

1. 0.47, 0.094

2. $264

3. $500(1.03)^4$

4. 720, about 513.34

5. About 33.54

Lesson 3-3

1. 0.91, 0.0648

2. $204.75

3. $582.30

4. 2,460

Lesson 3-4

1. $3,240

2. $3,974.73

3. About 9.58, about 17,504.29

4. 0.790957

Lesson 3-5

1. a. 16 g
   b. 441 mg
   c. 8%

2. 64 g

3. 62.5%

4. 5 g

5. $M = \frac{k}{P}$

Lesson 3-6

1. 23%, 79.4%, 3.5%

2. $860.30

3. $3,519.32

4. $9,300.20

Lesson 4-1

1. ¾, 13/16

2. 0.6, 0.125

3. 60%, 75%

4. 13%, 5.8%

5. 15, 16, 17, 18, 19; 2, 3, 4, 5, 6, 7, 8

Lesson 4-2

1. 30, 120

2. $\frac{1}{9}$

3. $\frac{1}{4}$

4. $\frac{1}{6}$

5. $\frac{2}{3}$

Lesson 4-3

1. $\frac{1}{3}, \frac{4}{5}$

2. $\frac{5}{36}, \frac{21}{36}$

3. $\frac{13}{21}$

4. $-\frac{13}{9}$

Lesson 4-4

1. 6

2. $348,000

3. $107,000

4. 20

5. It depends on if you do or do not include $150,000: either 2 or 3.

Lesson 4-5

1. 1.2 1.3 1.3 1.7 1.7 1.7 2.1 2.3 2.4 2.4 2.4 2.5 2.5 2.8 3.1 3.7

2. 35.1

3. $\frac{43}{5} = 8.6$

4. 55

5. 1.2

6. 16

Lesson 4-6

1. 49.5
2. 191
3. 10.58, 121.95
4. 110

Lesson 4-7

1. \$1.00–\$1.99
2. 14
3. 0.12
4. About 8 square centimeters
5. About 83.8%

Lesson 4-8

1. 169.2°
2. 4.29
3. 0.02
4. $\frac{12}{n}$
5. $n = 9$
6. $n \approx 0.0077$

Lesson 4-9

1. Mean: about 25.9; median: 19; mode: 18; midrange: 53
2. \$3.85, about 51.7%
3. 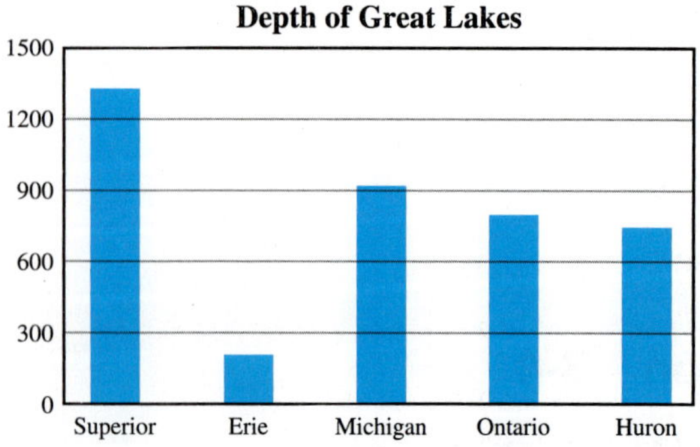

4. 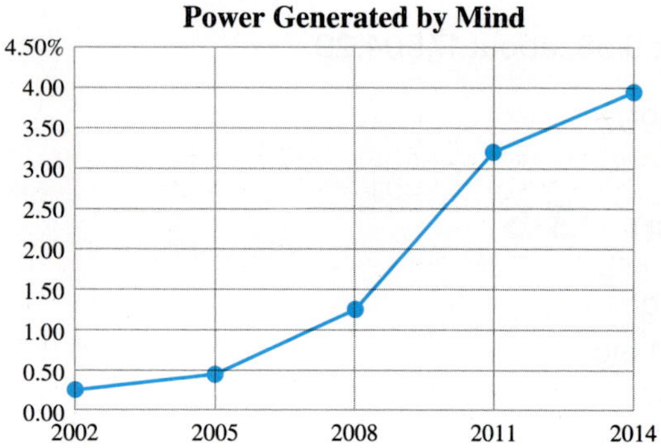

Index

S

T